GAMETES & SPORES

John Farley is professor of biology at Dalhousie University, Halifax, Nova Scotia. He is the author of *The Spontaneous Generation Controversy from Descartes to Oparin* (also published by Johns Hopkins).

JOHN FARLEY

GAMETES
& SPORES

Ideas about Sexual Reproduction
1750–1914

THE JOHNS HOPKINS UNIVERSITY PRESS
Baltimore and London

The Johns Hopkins University Press, Baltimore, Maryland 21218
The Johns Hopkins Press Ltd., London

Library of Congress Cataloging in Publication Data

Farley, John, 1936–
Gametes and spores.

Includes index.
1. Reproduction—Research—History.
2. Reproduction.   I.   Title.
QH471.F37   574.1′6   82-87
ISBN 0-8018-2738-8   AACR2

To the late George Lubinsky,
who first awakened my interest in the history of science

Contents

Acknowledgments

The idea of writing a "sex book" came from my first-born, Gael, who suggested that, having written a book on spontaneous generation, the original title of which was "Life without Parents," I should write something on "life with parents." It seemed an excellent idea, for while much has been written about eighteenth-century reproductive theories, the curtains have remained discreetly closed over nineteenth-century views on the subject. As one Victorian gentleman wrote in the *Edinburgh Review*, "It is better [in] every way that what cannot be spoken and ought not to have been written should not be written."

The initial work on the present volume was completed during a second sabbatical year in the Department of the History of Science at Harvard University. Once again I thank Everett Mendelsohn and other members of that department for making that and other visits so rewarding and enjoyable. I thank also various librarians for their help in my endeavors: those at Woods Hole, where some of Frank Lillie's papers are housed; Elspeth Simpson at the University of Glasgow archives, where I read the Frederick Bower collection; the staff of the Gray Herbarium Library at Harvard University; and, above all, Mary Keeler and others at that mecca of all biological historians, the Museum of Comparative Zoology at Harvard.

I owe an enormous debt of gratitude to Frederick Churchill of Indiana University, not only for reading my original manuscript, but also for sending me the first draft of his magnificent "Sex and the Single Organism" (*Studies in History of Biology 3* [Baltimore: The Johns Hopkins University Press, 1979], pp. 139–77). Readers of Churchill's article will recognize that one of the major themes in this book—that sex lost all biological significance in the nineteenth century—was first clearly elucidated in that article. I had come to much the same conclusion during my initial readings on plant reproduction, but until receiving Churchill's article I was not aware that the conclusion applied also to animal reproduction and that it would come to have such an important place in the entire story. Without Churchill's pioneering work, I doubt that this book would have taken the form that it has.

I thank also many other individuals who have contributed to the book: Alice Baxter, for allowing me to utilize our article "Mendel and Meiosis," which appeared in the *Journal of the History of Biology* 12 (1979); Tina Simmons, for introducing me to the vast array of feminist literature dealing particularly with the role of women in Victorian society; Tony Chapman, for acting as my botanical guide; Anne von Maltzahn, for being such an extraordinarily fine teacher of German, and Anja Pearre, for translating many German passages I could not handle; Barry Fox, for his invaluable help in proofreading and in the use of the English language in general; Peter Buck, for many enjoyable hours in Boston, for asking questions I could never answer, and for suggesting a way to end this book; Goldie Gibson, for her painstaking art work; Mary Primrose and Joe Harvey, for their photography; and typists too numerous to mention. To all, my deepest thanks.

I acknowledge also the financial assistance of the Canada Council (now unfortunately partly renamed the Social Sciences and Humanities Research Council of Canada!), for both a Leave Fellowship and the subsequent grants that enabled me to visit the Harvard libraries. During the final stages of my work I received generous support from the Associated Medical Services, Inc., and the Hannah Institute for the History of Medicine, and to these organizations my thanks are also extended.

In the end, however, my greatest thanks must go to the plants and animals themselves for managing to hide their reproductive secrets. They could simply have split in two, but by not doing so very often, they baffled the best minds for centuries. After all, what can be more mystifying than a tiny spermatic animalcule, a beautifully architectured pollen grain, or a simple grain of dust liberated from a moss capsule?

GAMETES & SPORES

Introduction

Man has long considered the ability to reproduce to be one of the fundamental attributes of living things, and since antiquity, much effort has been directed toward understanding the nature of reproduction. Much of the work has necessarily been speculative, but much of it has been observational and even experimental. The present study begins with mid-eighteenth-century views of the reproductive process. By this time, sexual reproduction was assumed to be universal and Carl Linnaeus had developed a classificatory scheme based on the ubiquity of sex organs.

Whether or not sexual reproduction is universal is one of the puzzles examined in this book. By the early nineteenth century the assumption seemed more and more unlikely. Some of the lower animals appeared only to split in two or to bud; moreover, French botanists had concluded that the mysterious powder released from moss capsules and from the underside of fern leaves consisted not of seeds, as Linnaeus had believed, nor of pollen grains, but of asexual spores. In the middle of the century, however, inspired by the ideas and philosophy of Matthias Schleiden, German botanists, made the amazing discovery that all vascular plants liberate both gametes and spores. They illustrated, in the words of Wilhelm Hofmeister, an alternation of generations, a *Generationswechsel,* between a spore-forming stage and a gamete-forming stage. Furthermore, increasing knowledge of the marine invertebrates and insects had revealed that they, too, often multiplied by asexual means, and some even displayed what the Danish naturalist Japetus Steenstrup also called a *Generationswechsel.* Asexual reproduction, therefore, was shown to be far more widespread and more significant than had been believed in the previous century.

What was the nature of the substances liberated by plant and animal sex organs? That question, too, was debated at length and forms another basic puzzle examined in this book. Eventually, seventeenth-century naturalists

1

assumed that ovaries produced "eggs," somewhat like the eggs of birds, with a nutritive part and a part from which the new being arose. They assumed that the testes liberated only a seminal fluid, a belief whose accuracy was first challenged by the amazing discovery, in the 1670s, that this fluid teemed with tiny "animalcules." What were these small animals? Were they merely parasites of the testis, as most believed, or did they play an essential role in the sexual process? And if they were indeed parasites, how did they arise in the testis? Did they lay eggs, like other organisms, or were they products of spontaneous generation? The cell theory of Schleiden and Theodor Schwann provided the turning point in this particular puzzle. The "animalcules," or spermatozoa, came to be seen as cells derived from cells of the testis, and eggs were seen to contain "ova," cellular products of the ovaries. Antherozoids (sperm) and ova were found in the antheridia and archegonia of the lower plants, and they, too, after much argument, were called cells. Likewise, the brilliant work of Hofmeister revealed that the embryo sac and pollen of seed plants were not, in reality, ova and sperm, but were much-reduced stages, formed from spores, that eventually produced ova and nonmotile sperm cells respectively. Plants, it appeared, reproduced mainly by means of spores, and this provided more justification for the nineteenth-century belief in the importance of asexual reproduction.

The greatest puzzle of all, however, involved the role played by the products of the sex organs. In the eighteenth century, an age when even the bewildering complexities of embryogenesis needed to be analyzed in mechanical terms, the egg was generally assumed to contain a preformed embryo on which the seminal fluid acted in order to initiate development. Only a few naturalists—including, quite understandably, the discoverers of sperm—believed that sperm contained the preformed embryo, which, to develop, needed to be emplanted in the nutritive egg. So powerful was the antagonism toward these "spermist" theories, so powerful was the belief that the female alone was responsible for procreation, that even when the necessity of sperm cells to the reproductive process was recognized in the mid nineteenth century, their actual role was assumed to be severely restricted. Even if sperm actually penetrated the egg—and that question provided the focus of the Great Debate (see Chapter 2, below)—their role, once they were inside the egg, was seen to be one of "rejuvenation" and "stimulation," although many also assumed that they somehow passed male characteristics into the egg. In the nineteenth century, therefore, sperm were considered analogous in function to rain water and sunlight, and ova were considered to differ from spores and buds only in the stimulus that was required for them to develop.

As a result of the close association made in the nineteenth century between ova and asexual cells, and the realization that asexual reproduction was a widespread phenomenon, the question then arose, what were the reasons for sexual reproduction in the first place? The answer, as Chapter 4 argues, reflects Victorian social attitudes toward sex. Sex was simply the means of procreation used by organisms that had, by the division of labor, placed the task of procreation into the hands (or, more appositely, ovaries) of a special individual, the female. Socially and biologically, the female existed solely to bear and raise offspring. The nineteenth-century biological attitude toward sex mirrored that century's social attitude toward women. The status of women was therefore a law of nature; to argue otherwise was to threaten the social and biological fabric of the race.

At the end of the nineteenth century these biological theories rapidly collapsed. New cytological techniques revealed that both the production of gametes and their subsequent fecundation were unique events, quite unlike the production of buds and spores. Discrete nuclear bodies (the chromosomes) were involved, and by the first decade of the twentieth century, a dramatically new interpretation of the purpose of sex came into focus. The process was not so much a procreative tool but was a means of introducing, by genetic recombinations, the variations in a population without which evolution, as we know it, would not have occurred. The male's role was seen to be as important to this process as that of the female, and the act of reproduction was seen to be distinct from the act of gamete formation and fecundation. That distinction ushered in the "sexual revolution" of the twentieth century; whether the biological distinction had any impact on the social distinction is not clear, and this book does not speculate about the matter. Certainly, the new scientific theory preceded the social revolution, and educated women would have read that the essence of sexual reproduction lay in recombination, not in procreation. At that time, particularly in America, women were pursuing a higher education, many in the field of biology. In addition, biologists as influential as Thomas Hunt Morgan and Jacques Loeb taught at Bryn Mawr, one of the leading women's colleges in the United States.

For all intents and purposes, this study ends with the year 1914. By that year Morgan's *Heredity and Sex* and Frank Lillie's famous paper on fertilization had appeared. The latter's theory appears to have been the first to reconcile successfully the existence of sperm with the puzzling phenomenon of parthenogenesis—another problem that weaves its way through the present text. As explained in the Postscript, this book ends with 1914 because what our modern biology students learn about sex is essentially the same as what was known then. The Postscript shows briefly how modern views differ from what is actually taught to present-day beginning students and suggests that this seventy-year time-lag is inexcusable.

Such, then, are the bare outlines of this book. Naturally, many other facets of this complex problem are discussed. Time is spent on plants, in part because historians of biology seem so indifferent to post-Linnaean botany and in part because the life cycles and reproduction processes of plants are more complex than those of most animals. Then again, many facets of sexual reproduction have been omitted. For example, no mention has been made of ovulation, pregnancy, the estrous and menstrual cycles, and secondary sexual characteristics—the area that today comes under the aegis of endocrinology.

This book deals basically with eggs, spermatozoa, and spores—what they are, what they do, how they do it, and why they do it.

ONE

The Universality of Sex

THE INCIDENCE OF SEXUAL REPRODUCTION

According to modern estimates, 99.9 percent of all species reproduce sexually. This figure would come as no surprise to a reincarnated eighteenth-century naturalist, for he, too, believed that sex was virtually universal. His belief was based, quite simply, on the assumption that the anatomy, life histories, reproduction, and development of flowering plants and vertebrate animals provided the model for all other plants and animals. "A natural instinct," wrote Carl Linnaeus, "tells us to know first the objects which are closest to us, and the smaller ones last, for example: man, quadrupeds, birds, fishes, insects, mites; or first the larger plants and last the tiny mosses."[1] From such a perspective, not only was sexual reproduction assumed to be universal but so were sex organs. The universality of sex presupposed the universality of sex organs: testes, ovaries, stamens, and pistils.

The assumed ubiquity of sex organs enabled Linnaeus to build around them a complete classificatory scheme, which W. Watson described as "the master-piece of the most complete naturalist the world has seen."[2] The fact that Linnaeus's scheme swept all of Europe and North America indicates just how universal sexual reproduction was assumed to be. Even William Withering, chief physician of the Birmingham General Hospital and one of his mildly critical English followers, admitted in 1776 that "the system of Linnaeus is now very universally adopted; and though confessedly imperfect, it approaches so near to perfection, that we may perhaps never expect to see any other improvements than such as will be founded upon this plan."[3]

Linnaeus, born in 1707 and named after the great Swedish warrior-king Charles XII, enrolled at the University of Uppsala in 1728 and, as was the tradition in those days, traveled to Holland to complete his studies. By the time he left Holland in 1738, Linnaeus had not only gained a medical degree

Figure 1.1.   Carl Linnaeus. Reproduced by permission of the Museum of Comparative Zoology, Harvard University.

from Harderwijk but had also published fourteen papers, among them
Systema Naturae (1735), *Fundamenta Botanica* (1736), and *Genera Plantarum* (1737). Although all of these small tracts were later expanded into
major works, they contain the basic classificatory scheme that Linnaeus
adhered to for the rest of his life. Twenty-three of his twenty-four plant classes

were categorized on the basis of the number, length, position, and distinctness of the stamens.[4]

In his *Philosophia Botanica* (1751), Linnaeus succinctly described the nature of these sex organs in a series of aphorisms (Fig. 1.2):

III. The *stamina* are those parts of a flower appropriated to the preparation of the *pollen* or fecundating dust, and consist of the *filamentum,* the *anthera,* and the *pollen.*

12. The *pollen,* or impregnating dust, is that fine powder contained within the *antherae,* or tops of the *stamina,* and dispersed, when ripe, upon the female organ, for impregnating the same.

13. The germen, or seed bud, is the base or lower part of the *pistillium,* containing the rudiments of the unique ripe fruit, or seed, in the flowering state of the plant.

15. The *stigma,* the summit, or top of the style, is that part which receives the fertilizing dust of the *antherae,* and transfers its *effluvia,* through the style, into the middle of the *germen* or seed bud.

Thus, he concluded a few pages later, "every fructification has the *anthera, stigma,* and seed . . . [and] every vegetable, without exception, is furnished with flower and fruit."

Linnaeus also drew analogies between the reproductive organs in plants and those in the higher animals. As he noted in one aphorism, "The *calyx* then is the marriage bed, the *corolla* the curtains, the filaments the spermatic vessels, the *antherae* the testicles, the dust the male sperm, the *stigma* the extremity of the female organ, the *style* the *vagina,* the *germen* the ovary, the *pericardium* the ovary impregnated, the seeds the *ovula* or eggs."[5]

There were, however, plants in which the flowers, if present at all, were certainly not readily visible. Such plants—the algae, fungi, mosses, liver-

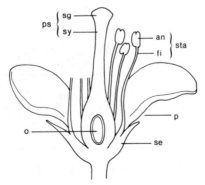

FIGURE 1.2. The sex organs of plants according to Linnaeus: *an,* anther; *fi,* filament; *o,* ovary; *p,* petal; *ps,* pistil; *se,* sepal; *sg,* stigma; *sta,* stamen; *sy,* style.

worts, ferns, and fern allies—were placed by Linnaeus in his twenty-fourth class, the Cryptogamia, "the fructifying bodies of which," he wrote, "remove themselves from our sight."[6] They enjoy their amorous relationships in a clandestine manner: *Nuptiae clam instituuntur* ("Secretly a union is arranged").

It was more than a century later that these peculiar plants revealed their secrets to botanists, and although Linnaeus was quite certain that sex organs existed, remarking in 1735 that "the aid of a lens teaches this," he was unsure as to the whereabouts of these organs.[7] He was also unsure of the nature of the dust liberated from the moss capsules and the "orbs," or "globes," lying beneath the leaves of the fern (Figs. 1.3 and 1.4).

In the spring, moss plants sometimes produce a small capsule carried at the end of a narrow stalk. When ripe, this capsule liberates a fine powder, as do

FIGURE 1.3.   The capsules of mosses, which release a fine dust. Courtesy of M. J. Harvey.

FIGURE 1.4. The "orbs," or "globes," of ferns, which also release a fine dust. Courtesy of M. Primrose.

the small round bodies usually located on the underside of fern leaves. Understandably, Linnaeus was quite confused over the nature of this powder. Did it consist of pollen grains that presumably fecundated the female organ situated somewhere else on the plant, or did it consist of seeds that germinated to produce a new moss or fern plant? Initially, Linnaeus assumed that the powder was pollen grains, remarking in a 1737 letter to Albrecht van Haller that "this powder, seen under a microscope, exactly agrees with the dust of the anthers in the other plants." But one month later he admitted: "[I know] nothing about the imperfect tribes of plants and must confess my ignorance whether what I see is seed, or dust of the anthers."[8] By 1751, however, Linnaeus had accepted that the powder of the fern consisted of true seeds. By the sixth edition of his *Genera Plantarum,* published in 1764, he had similarly concluded that in mosses, "what we have called anthers, should perhaps be called capsules, and their pollen true seeds, since in *Bauxbaumia* and others we have seen inside the operculum true pollen-producing anthers suspended from their own filaments."[9]

The view that moss capsules were petalless anther-bearing flowers that eventually produced a seed-bearing "fruit" came to be widely accepted during the latter part of the eighteenth century. One of the first clear pronouncements to this effect had come from the English apothecary John Hill, who gained lasting fame and unpopularity by satirizing the Royal Society after having been excluded from its illustrious circle. In his text of 1759[10] he reported having cut the capsule of the "Swan's Neck Bryum" longitudinally in half and having seen there a central seed vessel and a crown of anthers

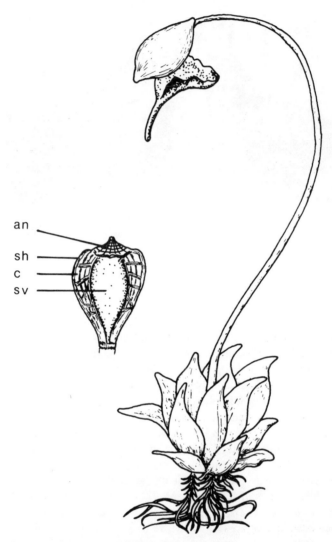

an
sh
c
sv

FIGURE 1.5. The "Swans's Neck Bryum" (after Hill). A capsule cut in half reveals both a seed vessel and anthers: *an*, anthers; *c*, cavity; *sv*, seed vessel; *sh*, shell.

loaded with pollen grains (Fig. 1.5). Johann Hedwig, a Chemnitz-based physician, likewise believed that the capsule of mosses was a seed-bearing and seed-liberating "fruit," but assumed that the actual sex organs (anthers and pistilla) were carried on the tips of the moss leaf and not inside the capsules as Linnaeus and Hill had assumed. This view was further substantiated when Hedwig discovered that the moss "seeds" germinated to produce a filamentous body, the "cotyledons."[11]

The utter simplicity of the Linnaean system carried all before it. By 1760 the system had become firmly established in England, where it satisfied the needs of practical horticulturists. "I had the felicity of taking the lead in introducing the Linnaean system and language to my countrymen by a course of public lectures," remarked Thomas Martyn, professor of botany at Cambridge.[12] Martyn's *Plantae Cantebrigiensis* of 1763 and William Hudson's *Flora Anglica* of 1762 became the standard botanical works of that period, representing the final rout of John Ray's system, which had dominated English botany since its inception in 1682. Despite Martyn's early recognition of the Linnaean system, however, it was mainly due to the work of Joseph Banks and the growth of Kew Gardens that British botanists accepted the Linnaean tradition and classification. The late 1700s ushered in a century of "imperial botany." The emphasis lay in collecting plants from the new colonies, in classifying them, in discovering plants suitable for cultivation in these faraway lands, and in training gardeners to work in the colonies. Kew Gardens became the focal point of this effort. Lying adjacent to the Royal Richmond Gardens and a property of the Carpel family, the gardens had been leased by the Prince of Wales in 1740. After his death in 1751, they had been joined with the Richmond Gardens by his widow, and after her death in 1772, Joseph Banks, then president of the Royal Society, had become scientific adviser to the gardens. Under his influence, Kew became the "centre of scientific intercourse between the mother country and the colonies."

The enthusiasm for Linnaeus in England is also indicated by the success of the Linnaean Society, which was founded in 1788 chiefly through the efforts of its first president, James Edward Smith, who had earlier purchased the Linnaean collection from Linnaeus's mother. But perhaps Erasmus Darwin's poem *The Loves of the Plants* best indicates the strength of the Linnaean cult in England:

With honey'd lips enamour'd Woodbines meet,
Clasp with fond arms, and mix their kisses sweet.   20

The fair Osmunda [a fern] seeks the silent dell,
The ivy canopy, and dripping cell;
There hid in shades *clandestine* rites approves,   95
Till the green progeny betrays her loves.[13]

Thus, it was believed, all plants arose from seeds, and all animals from eggs. Sex was universal: "From the general law of Nature there can be no considerable deviations; no being owes its formation to chance, and all are probably produced by a similar mode of generation, depending on the concourse of the two sexes, the minutest insect, as well as the elephant, the smallest moss, as well as the most stately and elevated oak."[14]

Such assumptions did not go unchallenged, however, for in the eighteenth century, French botanists began seriously to question the validity of the Linnaean system and the belief that sex was universal in plants. They, unlike the

Linnaean botanists, demanded that classification be more than a convenient means of recognizing plants; that it should be a "natural system," reflecting real natural affinities. To Linnaeus, on the other hand, only his genera and species had any real existence in nature. Their combinations into classes and orders, in which sex played such an important role, was an arbitrary device by which to gain ready access to the well-known and well-established natural genera. The French, however, demanded that all taxa be natural. The pioneers of this natural system were the five de Jussieu brothers, whose views dominated French plant taxonomy for well over a century. As a result of their efforts, classificatory schemes developed which denied both that the cryptogams were a natural grouping and also that they were sexual forms. In developing their taxonomic schemes, these French botanists first seriously challenged that basic eighteenth-century belief that sex was all-encompassing and all-important.

The eldest of these brothers, Antoine de Jussieu, born in 1686, was the first among them to attempt to construct a natural system of classification. In 1708 he took up a post in the Jardin de Roi, succeeding Joseph Pitton Tournefort to the chair of pharmaceutical botany in 1710. Tournefort, in his *Elémens de botanique* of 1694, had argued that it was not possible to find seeds in all plants, but had acknowledged, nevertheless, "that the views of those who believe all plants have seeds is founded on very reasonable conjectures." Unlike the later Linnaean botanists, however, Tournefort did not assume that the existence of seeds demanded also the existence of sex organs. The ferns, for example, were placed in the sixteenth class of his system and were characterized by the possession of seeds but the absence of flowers. Mosses, on the other hand, were placed in a seventeenth class, having neither seeds nor flowers, although Tournefort realized that moss capsules liberated a fine powder "which seems to take the place of seeds."[15] Curiously, a much-expanded version of this text was reprinted one hundred years later as part of the wholesale French attack on the Linnaean system.

Antoine de Jussieu published little, but instead directed his main efforts toward training students, among whom was his brother Bernard. Bernard, it seems, became an extraordinary teacher, whose pupils included Michel Adanson, the Richard brothers, and his own nephew, Antoine Laurent de Jussieu. Like his elder brother, Bernard wrote very little, but became an active protagonist of a natural classification system. He was charged with arranging the famous Trianon gardens at the Palace of Versailles, an experience which led him to devise the so-called Trianon system of 1759.

In 1763, using the sources in the de Jussieus' library, Adanson wrote his *Familles des plantes*, but he failed to persuade the de Jussieu brothers to adopt his system in the Trianon gardens. He attacked Linnaeus's claim that all plants, by necessity, bore stamens and pistils, and he challenged the classificatory scheme that rested on this belief. It was not feasible, he argued, to classify plants by reference to only one or a few principles; one had to consider the root, stem, leaves, flowers, fruit, and all parts and qualities together in order to arrive at "natural and invariable families." Continuing his argument, he wrote:

> Newton wished to reduce all physics to attraction; Linnaeus all plants only to the conessance of stamens or to fructification. Whereas it is necessary to collect together all parts in botany in order to have the true principles or the true system of this science. In the same way it is necessary to consider not only quality, such as attraction of mass, but the totality of mechanical principles as the basis for everything in Physics.[16]

Adanson believed that some plants, such as mushrooms and the common brown algae *Fucus,* were completely asexual—that is, had no sex organs—while mosses, liverworts, and ferns were sexual plants having the requisite sex organs.

After the death of Bernard, Antoine Laurent de Jussieu became professor of botany at the Muséum d'Histoire Naturelle and in 1789 published his *Genera Plantarum.* His scheme, based on the Trianon system, included all the cryptogams in the Acotyledones, a group lacking the seed cotyledons that were present in his other two divisions, the Monocotyledones and the Dicotyledones. Thus, like Linnaeus before him, he separated the lower plants from the others by negative characteristics.

Nevertheless, although a broadening of the characters upon which the plant classes were based had occurred, de Jussieu still believed that the Acotyledones were sexual. "The sexual organs in some are least or insufficiently conspicuous," he wrote; "in others they are most insignificant or uncertain, [and] in others they are visible."[17] He still regarded the capsule of the moss as the fructifying organ, bearing as it did all the sex organs, and he believed the same was true of the structures on the underside of the fern leaves.

French botanists expanded their work on natural classification during the first half of the nineteenth century. It came to be accepted by them that the so-called seeds, liberated by some or all of the Acotyledones, were, in reality, asexual spores and that some or all of Linnaeus's cryptogamic plants did not reproduce sexually.

The idea that the "seeds" of cryptogams lacked an embryo and were "gemmae" (amorphous bodies that became detached from the mother plant to form a new plant) rather than seeds was first postulated by Joseph Gaertner, a pupil of Albrecht van Haller and a widely traveled and acclaimed naturalist.[18] His findings received little attention in Germany but not surprisingly were well received in Paris. More and more French botanists acquiesced in that view.

In 1813 Augustin-Pyramus de Candolle, for example, published his

Théorie élémentaire de la botanique, in which the Linnaean cryptogams were broken up into various divergent taxa. Born in Geneva in 1778, de Candolle had spent many years studying in Paris before being appointed to the chair of botany at Montpelier. A student of de Jussieu's natural methods, he classified the plants on the basis of a hierarchy of characters, with the embryo and the sex organs being the most important. Algae, fungi, and lichens, which he assumed lacked sex organs, were placed with the sexual mosses and liverworts in the "cellular," or acotyledonous, plant category, while the ferns and their allies, on the basis of their vascular tissues, were placed in the vascular, or cotyledonous, group.[19] Achille Richard, however, went further. "We believe," he wrote, "that plants designated by the name Cryptogamous are entirely lacking sexual organs and that nothing in them can reasonably be compared to similar parts in phanerogams."[20]

Despite the views of Richard, however, the basic question—whether the acotyledonous plants, now divided by the French into those with vascular tissue (the ferns and their allies) and those with "cellular" tissue (algae, fungi, lichens, liverworts, and mosses), were sexual beings bearing sex organs—had not been resolved. The problem was discussed at length by Adrien de Jussieu in his *Cours élémentaire d'histoire naturelle,* which was published in 1843 and used as an elementary text in French colleges. Adrien, son of Bernard, was clearly unsure. If sex organs do exist in the acotyledonous plants, he concluded, "it is evident that the contents are quite different from that described in phanerogamic plants."[21]

The problem that plagued these botanists, and continued to plague them until the cell theory became established in the 1840s, lay with the criterion used to determine sex. In the first part of the nineteenth century, as in the eighteenth, to be sexual was to possess sex organs, and sex organs were assumed to be anther- or pistillike. Couched in these terms, the question whether the cryptogams were sexual became unanswerable. By the early nineteenth century many authors had described organs on these plants and had termed them *anther*-idia and *pistil*-lidia, assuming them to be analogous to the anthers and pistils of the higher plants. But as everyone realized, these organs were *not* identical in structure to the sex organs of the phanerogams. The question was, were these organs sufficiently similar to anthers and pistils to be considered sex organs? Linnaean botanists were sure that they were, but French botanists gave both positive and negative answers.

In the early decades of the nineteenth century, the Linnaean system began to crumble in England as well. A long line of taxonomists appeared opposed to the artificial system, and they were to argue that the lower plants were asexual, spore-bearing forms. According to Reynolds Green, it was Robert Brown who "quietly and unobtrusively sapped the foundation of the Linnaean system."[22] Brown's *Prodromus Florae Novae Hollandiae* of 1810 was the first natural system published in England. Sent out to New Holland by Banks in 1801, Brown had collected nearly 4,000 species—three-quarters of which were new to science—before returning to England and being appointed secretary of the Linnaean Society.

Despite Brown's influence, however, the greatest reform in the classification of the English flora was due to the work of John Lindley. Initially an assistant to Robert Brown, Lindley became professor of botany at the newly established University College of London in 1829. In that year he published *Synopsis of British Flora,* the first attempt to classify England's native flora according to a natural method. The Linnaean system, Lindley wrote, "certainly does not now tend to the advancement of science, or to an accurate knowledge of things themselves." Such a system, he added, "skims only the surface of things and leaves the student in the fancied possession of a sort of information which it is easy enough to obtain, but which is of little value when acquired." A natural system, on the other hand, "requires a minute investigation of every part and every property known to exist in plants."[23]

Lindley divided the plant kingdom into two classes, the Cellulares and the Vasculares. The former, synonymous with Linnaeus's cryptogams, were, according to Lindley, typified by the absence of vascular tissue, cotyledons, and sex organs. A year later he briefly presented the criteria on which his system was based. Physiological characters—"that is to say [those] which depend upon differences of internal anatomical structure"—were of prime importance, he noted.[24]

Throughout the nineteenth century Lindley continued to argue for the asexual nature of the lower plants. In his magnum opus, *The Vegetable Kingdom,* first published in 1845, he subdivided his cellulares, or asexual flowerless plants, into the Thallogens and Acrogens, the former being without, and the latter with, stems and leaves. "It is true," he wrote, "that such names as Antheridia and Pistillidia are met with in the writings of Cryptogamic Botanists . . . but these are theoretical expressions and unconnected with any proof of the parts to which they are applied performing the office of anthers and pistils."[25] In the mosses, ferns, and their allies, he noted, "sexes are wholly missing; that is to say, nothing can be found which resembles the anthers and pistils of flowering plants. There is no evidence to show that any one order of Acrogens possesses organs which require to be fertilized the one by the other to effect the generation of seeds. Hence, these reproductive bodies of Acrogens which are analogous to seeds are called spores."[26]

By the second quarter of the nineteenth century most European and North American botanists had rejected the Linnaean system and no longer believed that all plants bore sex organs. Mosses and ferns, they concluded, did not bear anthers and pistils. Instead, they reproduced by means of spores liberated from sporangia lying inside the moss capsule and on the underside of fern leaves. As William J. Hooker, professor of botany at Glasgow University, remarked in his *Flora Scotia* (1821), "There are no sexes as in the phanerogamous plants and, consequently, no true seed, which can only be produced through their co-operation."[27]

THE NATURE OF SEXUAL REPRODUCTION

What was the nature of the sexual cooperation to which Hooker referred? Indeed, was any such cooperation obligatory? Like the problem of the distribution of sex, these questions were subject to various and conflicting answers.

By the end of the seventeenth century, William Harvey's famous concept—that "all embryos are procreated from some primordium or conception," and that while an egg is "a conception exposed beyond the body of the parent, a conception is an egg remaining within the body of the parent until the foetus has acquired the requisite perfection"—had gained general acceptance. By 1679 it could be stated in the pages of the prestigious *Journal des Sçavans* that "the view of the formation of man as well as all the other animals by means of eggs is so common at present that there are almost no philosophers who do not admit it today."[28] According to Regner de Graaf and others, these eggs arose in the "female testes" or ovaries. But these eggs were not cellular ova in our sense of the word; they were, essentially, chicken eggs without a shell. "We shall call these vesicles *ova*," wrote de Graaf in 1672, "on account of the exact similitude which they exhibit to the eggs contained in the ovaries of birds."[29] Likewise, sexual plants were assumed to arise from seeds. But the nature of these eggs and seeds and the role played by the male semen and pollen became subjects of controversy.

By the end of the seventeenth century the mechanical philosophy had triumphed over its rival philosophies. According to the Cartesian proponents of this philosophy, all psychic, "occult," and vital characteristics must be excluded from natural science. The physical and biological worlds must be understood by reference only to inert matter and motion. Mechanism, it could be said in 1693, had become "a constant maxim among modern philosophers."[30] But while it made sense to compare muscles and bones to pulleys and levers, lungs to bellows, blood vessels to hydraulic pipes, and the heart to a pump, it made little sense to describe the process of development in these terms. It seemed inconceivable that a complex and highly organized living being could ever be produced solely by inert matter in motion. "All the laws of motion which are as yet discovered," wrote George Garden in 1691, "can give but a lame account of the forming of a plant or animal. We see how wretchedly Descartes came off."[31]

Since Cartesian mechanists denied the validity of any explanation of development which incorporated vital and directive forces, they came almost inevitably to the doctrine of germ preexistence. The germ, they argued, must contain within it a fully formed and miniature embryo that simply grew or "evolved" during development. Such preformed embryos, it was further assumed, could not be a product of the parent, but must have been created by God at the beginning. Every germ of every individual that did exist, exists now, and will exist in the future was created by God in the act of creation, and every germ has preexisted since that miraculous day. By that time, naturalists believed that all organisms developed out of the universally occurring egg, and thus most of them assumed that the egg was the site of the preexisting embryo. They were, in other words, "ovists." Using this argument, they were able to "save the phenomenon" by ascribing the mysteries of development to the mere mechanical unfolding and growth of preexisting parts within the egg.

Although the actual originator of the preexistence theory cannot be named with certainty, the most precise formulation of the theory was devised by Nicolas Malebranche in 1673. From the premise that "the diminutive part of matter which is hidden from our eyes is capable of containing a world in which may be hid as many things as appear in this great world in which we live," and from the premise that "we have evident and mathematical demonstration of the divisibility of matter *in infinitum,* and that is enough to persuade us there may be animals, still less and less than others *in infinitum,*" Malebranche presented his preexistence theory of generation both for plants and for animals:

> We may with some sort of certainty affirm, that all trees lie in miniature in the cicatride of their seed. Nor does it seem less reasonable to think that there are infinite trees concealed in a single cicatride since it not only contains the future tree whereof it is the seed, but also abundance of other seeds, which may all include in them new trees still, and new seeds of trees... and thus *in infinitum*....
>
> We ought to think, that all the bodies of men and of beasts, which shall be born or produced till the end of the world, were possibly created from the beginning of it. I would say, that the females of the original creatures were for ought we know, created together, with all those of the same species which have been, or shall be, begotten or procreated whilst the world stands.[32]

These arguments made rather obscure the role of the pollen and the "frothy white liquid" emitted by the male animal. Ovists denied any material role for either, assuming instead that both contained a "germinative spirit" that stimulated the egg to develop.

Hieronymous Fabricius and William Harvey could find no trace of semen in the uterus after copulation and concluded thereby that a nonmaterial seminal spirit was involved. "The woman after contact with the spermatic fluid *in coitu,*" Harvey wrote, "seems to receive influence, and to become fecundated without the cooperation of any sensible corporeal agent, in the same way as iron touched by the magnet is endowed with its powers and can attract other iron to itself." Fabricius spoke of an "irradiant spiritous substance,"[33] while Jan Swammerdam alluded to "the fertilizing aura of the male seed."[34] Such views were perfectly compatible, of course, with ovist theories of preexistence.

In November 1677, Antoni van Leeuwenhoek reported in a letter to Nehemiah Grew, secretary to the Royal Society, that the semen of a healthy male "immediately after ejaculation" contained a large number of small animalcules, a million of which "would not equal in size a large grain of sand" (Fig. 1.7). Their bodies, he reported, were rounded in front and pointed behind, carrying a long thin tail like a "small earth-nut with a long tail."[35] Over the next few years, urged on by Grew, Leeuwenhoek investigated the nature and characteristics of these animalcules and speculated on their role in reproduction.

That they were truly animals, analogous to infusorians, was intuitively obvious to Leeuwenhoek. Not only were they "composed of such a multitude

FIGURE 1.6.   Antoni van Leeuwenhoek. Reproduced by permission of the Museum of Comparative Zoology, Harvard University.

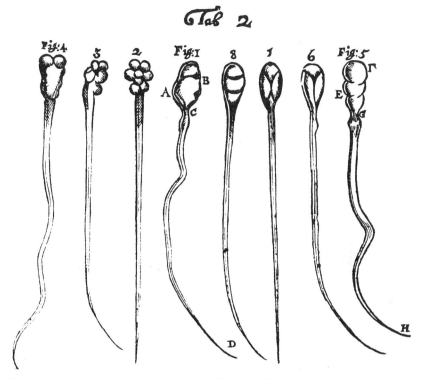

FIGURE 1.7. Leeuwenhoek's drawings of spermatic animalcules. From Antoni Leeuwenhoek, "The observations of Mr. Antoni Leeuwenhoek, on animalcules engendered in the semen," *Phil. Trans. Roy. Soc.*, 1679.

of parts as compose our bodies," but of necessity they had to arise from fertilized eggs.[36] In 1680 Leeuwenhoek described the appearance of "irregular particles" in a rat testis, and assumed that these were immature animalcules arising from eggs. But whence did the eggs arise? Initially, he believed that they were disseminated through the air, but in a letter to Robert Hooke he postulated that the eggs of the animalcules might be transmitted via the father's semen:

> Shall we imagine that the seed of these animalcules is already existent, even during the moment of conception, and that this semen keeps lying in the man's testicles till he has attained the age of fourteen, fifteen, or sixteen years, and that the animalcules do not come to life or are full grown till that time, and that then there is a possibility of generation?[37]

Given Leeuwenhoek's acceptance of this explanation of the source of animalcule eggs, it is not surprising to find him eventually subscribing to a spermist view of preexistence. He had already criticized de Graaf's discovery of the mammalian egg within the ovary, and in doing so had written: "It is exclusively the male semen that forms the foetus and . . . all that the women may contribute only serves to receive the semen and feed it."[38] Five years later he modified this claim by arguing that a human "originates not from an egg but from an animalcule that is found in the male semen."[39]

It is, to Nicolas Hartsoeker, however, that we owe the most explicit state-ment in support of sperm preexistence. Discovering the animalcules indepen-dently in March 1678, he argued that they were joined to the egg by an umbilical vessel in the tail and contained an infinite number of other animal-cules, so that the first male of each species "had been created with all those of the same species that have been engendered and will be engendered until the end of time." Similarly, in plants the pollen grains contained the rudiments of the entire plant, in which also "there are new seeds which contain new germs, and these new germs contain new plants with their seeds, and thus to infin-ity."[40] We owe to Hartsoeker also that most extraordinary claim that the sperm actually contained a fully formed miniature adult coiled up within—the famous homunculus (Fig. 1.8). It should be pointed out, however, that a belief in spermism did not also imply a belief in the existence of a physical homunculus; the spermists argued for the preexistence of parts, not necessar-ily the complete organism.

"Spermism," or "animalculism," and its botanical equivalent, "pol-lenism," gained some support during the early years of the eighteenth cen-tury. It was adhered to by the physician Nicholas Andry de Bois-Regard, by William Derham in his text on natural theology, and by a contributor to the *Lexicon Technicum*.[41] Samuel Moreland argued in 1702 that the pollen grain, or "farina" as it was called, "is a congeries of seminal plants, one of which must be convey'd into every ovum before it can become prolifick,"[42] a view to which the French apothecary Claude Geoffroy subscribed in 1711. As Geoffroy wrote, either the pollen reaches the young fruit and excites "a fermentation capable of developing the young plant enclosed in the embryo of the seed," or the pollen itself contains "the first germ of the plant."[43] Geof-froy supported the latter view, and whereas support for spermism faded quite rapidly, pollenism continued to attract some support. Much of it came from the anti-Linnaeans, who tended to attack both Linnaeus's theoretical beliefs and his classificatory scheme. John Hill, for example, argued that the plant embryo is contained in the farina and that "nature provides for the reception of the embryo, thus formed in the Farina, a peculiar case, the seed, which receives it naked, and gives [it] the first means of limited growth."[44] This doctrine was not easily dismissed. At the end of the century Johann Hedwig still supported the pollenist theory.

In the learned world, animalculism never enjoyed the popularity that ovism did. In an age dominated by theories of preexistence and the outlook that nature appeared to do nothing without a purpose, this was not surprising. Spermism implied an enormous wastage, for only a minute fraction of the

FIGURE 1.8. The preexistence of sperm: Nicolas Hartsoeker's homunculus. From Nicolas Hartsoeker, *Essai de dioprique* (Paris, 1694).

little homunculi found an egg in which to develop. The same criticisms applied even more to plants, where the pollen was often scattered to the four winds. William Smellie of Edinburgh made much of this in his argument that pollen had no role whatsoever in reproduction. Accidental causes were, he wrote, "repugnant to sound philosophy"; the spreading of pollen, if, indeed, it played a role in generation, would introduce "universal anarchy." "Instead of a regular succession of marked species, the Earth would be covered with monstrous productions, which no botanist could either recognize or unravel."[45] Usually assuming that the sperm were accidental parasites of the testis, ovists restricted the stimulating role in fecundation to the seminal fluid.

How the sperm found their way into the testis was a mystery to them. Presumably they produced eggs, as did all organisms, but did the eggs pass out with the urine to be picked up again in food and drink or even from the air? Or were the eggs of the sperm parasites passed from father to embryonic son during copulation? The problem was not an easy one to solve.

Animalculism seems to have gained some popular support, however. The widely read "Aristotle series," a compendium of information drawn from folklore, myths, ancient medicine, and Greek philosophy, and a source of much sexual information throughout the late seventeenth and the eighteenth centuries, clearly adhered to it. *Aristotle's Compleat and Experienc'd Midwife* drew the oft-quoted analogy between the male "seed" and the seeds of plants. "For as the seed of plants can produce no fruits, nor spring unless sown in ground proper to waxen and excite their vegetative virtue, so likewise the seed of man, though potentially containing all the parts of a child, would never produce so admirable an effect, if it were not cast in to the fruitful field of Nature, the womb."[46] Man, quite literally, "sowed his oats." Laurence Sterne, writing of the birth of Tristram Shandy, made reference to animalculism in this delightful satire:

> The HOMUNCULUS, Sir, in however low and ludicrous a light he may appear, in this age of levity, to the eye of folly or prejudice:—to the eye of reason in scientifick research, he stands confess'd—a Being guarded and circumscribed with rights:—The minutest philosophers, who, by the bye, have the most enlarged understandings, (their souls being inversely as their enquiries), shew us incontestably, that the HOMUNCULUS is created by the same hand,—engender'd in the same course of nature,—endow'd with the same locomotive powers and faculties with us:—That he consists, as we do, of skin, hair, fat, flesh, veins, arteries, ligaments, nerves, cartilages, bones, marrow, brains, glands, genitals, humours, and articulations;—is a Being of as much activity,—and, in all senses of the word, as much and as truly our fellow-creature as my Lord Chancellor of England.—He may be benefited, he may be injured,—he may obtain redress;—in a word, he has all the claims and rights of humanity, which *Tully, Puffendorf,* or the best ethick writers allow to arise out of that state and relation.
>
> Now, dear Sir, what if any accident had befallen him in his way alone?—or that, through terror of it, natural to so young a traveller, my little Gentleman had got to his journey's end miserably spent;—his muscular strength and virility worn down to a thread:—his own animal spirits ruffled beyond description,—and that in this sad disorder'd state of nerves, he had lain down a prey to sudden starts, or a series of melancholy dreams and fancies for nine long, long months together.—I tremble to think what a foundation had been laid for a thousand weaknesses both of body and mind, which no skill of the physician or the philosopher could ever afterwards have set thoroughly to rights.

Alas, Tristram's uncle exclaimed, "my Tristram's misfortunes began nine months before ever he came into the world."[47]

Thus, by the early eighteenth century, although they shared the belief that development involved the simple mechanical unfolding of preformed parts, the naturalists were divided over the question of the site of the preformed embryo. Most were "ovists" and assumed that the embryo was in the egg, while others assumed it was situated in the sperm. But a few, like Linnaeus, denied that the complete embryo existed in either the egg or the sperm.

OPPOSITION TO OVISM AND ANIMALCULISM

Linnaeus was utterly opposed to any form of preexistence theory and the implications that the products of one sex organ could occasionally develop without the influence of the other. "That the offspring proceeds not from the egg alone, nor from the male sperm alone," in plants as well as animals, appeared undeniable to Linnaeus, given "the consideration of mules, the reason of the thing, and the structure of the parts."[48] Both male and female sex organs were a universal property of all plants and all animals, according to Linnaeus, and, therefore, both must always be essential for procreation to take place. The Creator does nothing in vain.

> That vegetable generation is performed by the falling of the dust of the *antherae* upon the moist *stigmata,* where the particles burst and shed their seminal virtue, which is absorbed by the moisture of the *stigmata,* is confirmed by our sight, by their proportion, place, time, rains, culture of palm trees, nodding, sunk, and syngenesious flowers; nay, by the genuine consideration of all sorts of flowers.[49]

But Linnaeus was at a loss to explain how this fecundation occurred. The basis for his opposition to preexistence theories was simply that experience led him to no other conclusion than that both sexes contributed to the next generation. He quoted the case of the mule, crosses between two breeds of dogs, and the case of a certain Negro who "got a wench with child" and was "delivered of a boy, who was in colour altogether like the mother, except the *penis,* which was black, a sufficient indication who the father was."[50]

In 1759 the Imperial Academy of St. Petersburg offered one hundred ducats for the best dissertation on the sexes of plants. Linnaeus's essay, *Disquisitio de sexu Plantarum,* received the prize on September 6, 1760. In this essay, Linnaeus offered a few experiments in support of his claim that both sexes contributed materially to the next generation. He grew some hemp (*Cannabis*), for example, and noted that if the male plants were removed, the seed buds failed to develop, remaining "brown, compressed, membranaceous, and dry, not exhibiting any appearance of cotyledons or pulp."[51] He also discussed the production of "hybrids or mule vegetables," claiming that the hybrid's leaves resembled those of the father, while the internal parts, or medulla, resembled the mother.[52]

Linnaeus, following the writings of Theophrastus and Cesalpino, believed that plants consisted of two different substances: an inner medulla, in which the essence of the plant resided; and an outer cortex, consisting of epidermis, bark, and wood. Likewise, the flower consisted of the two elements: the pistils and ovaries, being made up of the medullary substance; and the stamens and pollen, comprising the cortex. Since "the medulla exists naked in the germen, it cannot support itself, or make any further progress, without the assistance of the cortical substance which it has left; it must therefore receive this assistance by some means or other, and in fact, does receive it from the stamina and their pollen."[53] Thus, on theoretical grounds, fecundation involved the material interaction of both medullary and cortical substances, of both ovary and pollen. The offspring from the sexual act therefore contained medullary substances from the female parent and cortical substances from the male parent; the offspring did not exist totally preformed in either the pollen or the ovary.

Linnaeus was not alone in opposing preformation theories. In the mid eighteenth century, Cartesianism collapsed in France as Newtonianism became increasingly popular. Newtonianism, unlike Cartesianism, denied that matter was entirely passive. Forces existed in nature which bestowed on matter a dynamic attribute that was inconceivable to the Cartesian mechanists. These new and dynamic Newtonian concepts were first formulated in France by Pierre-Louis de Maupertuis and Georges-Louis de Buffon.

It remains puzzling, wrote Maupertuis, how theories of preexistence "make natural science more clearly understood than if new productions were admitted."[54] Converted to Newtonianism during a visit to London in 1728, Maupertuis argued that the fetus must arise from the mixture of two seminal fluids. According to his thinking, no preformation theory could explain how parental characters are passed to the offspring, how monsters sometimes arise, and how offspring are sometimes malformed as a result of a mother's imagination. The mechanism may well be obscure, but as he wrote with reference to Newtonian attractive forces:

> Why should not a cohesive force, if it exists in Nature, have a role in the formation of animal bodies? If there are, in each of the seminal seeds, particles predetermined to form the heart, the head, the entrails, the arms and the legs, if these particular particles had a special attraction for those which are to be their immediate neighbours in the animal body, this would lead to the formation of the fetus.[55]

Buffon, writing in the middle of the eighteenth century, could not accept preformation theories either, and rejected "from philosophy all opinion which leads necessarily to the idea of the actual existence of geometric or arithmetic infinity." Buffon's view of nature was essentially dynamic, incorporating a constant cycling of "organic molecules" that were released from an organism at death and re-formed again into a new organism. All beings, he wrote, are formed by the grouping of these organic molecules; "reproduction and generation are therefore nothing but a change of form which comes about simply by the addition of these similar parts."[56]

Form was imprinted on this huge and indestructible mass of organic parts by the "moule interieur," which Buffon likened to the gravitational force. It was this "mold" that controlled generation, development, and growth. In early life the organic particles were absorbed and employed in augmenting the different parts of the body; the force of the "moule interieur" assured that each part of the body would receive only those molecules it could use. After puberty, when growth stopped, the excess organic particles were passed to reservoirs—the testes and the ovaries. Therefore, the spermatic animalcules were not living animals at all, but were "the first gathering of these organic molecules. . . . They are perhaps even the organic particles which constitute the organized bodies of animals."[57] For an embryo to form, the organic particles of both seminal fluids had to mingle.

Buffon examined microscopically the seminal fluids from numerous animals and the seeds from plants. However, because he allowed this material to stand for a few days in corked vials containing water, contamination took place. As a result, his descriptions of sperm "metamorphosis," to which the sperm's lack of animality was attributed, were in reality descriptions of numerous protozoan contaminants within the water.[58]

Linnaeus's interpretation of the spermatic animalcules was very similar to Buffon's. Linnaeus first observed them in 1737 and in a later publication concluded that they "are not in the least animalcules which enjoy voluntary motion, but inert corpuscles which the innate heat keeps afloat, just like oily particles."[59]

The observations, experiments, and theoretical postulates of Buffon and Linnaeus were effectively challenged in the latter part of the eighteenth century by three prestigious naturalists: Lazzaro Spallanzani in Italy, Charles Bonnet in France, and Albrecht von Haller in Switzerland, all of whom came to subscribe to a form of ovist preexistence.

LATE-EIGHTEENTH-CENTURY OVISM

Of the three naturalists mentioned above, Spallanzani had the greatest impact on the reproduction debate. In particular, his mastery of experimental techniques was justifiably admired in France well into the nineteenth century. Spallanzani was an ovist. "It is very natural to believe," he wrote in reference to the amphibian egg, "that these different orders of fetuses, which appear every year in the ovaries, are not formed successively, but coexist from the beginning with the female, and that they are developed and only then rendered visible." Such eggs were, to Spallanzani, "tadpoles concentrated and folded up."[60] Thus, Spallanzani was concerned with the nature of the spermatic animalcules and particularly with showing that they were parasites that had no role in reproduction.

Spallanzani first attempted to prove the animality of the spermatic animalcules by comparing their behavior with recognized microscopic animals, the infusorians. After investigating the infusorians, he turned to the animalcules in semen. Contrary to Buffon, he noted that no change of form took place and that their tails were never lost. Thus, he realized, Buffon's description of such changes was due to infusorian contamination in the vials of decaying semen. Finally, after showing that the behavior of animalcules when exposed to changing temperatures and various chemicals was identical to that of the infusorians, he concluded: "We have a complex of proof both convincing and decisive, favoring the recognition of a true and rigorous animality for the spermatic animals."[61] The spermatic animals were clearly parasites, and as he proved on at least three occasions, were not necessary for fecundation. Seminal fluid devoid of animalcules or containing dead animalcules was capable of initiating the development of frog eggs.

By 1777 Spallanzani was no longer sure, as he seems to have been previously, that fecundation involved an *aura seminalis*. To investigate this premise experimentally, he placed toad semen in a watchglass and covered it with another watchglass, to the underside of which toad eggs were attached by means of mucus. Development did not take place until the eggs were allowed to fall into the semen. "These various facts," wrote Spallanzani, "concur to prove that fecundation in the fetid toad is not the effect of the *aura seminalis, but of the sensible part of the seed.*"[62] Later, in an addendum attached to the French translation of his *Dissertazione*, published in 1785, he reported how the capacity of semen to fertilize gradually diminished as the semen was passed successively through several layers of filter paper. In addition, he noted, "if the paper containing the residuum of the freshly filtered spermatified water is then pressed into pure water and applied to nonfertilized eggs, these will develop very well; this proves that the filtration removes from the spermatic water its fertilizing power."[63] Having proved earlier that samples of semen devoid of any animalcules were capable of initiating development of the tadpole, he necessarily concluded that the thick liquid retained on the filter paper, and not the spermatic animalcules, was responsible for initiating development. He had already concluded that "the spermatic fluid is the stimulating fluid, which, by penetrating the heart of the tadpole, excites more frequent and stronger pulsations and gives rise to a very tangible augmentation of parts and to that life which follows fecundation."[64] The spermatic animalcules were parasites of the semen which "are born, are nourished, and propagate in us and in animals, and which pass from generation to generation with sucking from the mother, in the uterus, or by means of the milk."[65]

In 1780 Spallanzani turned his considerable talents to the plants. Examining the "seeds" before and after fecundation, he noted that before the event, they were composed of a homogeneous, spongy, jellylike substance, while afterward their "lobes and plantules" appeared. Was this embryo transmitted by the pollen to the seed? The pollen, he reported, was filled with a globular liquid, and since no evidence of the embryo could be seen in them, he concluded that "the embryos are not concealed in the dust of the flowers."[66] But such a claim required proof, proof that came by denying the pollen access to the seeds. Removing immature anthers from eighty-four flowers of the genus *Ocymum,* he noted that in twenty-five of them the seeds appeared normal—that is, the lobes and plantules appeared—but in none of those tested did the seeds germinate. Thus, he concluded, although the dust of the flower is essential for complete development to occur, the embryo appears without its influence, and so the dust cannot be the carrier of the embryo.[67] In other genera, however—*Cucurbita, Cannabis,* etc.—the seeds from plants whose male organs had been removed not only appeared normal but actually developed. "Neither the embryo of this plant nor its fructification depend upon the dust of the stamens," Spallanzani concluded.[68] Rejecting as absurd any claim that such embryos, contained exclusively in the seed, could be organized out of homogeneous matter "by mechanical laws alone," he naturally concluded that ovist preexistence was as much proven for plants as for animals, and thus "a new and very luminous analogy exists between them."[69]

Spallanzani was unimpressed by Linnaeus's claims. Pouring scorn on "System makers," who insist that nature always and everywhere must abide by the same set of rules, and on the theory that pollen must *always* fecundate the egg, he clearly labeled Linnaeus as unqualified to deal with such matters. Linnaeus and his followers, he argued, were mere "nomenclators," "inventors of phrases," ignorant of, or indifferent to, the inner workings of plants. Calling for a change of approach to botany, Spallanzani argued that "natural bodies are not simple beings, but very complex. One can compare them to a clock consisting of many wheels enclosed in a case that conceals their size, their reciprocal action, and the influence of their springs." Thus, to gain the right concepts, it is necessary "to view the internal parts, to remove them from their box."[70]

Not surprisingly, the French school of anti-Linnaean botanists, as part of their general attack on the Linnaean system, also interpreted the sexual act in terms of a preformed embryo. But whether they also believed in its extreme version, preexistence, is not clear. Their theoretical views on fecundation were derived, in part at any rate, from the readily observed fact that a seed did contain within it the rudiments of a stem, root, and, above all, the leaves (i.e., the cotyledons). The question was whether these preformed parts appeared only after the entry of the pollen or whether they existed prior to that in the embryo sac.

Joseph Tournefort, predecessor of Antoine de Jussieu at the Jardin de Roi, whose text of 1694 was republished in 1797, appears to have been a full-fledged ovist. The germs of the plant and the eggs of the animal "seem to be only embryos," he noted, "where all the parts are contained in space so small that our imagination cannot conceive of it."[71] Likewise Adanson, in the middle of the eighteenth century, stated that "the embryo exists completely formed in the seed of plants which have not been fecundated, in the same manner that the fetus exists totally formed in the eggs of the frog and in those of the hen before fecundation, according to the observations of Malphighi, Haller, etc." Fecundation, Adanson believed, operated by a vapor, "une espèce d'esprit-volatil," which descended to the seed after the pollen dust had mixed with the fluid of the stigma, there to provide the first impetus for the seed to unfold or evolve.[72] Identical views were subscribed to by Antoine-Laurent de Jussieu, while Augustin-Pyramus de Candolle spoke of the pollen carrying "le liquide fecondateur," which fecundated the new being, "lui imprime le mouvement vital."[73]

The belief that the seed contained within it a preformed embryo led to the theory that spores and seeds were fundamentally different. Achille Richard, for example, who believed all cryptogamic plants to be asexual, noted that "a reproductive corpuscle of a fern or a mushroom, etc., if placed on the ground, will develop there, but it will not be, as in the embryo of a phanerogam, parts already formed, only reduced in some way to their rudimentary state."[74] In 1843, just before a series of fundamental discoveries on plant life cycles were reported (these will be discussed in Chapter 2), Adrien de Jussieu noted that although spores were analogous to seeds in that they developed into complete plants, they were quite different in that they were constructed of only a membrane surrounding a liquid. Similarly, he noted that "the reproductive organs of cryptogams differ from those of phanerogams, and that one recognizes ovaries in the former only if one defines it [the ovary] as any cavity enclosing bodies susceptible of developing into a plant similar to that which gives them birth: a definition so general that one will be forced to rank under it a certain number of different parts without any true relationships one with the other."[75]

It is worth noting in passing that the results of Spallanzani's experiments were used by those few naturalists who believed sex to be entirely absent in plants: Franz Schelver and his pupil, August Henschel, in Germany; and William Smellie in Scotland. Spallanzani, being an ovist, believed, and even proved experimentally, that the action of pollen was not *always* necessary. Smellie, on the other hand, claimed that Spallanzani's experiments, his work on aphids, and Trembley's discoveries regarding asexual budding in hydra indicated that some animals multiplied without sex. If that is true of animals, Smellie noted, "what should induce us to fancy that the oak or mushroom enjoy these distinguished privileges?" Spallanzani's experiments, he wrote, "have annihilated this beautiful Fabrick"—that is, the belief that sex is universal.[76]

Smellie's paper, published in the first edition of the *Encyclopaedia Britannica,* which he edited, was basically an attack on the Linnaean system: sex is not universal, plants are devoid of sex, and pollen grains do not play a role in generation. In a curious foretaste of things to come, Smellie noted an experiment by Dr. Hope, professor of botany at Edinburgh, who, crossing red-flowered male *Lychnis* with white-flowered females, obtained seeds that all germinated into red-flowered plants. Hope, a Linnaean, naturally assumed that this red color arose from the influence of the male. Smellie, on the other hand, argued that if indeed the pollen had any role to play in this regard, the offspring should have been a mixture of red and white. "To whatever cause, therefore, this change may be attributed," he concluded, "it can never be ascribed to anything analogous to generation." [77]

I have shown that much of the debate in the eighteenth century over the distribution of sexual reproduction and the role played by pollen in fecundation was intimately related to the clash between Linnaean and anti-Linnaean botanists. I have also shown that similar puzzles surrounded the nature of fecundation in animals. On one hand, Linnaeus and his followers, Buffon, Maupertuis, and others argued that both seminal fluid and pollen must contribute material to the forthcoming generation. On the other hand, the "ovists"—many of whom were anti-Linnaeans—argued that all the material for the new generation was contained in the egg; the male seminal fluid and pollen material was only a stimulant. Neither group saw any role for the spermatic animalcules; indeed, Buffon and Linnaeus denied that they were animate. However, mainly because of the work of Spallanzani, whose experimental expertise was so much admired, the French naturalists in particular came to believe that the twin semen theories of fecundation had been thoroughly undermined, and, thus, by the early nineteenth century, they had accepted his position: the parts were preformed in the egg; the male semen and pollen stimulated the egg to develop; and the spermatic animalcules, accidental parasites of the testes, had no role in generation. It should be pointed out that only the animalculists, whose theories were no longer viable by the end of the century, believed that the spermatic animalcules played any role in the reproductive processes.

In addition, French naturalists of the eighteenth century presented a thoroughly mechanistic interpretation of fecundation and development. Ovists spoke of the unfolding, or *l'évolution,* of preformed parts; Maupertuis spoke of predetermined particles, and Buffon spoke of organic molecules forming a complete new organism immediately after mingling with other organic particles at the time of fecundation. Such mechanistic views were not shared by late eighteenth and early nineteenth century German naturalists.

THE GERMAN APPROACH

The mechanistic philosophy never gained much favor among German naturalists. Through their influence the doctrines of preexistence, preformation, and *l'évolution* were replaced in the nineteenth century by epigenetic theories of development. Epigenesis teaches that development is not the mere mechanical unfolding of preformed parts, but is a process of simultaneous growth and differentiation from a completely homogeneous beginning. As Caspar Wolff, the foremost epigenesist of the eighteenth century, remarked in 1759, "Those who teach the system of predelineation do not explain generation but deny that it occurs."[78] Karl Ernst von Baer's *Entwicklungsgeschichte der Thiere,* published in 1828, provided observational data to substantiate this belief in epigenesis. In the first scholium of that work, von Baer examined the question "whether or not the whole embryo can be present with all its parts, but built so fine that the measurer and the microscope cannot reach it." Observation shows, he argued, that this is not the case—the younger the embryo, the less fine its features. A functioning muscle, for example, is made up of a series of very fine threads; in the developing muscle, however, these threads appear quite thick, and when they first appear they look more like unformed globules. In other words, he concluded, "the single organic elements, from which [the organs] arise, are more finely perfected the more developed the animal." Thus, since the early stages of development are characterized by the presence of large, formless globules, "still less can the whole embryo of the chick hide itself through being minute."[79] Long before the publication of von Baer's work, however, the doctrine of preformation had been rejected among German naturalists. As Lorenz Oken wrote in 1810, "The theory of preformation contradicts the laws of Nature's development."[80]

The theory of epigenesis implies also that each embryo is a totally new formation, originating afresh at each fecundation. Thus, instead of interpreting fecundation in terms of the stimulus of preformed parts, epigenesists tended to view fecundation in essentially chemical terms: two fluids come together, react, and produce a new being. The epigenesists, like the mechanical ovists, concluded that the spermatic animalcules played no role in the process; fecundation involved the activity of the seminal fluid.

Such chemical and dynamic interpretations of fecundation were fairly common among German naturalists, even in the eighteenth century. For example, Joseph Koelreuter, professor of natural history at Karlsruhe from 1764 until his death in 1806, described pollen grains as a collection of organic particles. Upon ripening, these particles liquify, and by the contraction of the hard shell surrounding the pollen, the liquid is squeezed out through "excretion canals." This "male fertilizing principle," he argued in 1761, comes together with the oily secretion of the stigma and passes to the ovaries. Koelreuter believed that the mixing of two fluid secretions was essential for fecundation to occur, but, unlike the French, he interpreted the reaction between the fluids in chemical terms. "From the union and commingling of these two materials," he wrote, "there arises another [material] of an intermediate sort," the actual combination of these two secretions being similar to "the union of an acid and an alkaline substance" from which "a third or intermediate salt originates."[81]

Similarly, Kurt Sprengel denied any kind of preexistence and even any contact between pollen and ovary, "by which supposition," he wrote, "we should explain the germination of the seed entirely on anatomical principles." Fructification, he urged, is a dynamic operation: "The pollen is brought into contact with the stigma and awakens in it, as well as in the germs, a new life."[82]

Again, the nature of this interaction was viewed in chemical terms. Fecundation was effected by the combination of hydrogen and carbon, Sprengel wrote, because "carbon and hydrogen are the essential component principles of the fecundating fluids of all vegetables."[83] From such a chemical viewpoint, Sprengel denied that sexual reproduction necessarily demands the existence of sex organs. "The preconceived opinion, that in all cryptogams, as in phanerogams, there must necessarily be found two distinct organs of fructification," he wrote, "imposes upon Nature a law, which human reason has established upon analogy only, without proof of a satisfactory induction." The vessels of ferns and the inner structure of the moss leaf are in themselves adequate places for the plant juices to be freed from oxygen, he argued, and thus "suffer the carbon and hydrogen, the chemical agents in the first formulation of the fruit, to enter into combination for this purpose."[84]

Historians of biology have long recorded that the mechanistic philosophy collapsed at the end of the eighteenth century and that theories of preexistence were replaced by more dynamic and vitalistic theories of epigenesis. But many French naturalists continued to believe in a mechanical, ovist preformation well into the nineteenth century. This suggests that the early nineteenth century move toward epigenetic theories of development and a more chemical interpretation of fecundation did not reflect a change in the biological paradigm. Rather, it reflected a change in the locus of scientific practitioners. After the Restoration in France, the center of scientific activity gradually shifted away from Paris, crossed the Rhine, and took root in the newly revitalized German universities.

"There is no people," wrote James Bryce in 1885, "which has given so much thought and pains to the development of its university system as the Germans have done—none which has profited so much by the services universities render—none where they play so large a part in the national life."[85] In the early years of the nineteenth century a group of German intellectuals, led by Wilhelm von Humboldt, initiated a reform movement directed toward the spiritual and philosophical rejuvenation of German culture. One of the foci of this movement was the German university, which throughout the eighteenth century had suffered from poor finances and declining enrollments. As a result of these reforms, the universities not only grew in prestige but came to possess a unique set of characteristics, quite unlike those of other Western countries.[86]

The goal of the German system, as set forth by the reformers, differed markedly from that of the university systems of France and Britain. Instead of defining the university as a pedagogical institution where existing knowledge is imparted to a group of potential lawyers, physicians, clergymen, and civil servants, the German reformers stressed the cultivation of human individuality. Within the primary university faculty—that of philosophy—students were encouraged to think independently and to undertake independent research. "However diligently he attends lectures, studies his textbooks, and however brilliant his examination results," one educator wrote, "we should say that he lacks something—nay, more, that he lacks the most important thing, the trial of his strength in independent research."

In this essentially antiutilitarian mold, the German student was necessarily led to cultivate pure learning. The natural sciences, having previously been taught only as auxiliaries to medicine and technology, were moved into the philosophy faculty, where they rested on their own intrinsic value.

Initially, such a philosophy of learning opposed any form of specialization. Instead, students were to be imbued with *Wissenschaftideologie*—the great organic unity of all knowledge. As the natural sciences gained a stronger and stronger place within the universities, however, and as *Wissenschaftideologie* collapsed, an era of specialization began. Still, the science that grew and was nourished there came to share many of the ideals of the earlier period. The emphasis still lay with research and pure learning. In addition, with the development of research laboratories, an explosion in laboratory-based science took place. As Steven Turner recently remarked, "The laboratory system gave institutional expression to the ethos of learning which had come to dominate the universities—a professional commitment to research and to the elaboration of research methods and a fundamental concern with training students in the techniques of scientific investigation."[87] Thus, the German university was not a school where the student received a set of materials upon which he was later examined; rather, it was a seat of research.

At the same time, the Prussian state gained a virtual monopoly over university appointments. In place of the usual pedagogical virtues by which earlier academics had gained appointments, the state began to demand disciplinary expertise. As one ministerial official remarked, no one would be appointed to a professorship until "he has written a solid book, a work which one can display and reap honour from, a work one can stand on."[88] Private study and research ability, not teaching, became the necessary qualifications for entry into the highly prestigious German academic community. As a result, much of nineteenth-century biology came to be dominated by German, laboratory-trained, professional scientists.

The science these professionals practiced had certain peculiarities. The German university professor, his assistants, and his graduate students, became totally divorced from their naturalist cousins both socially and intellectually. A new style of investigator was born, one who probed beneath the surface of things in an attempt to explore their underlying mechanics. As Thomas Henry Huxley, a later admirer of the new German science, remarked: "There is very little of the genuine naturalist in me. I never collect anything, and species work was a burden to me; what I cared for was the architectural and engineering part of the business."[89] New specialist journals sprang up in vast numbers, and contributors to them spoke only to their fellow professionals and in a language that appeared ever more unintelligible to the amateur naturalist. As David Allen puts it:

> The non-professional elite, accustomed hitherto to reading the raw scientific texts for their straightforward enlightenment, now found it increasingly impossible to keep up. The jargon was too dense and the accounts took too much specialized knowledge for granted in advance. For the first time people found themselves confronted with their own native tongue in a form that might just as well have been Urdu or Swahili.[90]

Natural history, the naming and classification of plants and animals, and all the traditional pursuits of the amateur naturalist became identified with old-fashioned science; they were no longer worth the attention of the consciously professional biologist. Locked in his laboratory, the new professional studied different things, asked different questions, and expected different answers than did his amateur peer. This trend generated a tragic and fundamental split in the biological sciences between the laboratory worker, with his microscope, and the naturalist, with his hand lens—a split that is still with us today.[91]

In the field of reproductive biology, the laboratory professional tended to concern himself with the mechanisms by which the semen initiated egg development. The naturalist, on the other hand, was more prone to be concerned with the mechanisms of inheritance—in particular, with how male characteristics are passed to the offspring. Very few nineteenth-century biologists demanded that both questions be answered.

TWO

The Great Debate
1821–1856

Twice during the nineteenth century the scientific laboratory became the center of debate over the nature of fertilization, or "fecundation" as it was then called: first, between 1821 and 1856, and second, during the final two decades of the century. The fact that both periods were immediately preceded by important technological innovations seems to support those who argue that the use of new techniques alone is sufficient to account for the appearance of new and more accurate theories in biology; that once acquainted with new tools, scientists necessarily come to hold the correct opinions. Adherents of this view of scientific progress would argue, perhaps, that the outcome of the 1821–56 debate over whether or not the sperm and pollen actually penetrate the egg was an inevitable consequence of the increased use of achromatic microscope lenses developed in the 1820s; that those with the requisite skills to operate these new instruments would see at once that the sperm and pollen did actually penetrate the egg and embryo sac respectively.

Clearly, however, the first debate—indeed, both debates—was far more complex than this simplistic interpretation of scientific progress would suggest. Participants on both sides of the controversy used the new microscopes; it is manifestly absurd to believe that those on the "right" side were more skillful than those on the "wrong" side.

More important, it is clear that the acts of sperm penetration and pollen penetration could not be observed readily, if at all, using the techniques that were available during most of the nineteenth century. With luck, in some animals one might see the sperm inside the egg investments lying in contact with the membrane surrounding the egg. But without nuclear stains, which were developed later, in the 1870s, it would have been almost impossible to see the actual entry of sperm nuclei in animals and plants. Those who claimed to have seen the sperm within the egg were probably in error. What they did see, presumably, were sperm lying within the egg investments either directly above or below the egg, and under the microscope, this juxtaposition gave the impression that the sperm were inside the egg itself.

As a result of these and other problems, which will be discussed later, the 1821-56 debate cannot be interpreted as the gradual discovery of "visual truths." Indeed, visual truths alone would more likely have led to the disavowel of sperm entry than to its acceptance. Neither can the debate be understood solely by reference to the new achromatic microscopes, although obviously, without this new tool there might not have been a debate in the first place. The new tool gave those microscopists who had sufficient skills and patience the ability to see the egg, sperm, and pollen grains during the period of fecundation. What these microscopists believed was happening during fecundation, however, involved their interpretation of a set of highly unreliable observations and experiments, and such interpretations reflected the scientists' theoretical presuppositions about a whole range of related phenomena.

In brief, four theoretical assumptions dominated the debate. First, the historical background of eighteenth-century preexistence theories had a profound effect on the debate at its inception. Second, the late 1830s saw the intrusion of the new cellular theory of organization into biological thought, an intrusion which drastically altered the nature of the debate. Third, the period under discussion witnessed the input of new physiological ideas into biological thought, ideas which demanded that chemical and physical principles alone be the basis of all physiological explanations, including that of fecundation. And finally, closely tied to the third assumption, a new breed of German-trained or German-inspired laboratory investigator entered into the debate, and new questions began to take priority over those that had been posed by amateur field naturalists and plant breeders.

Nevertheless, it is clear that the development of achromatic lenses, whose images were less clouded by chromatic aberrations, was the major factor behind the initiation of the Great Debate. By the end of the eighteenth century, microscope manufacturers were aware that spherical and chromatic aberrations were inherent in their products. As George Adams, a Fleet Street optician, remarked in his *Essays on the Microscope* (1787), "[Because of] the different refrangibility of rays of light, which frequently causes such deviation from truth in the appearance of things, . . . many have imagined themselves to have made surprising discoveries, and have communicated them as such to the world; when, in fact, they have been so many optical deceptions owing to the unequal refraction of the rays."[1]

At the same time, these manufacturers realized that one of the "optical deceptions," chromatic aberration, could be minimized by combining lenses that had different properties. Prior to the 1820s, such crown glass–flint glass combinations gave satisfactory images only at low magnifications. During the 1820s, however, high-power achromatic images were obtained when Giovan

Amici and others succeeded in manufacturing combined lenses of very short focal length. Then, with the work of J. J. Lister, the father of Joseph Lister, a theoretical basis for the design of such lenses was developed. According to Savile Bradbury, Lister's famous paper of 1830 "proved to be the turning point in the design of microscopic objectives; indeed, it may be said that as a result of this work the *design* of microscope objectives became possible, whereas before they could only be constructed on an empirical basis." Continental microscopists took the lead in applying the new lenses, using cheap, well-made instruments constructed in Austria, France, and the German states. Among the most prominent of these early pioneers were the teams of scientists working in Johannes Müller's laboratory in Berlin and in Jan Purkinje's Institute of Physiology at Breslau. These new professionals contrasted sharply with the amateur microscopists in England, who preferred the more costly and elaborate English products, which were intended for "the rich and influential amateurs whose view prevailed in the microscopical societies. These gentlemen, who regarded their costly and monumental instruments as precious toys rather than as the tools for work and study, amused themselves by resolving test objects."[2] Matthias Schleiden, too, poured scorn on the English instruments, remarking that the deficiency of microscopical botanical studies in England "can only be attributed to the defectiveness of their instruments."[3] Equally important, of course, was the lack of research laboratories in England and an attitude toward science that downplayed the importance of research activity.

Using the new lenses, Jan Purkinje's laboratory revealed much about the nature of the ovarian egg. In the eighteenth century the egg, of which the chicken egg was the model, was assumed to be akin to the "proper egg" of Aristotle. It contained both a part from which the embryo was engendered and a part that was responsible for supplying this embryo with food. By the nineteenth century, however, no consensus had emerged as to where these parts were located within the egg. Was the chicken engendered in the yolk or in the white, or even, as Fabricius claimed, in the chalazae? Examining the blastodisc (or cicatricula as it was then called), Purkinje noted a "teat-like whitish hillock." Teasing the hillock apart with a needle, he "was filled with no little wonder, as a most exquisite vesicle appeared." He termed it the *vesicula germinativa,* imagining it to be "tenanted by the female germ from which the chick in its turn would develop." A few years later Karl Ernst von Baer, using only a hand lens, discovered "opaque globules" in the fallopian tubes of mammals, and opening up the so-called de Graafian "ovum," saw there these same globules, "so plainly that a blind man could scarcely deny it." Von Baer likened the de Graafian "ovum" (or better, follicle) to the chicken egg, inside of which there was another ovum—the true "foetal ovum" from which the new mammal arose. The latter, von Baer argued, was equivalent to the germinal vesicle of the chicken egg.[4]

THE DEBATE BEGINS

Also using the new Amici lenses, the Frenchmen Jean-Baptiste Dumas, his brother-in-law Adolphe Brongniart, and the Swiss scientist Jean-Louis Prevost initiated the great debate over fecundation by claiming, contrary to popular opinion, that the sperm and pollen particles actually penetrated the egg to contribute materially to the new generation.

Dumas, who was later to gain fame as an organic chemist, had emigrated to Geneva in 1817 when political turmoil erupted in his native town. In Geneva he studied pharmacy and lived in the house of his patron, the pharmacist Le Royer. In addition, he and several of his fellow students were given the opportunity to carry out experiments in the laboratory of Le Royer's pharmaceutical firm. At this time Dumas cultivated the friendship of the physician J.-L. Prevost (a relative of Le Royer's), who had returned to Geneva after completing medical studies in Paris, Edinburgh, and Dublin.[5] Dumas and Prevost collaborated in a series of studies on the role of sperm, publishing their first paper in 1821. Sperm, they claimed, were not parasitic animalcules of the testis but, "whatever be their intimate nature . . . are the result of a true secretory action and the active principle of fecundation."[6]

In 1822 Dumas moved to Paris and took up a position at the Ecole Polytechnique, where he was to move in the illustrious circle dominated by such giants as Arago, Leplace, Cuvier, Ampère, and the geologist Alexandre Brongniart, whose daughter he married in 1826. In 1824, with two of his peers, Victor Audoin and the botanist Adolphe Brongniart, he founded the *Annales des Sciences Naturelles,* the early volumes of which contain the reports of the work done on animal fecundation by Dumas and Prevost in Geneva, and on plant fecundation by Adolphe Brongniart in Paris. Unfortunately for them, however, by claiming in these papers that the sperm and pollen particles actually penetrated the egg, they not only were taking issue with current thinking, but worse, appeared to be resurrecting an outmoded theory—that of eighteenth-century animalculism.

As the previous chapter indicates, only the animalculists and pollenists of the seventeenth and eighteenth centuries believed that the sperm and pollen grains played any role in generation. According to them, the sperm and pollen contained a preformed embryo which needed to be implanted inside the egg in order for development to begin. These theories were viewed as historical relics by naturalists of the early nineteenth century. By then fecundation was seen as either the dynamic interaction of male and female fluids or the interaction of a male fluid with a preformed egg. Both theories explicitly denied that spermatic animalcules or pollen particles had any role in the reproductive process.

Theories stressing the interaction of male and female fluids were particularly widespread in Germany, where dynamic epigenetic views of develop-

ment had replaced the older mechanistic concepts. Such views were a continuation of theories proposed by Joseph Koelreuter and Konrad Sprengel in
the latter part of the eighteenth century and adhered to by Karl Friedrich
Gaertner in the early part of the nineteenth. Although Gaertner's monumental
Versuche und Beobachtungen über die Bastardzeugung did not appear until
1849, he began to publish on the subject of plant sexuality in the 1820s.
Fecundation, he argued, was brought about when "the principal material, the
liquid contained in the pollen, reaches the ovules after being combined with
the liquid material secreted on the stigma, so as to give birth there to the
embryo." In complete antithesis to theories of preexistence, he argued also
that "the embryo does not pre-exist in the ovules but on the contrary is the
product of fecundation."[7]

On the other hand, although they refuted extreme preexistence theories,
many French naturalists still maintained the essentially eighteenth-century,
mechanistic view that fecundation involved the stimulating effect of male
fluid on a preformed egg. Those who held to these views included such
prominent figures as Georges Cuvier, Charles Brisseau de Mirbel, Alphonse
de Candolle, and Achille Richard.

Writing in 1806 in his *Histoire naturelle,* prior to the discovery of the
pollen tube, de Mirbel spoke of germs preexisting at the time of fecundation
and described fecundation as a stimulation by the subtle liquid contained in
the pollen. Likewise, in 1835, de Candolle spoke of a "fecundating fluid, *an
aura seminalis,* " while in 1828 Richard described fecundation as a phenomenon "in which the embryo, still in a rudimentary state, receives and preserves
the vivifying principle of life." To Richard, the pollen tube was a device
whereby the vivifying liquid was passed to the rudimentary embryo, and the
granules played no material role in the new generation.[8]

Prevost and Dumas shared these mechanistic assumptions. As Elizabeth
Gasking has pointed out, they were little influenced by the newest epigenetic
and Germanic views, but represented "the continuity between the eighteenth
and nineteenth centuries."[9] Their first paper on generation, published in the
Annales in 1824, opened with an explicit statement in support of the
mechanistic outlook on life:

> There occur in animals two sorts of phenomena which it is impossible to confuse
> with each other. There are intellectual phenomena, whose manifestation pre-
> supposes an immaterial principle. . . . There are bodily phenomena which seem
> to be susceptible of a purely physical explanation, since we see in them only an
> elaboration of already existing material without any creation, and since in many
> cases, when the action is of a simple enough nature, it is easy to demonstrate the
> series of chemical and physical effects which determine the result.[10]

Physiology, they admitted, had yet to find a home in the exact sciences. "It
is," they wrote, "a magnificent palace whose buildings can only begin when
one has found the stones which must serve as foundations," and the problem

of generation, "enveloppé du voile le plus épais," was clearly one that demanded a causal as opposed to a descriptive explanation. Thus, unlike the Germans of this period, Prevost and Dumas had no metaphysical aversion to theories of preexistence. "It seems easier to imagine a time," they wrote in reference to Bonnet's theory of *emboîtement,* "when nature by labor gave birth, as it were, all at once to the whole of creation, present and future, than to imagine a continual activity, which is repugnant to our faiblesse." Nevertheless, they believed that it was necessary to discard such theories on the grounds that they represented "a gratuitous hypothesis" without proof. The problem of generation, like the chemical sciences, they argued, demanded for its solution a rigorous experimental approach.[11]

But an experimental approach to the vexing problem of fecundation can lead to some perplexing results. Nineteenth-century microscopists were not aware that in many cases freshly released eggs and sperm are incapable of fecundation. Eggs must first reach a certain stage of maturity, which often is not attained until sometime after their release from the ovary. Sperm, too, particularly in those animals in which fertilization is internal, have to pass through a "capacitation" period before they are capable of entering the egg. In mammals this waiting period often extends to several hours, with the result that only sperm taken from the uterine tract sometime after coitus are capable of fecundating. Freshly released mammalian sperm or sperm obtained directly from the male genital organs are simply incapable of penetrating the egg. Ignorance of this fact prevented any successful *in vitro* fertilization of mammalian eggs until the mid-1950s. Although no proof of a similar capacitation period exists for invertebrates, it still cannot be ruled out.[12]

Nineteenth-century microsopists were also not aware that pricking frog eggs with a needle—something that Spallanzani probably did—can induce many eggs to develop as far as the gastrula stage. In addition, cells in the embryo sac of many plants, including the pumpkin with which Spallanzani worked, can become egglike and develop in the absence of fertilization. It is not surprising, therefore, that these experiments on sperm and pollen were so contradictory and inconclusive and that a solution to the problem by experimental means was virtually unattainable in the early nineteenth century. It simply was not possible to refute Spallanzani's experiments, as Prevost and Dumas attempted to do, with the knowledge that was then available.

Prevost and Dumas wrote their first paper on the male liquid. Examining the "agent fecondateur" from many vertebrates and invertebrates, they noted the universal existence of animalcules in mature forms, but their absence in the immature and aged. These results, coupled with the well-known absence of animalcules in the sterile mule, convinced them that "there exists an intimate relation between their presence in the organs and the fecundating capacity of the animal."[13]

Their second paper, published in the same year, dealt with fertilization in the frog. Many of the experiments they described were modeled after

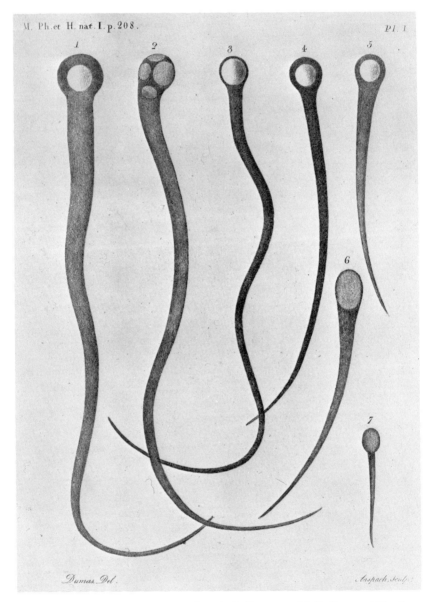

FIGURE 2.1.   The male animalcules. From J.-L. Prevost and J.-B. Dumas, "Essai sur les animalcules spermatiques de divers animaux," *Mém. Soc. Phys. Hist. Nat. Genève* 1 (1821). Reproduced by permission of the Museum of Comparative Zoology, Harvard University.

Spallanzani's work of the previous century and were designed to show that the sperm could and did come into actual contact with the egg. They argued that since the layer of mucus surrounding the egg became red and swollen when placed in a mixture of blood and water, the mucus itself was capable of absorbing water and solids of a limited size. They also noted the presence of a few sperm in the egg mucus and concluded thereby not only that the mucus could absorb small particles but that the sperm themselves were capable of penetrating the mucus and coming into intimate contact with the egg. An additional series of experiments convinced them that semen which did not contain motile sperm was incapable of fecundating the egg. The most famous of these experiments, and the one that drew the strongest criticism, involved the passage of semen through five filters and the production of a sperm-free filtrate that was incapable of fertilizing the egg.[14]

Having proved to their own satisfaction that the sperm needed to be brought into actual contact with the egg for fecundation to occur, Prevost and Dumas also claimed on rather insecure grounds that "it is infinitely probable that the number of animalcules employed corresponds to the number of fetuses developed," and that "the action of these animalcules is individual not collective."[15] In this paper of 1825 and in one dated 1827 they presented their individual and somewhat different hypotheses on the nature of fecundation, and it is abundantly clear from their arguments why their work was dismissed by many as akin to the outmoded doctrine of animalculism.

Prevost noted the existence of a "small elongated shadow" running from the periphery to the center of the blastodisc of a nonfecundated egg; in the fecundated egg, this "shadow" contained a "trait central," which called to mind the spermatic animalcule. Prevost argued that this trait disappeared shortly after fecundation and formed no part of the embryo. On the contrary, the embryo arose as a result of the action the sperm exercised on the blastodisc. "Neither one nor the other of these agents," he claimed, "formed a part of the being which is created." He concluded his paper with a classic epigenetic statement, that the sperm and blastodisc "will only give rise to the first of successive acts as a result of which this being will be created."[16]

Dumas, however, argued differently, and because of the more prominent position he held in French scientific circles, it is not surprising that the work of both men should have become associated with the views that he expressed. "The spermatic animalcule," Dumas wrote, "penetrates into the ovule to graft itself on the cellular-vascular membrane." In words that seemed reminiscent of a modified form of animalculism, he continued: "The development of the foetus, observed with care, shows us that the animalcule is nothing other than the rudiment of the nervous system, and that the membranous lamina on which it is implanted forms . . . all the other organs of the foetus."[17]

On the surface, the difference between the two men's hypotheses seemed to be trivial, resting as it did on the fate of the sperm once it entered the egg.

But given the historical background of the debate, the two hypotheses could be interpreted as fundamentally contradictory. Prevost, like the German developmentalists, was arguing that development was a special property of organic bodies, and that the sperm merely acted as "le principe d'action." Dumas, like the animalculists, was arguing that the spermatic animals actually contributed materially to the new generation.

At about the same time, Adolphe Brongniart, son of the famous French geologist Alexandre Brongniart, published an essay on generation in phanerogamic plants which won him the Académie des Sciences 1827 prize in experimental physiology. In this essay he argued—also in opposition to conventional views—that the granules of the pollen, and not the liquid contents, were the active agents of fecundation. Analogous to spermatic animalcules, these pollen granules were termed by him "spermatic granules." Brongniart noted that after landing on the stigma, the pollen grains produced a "vésicule allongée, plus ou moins tubuleuse," the famous pollen tube, which then grew down to penetrate the spaces between the "utricles" of the stigma and by means of which the spermatic granules were carried to the interior of the stigma (Fig. 2.2). These granules, he wrote, "do not penetrate into the stigma neither by an insensible transudation across the pollen membrane nor by the sudden rupture of the pollen and by the emission of the granules to the stigma surface, but by means of a tubular and membranous appendix."[18] Once inside the stigma, he argued, the granules are released into the fluid of the stigma, find their way down the style and enter the ovule, where they pass through the micropyle and become absorbed by the embryo sac. Pointing out that the process was very similar to the act of conjugation seen in some algae, whereby the contents of one utricle migrate across to be absorbed by the

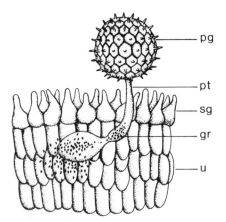

FIGURE 2.2. The pollen grain, pollen tube, and "spermatic granules" (after Brongniart): *gr*, spermatic granules; *pg*, pollen grain; *pt*, pollen tube; *sg*, stigma; *u*, utricles.

contents of a neighboring utricle, he concluded: "We can say that generation in the vegetables consists essentially in the union, that is in the combination of one or more granules furnished by a special organ with one or more granules furnished by another organ, in a special cavity of the latter organ."[19] Brongniart maintained that this meeting of granules validated the theory of epigenesis, since no embryo existed prior to fecundation, and since the pollen fluid did not serve merely to vivify a preformed embryo. He undermined his case severely, however, by mentioning the work of some eighteenth-century pollenists and by stressing the parallel between his views and those of Dumas: that whereas in the animal kingdom a single sperm forms a special organ (the nervous system), in plants, where organization is less determined, an indeterminate number of spermatic granules are required.[20]

Quite understandably the views of Prevost, Dumas, and Brongniart were severely criticized. Not only were their theories at odds with all current ideas on fecundation, not only had they revived, in the words of the Scottish physiologist Allen Thomson, "an old and fanciful notion that the animalcule forms the rudiment of the new being,"[21] but in asserting that spermatic animalcules played a central role in reproduction, they had also necessarily implied that these animalcules were neither organisms nor parasites.

In these early years of the nineteenth century, however, spermatic animalcules were clearly considered to be parasitic organisms specific to the male testis—of this there could be little doubt. Indeed, at that time the theory of parasites was more firmly grounded than it had been at the time of Spallanzani, who had concluded that they must be organisms, and thus parasites, on the basis of behavioral comparisons made with microscopical infusorians. But whereas ovist preformationists such as Spallanzani had always encountered difficulties when explaining how the eggs of these parasitic animalcules (eighteenth-century biologists assumed that parasites reproduced by eggs) passed from male testis to male testis, no such problem existed for many of those who held to the epigenetic concept of development. To them, and particularly to those influenced by *Naturphilosophie,* all distributional problems had been overcome by the assumption that spermatic animalcules were spontaneously generated within the testes of host organisms. As Karl Burdach argued in 1826 in his multivolume treatise on physiology, spermatic animalcules "must be considered, the same as Entozoa [parasitic worms], as the products of an organic substance which decomposes in the interior of a living organism. . . . Thus the spermatic animals appear to me to be the entozoans of the semen."[22] In the heyday of the theory of spontaneous generation the existence of spermatic animalcules in the testis was as understandable as the presence of intestinal worms in the gut; both parasites were generated from living animal tissue. The spermatic animalcules were, according to Johann Blumenbach, "animalculae of a stagnant animal fluid,"[23] and were termed "spermatozoa"—animals of the semen—by Karl Ernst von Baer. Indeed, being superficially similar to the small, tailed organisms, or cercariae, now known to be larval stages of parasitic flukes, these spermatozoa were placed in the infusorial genus *Cercariae*.

The theory of parasites was further strengthened when some microscopists described the existence of organs within the spermatozoa. In 1839, for example, Gustav Valentin, professor of physiology at the University of Berne, presented an account of the structure of bear sperm in which he reported the presence of a mouth, an anus, and fine "Blasen" (Fig. 2.3). The latter, he speculated, were either digestive vesicles, excretory organs, or a twisted gut. A year later, at the same institute, Friedrich Gerber described similar structures in the sperm of the guinea pig and even described the existence of sex organs in them (Fig. 2.4). Identical views were held by Christian Ehrenburg in Germany and by Antoine Dugès and Felix Pouchet in France.[24]

Prevost and Dumas needed to show that sperm were not parasitic organisms, but neither of them addressed this particular problem. Instead, they stressed the intimate relation between the presence of animalcules and the

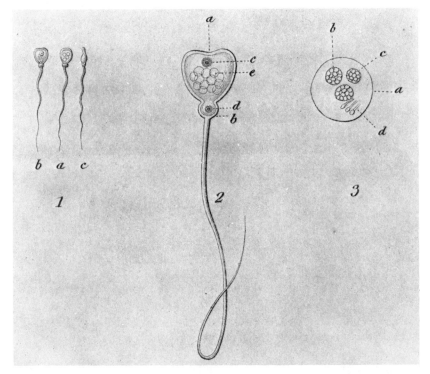

FIGURE 2.3.   The spermatozoa of the bear: (*1*) dorsal, ventral, and side views; (*2*) ventral view under high power (*c,* mouth; *d,* anus; *e,* interior bladders); (*3*) a single germ receptacle (*c,* egg yolk; *d,* immature sperm). From Gustav Valentin, "Über die spermatozoen des Bären," *Nova. Acta. Leopoldina* 19 (1839). Reproduced by permission of the Museum of Comparative Zoology, Harvard University.

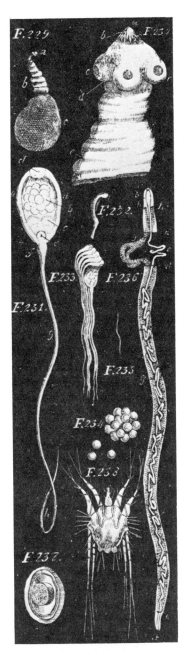

FIGURE 2.4.   Sperm of the guinea pig. From Friedrich Gerber, *Handbuch der allgemeinen Anatomie des Menschen* (Bern, 1840). The sperm is shown with other parasites (tapeworms, nematodes, etc.) that Gerber copied from Johann Bremser, *Über lebende Würmer in lebenden Menschen* (Vienna, 1819). Reproduced by permission of the Museum of Comparative Zoology, Harvard University.

fecundating capacity of the animal; the common structure of these animal-cules; their absence in immature, old, and sterile animals; and their common mode of formation by the secretory activity of the only constant and essential organ in reproduction. However, such arguments become convincing only if one rejects the possibility that sperm are parasitic animals. The French, unlike many German biologists of that era, were not favorably disposed to the doctrine of spontaneous generation, and so the formation of sperm within the testis may have been a strong argument against the parasite theory as far as they were concerned.[25] But if sperm were not parasites, what could they be? No answer was given.

Interpretation of the critical experiment involving the passage of seminal fluid through five filters depended on the answer to this question. Spallanzani, Prevost, and Dumas showed that only the seminal residue initiated egg de-velopment, while the filtrate had no effect. But whereas Spallanzani, who had shown that sperm were parasitic organisms, concluded that the fluid portion of the residue was responsible for initiating egg development, Prevost and Dumas, who had argued that sperm were the active agents of fecundation and were not parasites, naturally claimed that the spermatic portion of the residue was the fecundating agent. The value of this experiment hinged on the parasite question; with disagreement the experiment loses significance. As Johannes Müller correctly noted, "The fluid of the semen cannot be obtained separate from its components and consequently its peculiar properties cannot be ascer-tained." In the same vein, Thomson, in his acrid criticism of Prevost and Dumas' views, suggested that the experiments needed "repetition and some modifications, for other ingredients, besides the animalcules of the seminal fluid, might be retained on the filter."[26]

Nevertheless, subtle changes did occur in the 1830s. Microscopists did not deny the parasitic nature of sperm, but more and more they suggested that sperm had an important role to play in fecundation. None, of course, believed that that role was a material one. Pouchet, for example, claimed that sperm were parasites of the testis and were engendered by spontaneous generation, but he also believed that they were essential to fecundation. Bory de Saint-Vincent argued that sperm were parasites that, "by their continued agitation . . . contribute to the mixing of all chemical elements," a view to which Richard Owen subscribed. Allen Thomson had similar ideas. Although the nature of fecundation seemed a complete mystery to him, he argued that "neither experiment nor observation enables us to form the most distant conjecture what the nature of that action may be," and although he believed that sperm were best classified as parasitic animalcules related to cercarial infusorians, he nevertheless admitted that they seemed to be "intimately connected with the integrity of its [the semen's] fecundating property, if not as some are inclined to hold, the essential cause of it."[27]

Others, such as Gottfried Treviranus and Felix Dujardin, denied that sperm were parasites. "The more one studies the zoosperms or supposed spermatic animals," Dujardin remarked in 1837, the more one is convinced that they are not animals strictly speaking, not beings produced from an egg or sperm, as zoophytes are, capable of feeding, growth and reproduction. By using the best microscopes and comparing these bodies in different animal classes, one is led to think, on the contrary, that these zoo-sperms are simply a product, a derivation of the internal layer of seminiferous tubules; not a secretion, but a progressively formed product, a product conserving a sort of vitality necessary to contribute to the formation of the embryo.[28]

Thus, by the 1830s the nature and significance of sperm was once again a source of puzzlement. The view that sperm were simply parasites had received support from those who described complex organs within them. On the other hand, this view had been somewhat undermined by the decreasing popularity of the spontaneous-generation hypothesis. If sperm were parasites, how could they come to be present in the testes except by a spontaneous generation? In addition, their universality seemed to many to suggest a necessary function for them in fecundation. This in turn undermined the parasite theory, since parasites were generally regarded as a divine scourge and not as a necessary attribute of the body.[29] Despite this rising confusion, however, the views of Prevost and Dumas were unique in attributing to sperm a material role in fecundation. Thus their views were largely ignored, lumped with antiquated animalculist theories that were of historical interest only, and the discussion that did occur was usually limited to a criticism of the two men's experimental work.

Given this confused picture, one can appreciate the rather hopeless remarks of Peter Roget in his Bridgewater treatise of 1839: "No conceivable combinations of mechanical, or of chemical powers, bear the slightest resemblence, or the most remote analogy, to organic reproduction, or can afford the least clue to the solution of this dark and hopeless enigma."[30]

THE IMPACT OF THE CELL THEORY

By the 1840s the indifference to the claims of Prevost and Dumas changed quite abruptly, particularly in Germany, and the whole topic of fecundation became the focus of a stormy debate. The turning point came between 1838 and 1842 with the appearance of Matthias Schleiden's and Theodor Schwann's cell theory and Schleiden's equally significant "pollen theory." The cell theory rested on the assumption that, in the words of Schleiden, "every plant developed in any higher degree, is an aggregate of fully individualized, independent, separate beings, even the cells themselves," and more importantly, that "the most intimate connection of the two kingdoms of organic nature" is "the similarity in the laws of development of the elementary parts of animals and plants."[31] In other words, the key to understanding the basic similarity between animals and plants lay in studying their developmental histories, in discovering that all tissues, wherever they occur, originate only from cells. Using this approach, Schwann argued that the ovum de-

scribed by von Baer was probably a cell; the germinal vesicle reported by
Purkinje, a nucleus. And in 1838, using the same approach, Schleiden announced to a startled botanical community that the pollen tube actually grew
down into the embryo sac and implanted the new embryo—*not in the form of
a preexisting embryo, but as a cell.*[32]

Schleiden holds an important place in the history of German botany because he led a movement away from idealistic morphology toward a cellular,
developmental inductive approach. Curiously, however, his actual botanical
theories had little lasting impact, being almost universally discredited by the
late 1850s. Born in Hamburg in 1804, Schleiden received his doctorate of law
from the University of Heidelberg in 1826 and immediately thereafter set up a
legal practice in his home town. He found no satisfaction in this profession,
however, and after an unsuccessful suicide attempt, moved to the University
of Göttingen to study medicine. There he developed an interest in botany,
which was then an important part of medical training. Moving shortly thereafter to Berlin, Schleiden worked with the illustrious group of medical
physiologists in the laboratory of Johannes Müller. He gained his doctorate of
medicine in 1839 at the University of Jena and was appointed to the chair of
botany there. During these early years of his career he published his first
theory of pollen, his theory of cell development, and finally, in 1842, his
monumental *Grundzüge der wissenschaftlichen Botanik.*

There were many rational and scientific reasons why Schleiden should
have conceived of his pollen theory, which Louis Fuguier was later to call,
unfairly, "la théorie de Schleiden sur la préexistence des germes vegetaux."[33]
Indeed, Schleiden was not the only one to have proposed such a theory,
although he was alone in setting up a new classificatory scheme on the basis of
the theory—the first scheme, to my mind, which successfully linked together
the cryptogams and phanerogams.

As has been shown, the Linnaean system of classification, which had been
based on the assumption of universal sex in all plants, was in disarray by
1840. It had been replaced by French and German natural systems, all of
which stressed the asexual nature of the nonflowering plants, a theory that was
supported by the discovery that mosses, ferns, and their allies produced
spores, not seeds. In addition, microscopic examination of these spores had
revealed that they germinated to produce tubes exactly as pollen grains did.
De Mirbel, for example, described the production of such tubes from the
germinating spores of the liverwort *Marchantia* (Fig. 2.5) and stressed the
difference between these "séminules" and the fecundated seeds of
phanerogams: "Between the first utricle of *Marchantia* and the first utricle of
the phanerogams there is this notable difference: that that of *Marchantia* has
in it, as soon as it is formed, all the conditions necessary to form a complete
plant on the surface of the soil, while that of the phanerogams, to survive,
must begin development in the interior of the ovule."[34] Furthermore, Hugo
von Mohl reported that the mother cell of both spores and pollen separated

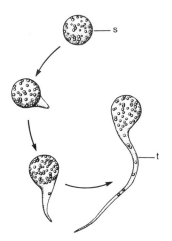

FIGURE 2.5. The germinating spore of the liverwort *Marchantia* (after de Mirbel): *s*, spore; *t*, tube.

into four parts. (This, of course, was the first description of what we now recognize to be a reduction division.) "After these researches," Franz Meyen noted in 1837, "the question should be asked: whether the sporangia of the higher Cryptogamia can be regarded as the same organ as the anthers in Phanerogamia, and the spores as the same organ as the pollen grains."[35]

Given this parallel between pollen grains and spores, it is hardly surprising that the claim was soon made that pollen grains are essentially spores that, instead of germinating in the soil, need to be implanted in the embryo sac in order for development to occur. The first to make this claim was the English naturalist William Valentine, who after arguing very surprisingly that "it is a well established fact that the embryo, or essential part of the seed, is derived from the pollen," went on to state: "We do not overstep the bounds of probability in supposing that in plants of a complicated organization there exists a necessity that the embryo should be protected by a nidus capable of imparting aliment. . . . Whilst in the *Cellulares* the process of their growth is so little complicated that the embryo requires no preparation to enable it to perform its functions."[36] According to Valentine, in other words, the lower plants are capable of reproduction "by the mere ejection of [their] pollen or sporules on the soil."

Schleiden took the same position. After examining the pollen tube of plants from approximately one hundred families, he remarked: "I have observed the entrance of the pollen-tube into the embryo-sac and the gradual conversion of its end directly into the embryo."[37] In the cryptogams, however, the spore or pollen is able to develop directly into the plant without the necessity of this implantation into the embryo sac (Fig. 2.6). "There can be no more striking analogy," he wrote in his text of 1842, "than the one presented between the

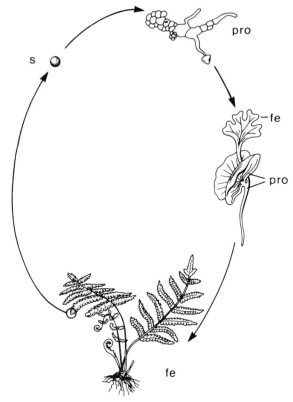

FIGURE 2.6.   The life cycle of the fern according to Schleiden. Only one type of reproductive cell is formed—the spore: *fe,* fern plant; *pro,* proembryo; *s,* spore.

germination of the spores of the Cryptogamic stem plants, and the behavior of the pollen-granules of the Phanerogamia upon the stigma, even as the history of development of the spores and the pollen granules, no less than their structure, are almost wholly identical."[38]

   With Schleiden's pollen theory, botany had turned a complete circle since the eighteenth century. At that time sexual reproduction had been viewed as the only mode of reproduction; it was assumed that reproduction required both male and female sex organs. With the collapse of the Linnaean system, however, nonflowering plants had come to be viewed as asexual forms, and the status and significance of sex had diminished. Schleiden's pollen theory refueled the debate. According to Schleiden, the only difference between sexual and asexual reproduction was the site where the reproductive cells developed: in cryptogams they developed in the soil; in phanerogams, in the embryo sac. "This, and only this," Schleiden remarked, "is the significance of the word 'sex' among plants." Sex was merely a special mode of asexual reproduction. It would be better, Schleiden claimed, to banish the word "sex" entirely from the plant kingdom than to continue "lame and unscientific" comparisons with the higher animals. He added, moreover, that if one chose to use the term "sex" at all in plants, one would have to reverse all previous notions. If the female organ was viewed as providing the material foundation of the new being, and the male part as merely promoting the development of the female-produced germ (a view to which most subscribed at that time), then, he argued, the anther would have to be regarded as the female ovarium, the pollen grain as the germ, and the embryo sac as the male organ.[39]

As we shall see in a later chapter, Schleiden's pollen theory enabled him to reclassify the plant kingdom and thus to remove the hiatus between asexual cryptogams and sexual phanerogams. Such a theory, he wrote, "explains the most widely differing facts from one natural law instead of from two. This simplification of the grounds of explanation is, however, one of the most important methodic claims of a sound natural philosophy."[40]

Schleiden's pollen theory (though not the classification scheme he derived from it) received a substantial degree of support. It was, de Mirbel and Edouard Spach remarked, "received quite coldly in Paris, but obtained an astonishing vogue in Germany."[41] Given the Germans' abhorrence of preexistence theories and animalculism, this was surprising. But fundamental differences existed between Schleiden's pollen theory and the older theory of pollenism. Pollenism implied the entry of a preformed *embryo* into the embryo sac and the subsequent unfolding or "*evolving*" of its parts. Schleiden's theory implied only the entry of a *cell,* which then developed in an epigenetic manner by exogenous or endogenous multiplication. The ascendancy of the cell theory in the 1840s, and the concept of the cell as the basic plant individual capable of developing into a wide range of cell types, thus removed the stigma of animalculism from Schleiden's pollen theory. The pollen contained cells; it was no longer considered the carrier of a preformed embryo. Plant development from this pollen was epigenetic, not *l'évolution* of preformed parts. The debate over the validity of Schleiden's claim was therefore fought among cell theorists on microscopic, not metaphysical, grounds, the basic question being whether the original embryonic *cell* existed in the embryo sac before or only after the entry of the pollen tube.

Perhaps the most influential German botanist to support the pollen theory was Stephan Endlicher, whose *Genera Plantarum* appeared in 1841. Arguing that the orientation of the embryo within the embryo sac precluded any possibility of its arising from a preexisting cell within the embryo sac, he concluded that the embryonic cell must be carried in from the outside in order to attain its full development within the embryo sac. Like Schleiden, he pointed out the remarkable similarities between spores and pollen grains and equated the sporangium and anthers with the animal ovary, and the spores and pollen with the animal egg.[42] Another botanist who accepted Schleiden's theory was

Heinrich Wydler, a professor at the University of Berne. "The anther," he wrote, "far from being the male organ, is on the contrary the female organ: it is the ovary. The grain of pollen is the germ of the new plant, the pollen tube becomes the embryo."[43]

In 1847 the Royal Netherlands Institute of Science and Literature proposed as a competition the examination of Schleiden's pollen theory. The prize from this competition was awarded to Hermann Schacht, a student of Schleiden's, whose prize-winning essay, *Entwicklungsgeschichte des Pflanzen-Embryon,* was eventually published in 1850. "The transformation of the extremity of the pollen tube to the embryo is so positively proven," he proclaimed, "[that] the absolute accuracy of Schleiden's theory of fecundation is beyond doubt."

> The grain of pollen therefore is *not* the fecundating organ . . . *but the egg of the plant.* Even less are the corpuscula of conifers . . . analogous to the cells existing in the embryo-sac of Orchids before fecundation which only the pollen tube makes capable of forming an embryo. Rather the pollen tube in *Taxus, Pinus and Abies* submerges itself in the corpusculum, expands itself in it, fills itself completely or partially with cells out of which the first cells of the embryo and its tube arise.[44]

In 1855, after the appearance of Theodor Deecke's work in support of Schleiden's theory, Schacht became even more dogmatic. Deecke's work, he claimed, "macht allem Streit ein Ende . . . aller Zweifel schwinden muss."[45]

Other German botanists, however, were not prepared to admit that "all doubt must disappear." Franz Meyen, professor of botany at Berlin from 1834 to 1840, clearly disagreed with Schleiden. In his *Neues System der Pflanzen-Physiologie,* published in 1839, he argued that the pollen tube did not convert into the embryo, although it contributed both dynamically and materially to the new generation.[46] But the most devastating blows against Schleiden's theory in Germany were to come from Wilhelm Hofmeister.

Hofmeister, a disciple of Schleiden's new scientific methodology, and, as we shall see, perhaps the single most significant figure in nineteenth-century botany, began publishing in 1847. In his first paper, which dealt with fecundation in the Oenothereen (evening primrose) "family," he described the appearance of numerous nuclei at the micropyle end of the embryo sac, one of which formed a pyriform-shaped cell. This cell—the germinal vesicle—and not those in the pollen tube, he argued, "is the true egg of the plant, the principle of the future embryo." To Hofmeister, the act of fecundation was one of chemical excitation:

> At the moment of fecundation, the germinal vesicle is separated from the end of the pollen tube by the embryo-sac membrane which remains intact. . . . Only through a double endosmosis can the liquid in the germinal vesicle unite with that in the pollen tube. . . . We do not regard fecundation, that is to say the excitation of the germinal vesicle to a specific development, other than as a production by the fluid which passes by exosmosis from the pollen tube into the embryo sac and from there into the germinal vesicle.[47]

Thus, he concluded rather cautiously, the "theory of Schleiden on the mode of formation of the embryo in phanerogams is entirely inadmissible for the family Oenothereen."[48] Two years later he was less circumspect. Arguing that the germinal vesicle, not the pollen cells, was the *Grundlage* of *all* phanerogams, he stated again that "the germinal vesicle which is to be fertilized remains as a completely enclosed cell. A direct entrance of any part of the contents of the pollen tube into the interior of the germinal vesicle is utterly impossible."[49]

The most vociferous opposition to Schleiden's theory came from the Paris-based French botanists, although even here some, such as August de Saint-Hilaire, seemed to favor Schleiden's views.[50] In 1839, however, de Mirbel and Spach produced a point-by-point refutation of Schleiden's views, and in a note of the same year, Brongniart also attacked Schleiden. From their observations, de Mirbel and Spach concluded, "the stamens were not yet mature when the primordial utricle approached the termination of its development."[51]

This debate over the role of the pollen had a considerable impact on the continuing investigation into the role of sperm in the animal kingdom, for proponents of the pollen theory tried to draw parallels with what seemed to be occurring in animals. One could, however, draw such parallels in diametrically opposite ways. One could argue, as Endlicher did, that the pollen grain is analogous to the egg and that in both cases the new organism is derived solely from the female organ. From this perspective, the products of the male organ served merely to stimulate the egg or pollen to develop. As has been mentioned, such a concept underplayed the significance of sex altogether by labeling sexual reproduction as a special form of asexual reproduction—one wherein eggs, pollen, and spores were basically identical.

This view reflected the commonly held early nineteenth century assumption (among German biologists at least) that all types of reproduction in animals were, as Johannes Müller argued, the result of a vital force that resided in the germ.[52] In higher animals this force resided only in the ova, but it existed in such an imperfect state that it needed to be acted upon by the semen. In germiparous forms, on the other hand, a portion of superfluous substance might separate from the parent animal to become the germ, which, because it contained sufficient vital force, could develop into several beings in the absence of any sexual influence. Compatible with this view, of course, was the belief that the semen was the essential stimulant and the sperm the parasites.

Schacht, on the other hand, pointed out that a remarkable analogy would exist between animal and plant reproduction if indeed the sperm did penetrate the egg as the pollen tube did. No one, however, went so far as to suggest that the sperm represented the future embryo. Writing in 1855, Schacht argued:

> While until now one has wished to ascribe to the male semen only a dynamical role, that is to say an actualizing [*befähigenden*] influence on the production of the embryo in the egg, now one must recognize in it a material contribution. . . .

> The act of fecundation loses the mystic darkness which has enveloped it so long, through a direct contribution of the pollen tube to the production of the plant embryo, in the same way the sperm contribute to the animal embryo.[53]

Schacht's argument was curiously illogical. Having earlier drawn a parallel between the pollen cell and an egg, describing both as containing the entire material of the next generation, he now turned to the animals and related the pollen cell to the male sperm. Moreover, having claimed that in plants, only the female cell (i.e., the pollen cell) contributed material to the new generation, he now claimed that both male and female cells contributed to the new animal generation. This, of course, suggested that whereas in plants sexual and asexual reproduction were fundamentally alike—the spore or pollen cell differing only in the stimulus required for development—in animals sexual reproduction was unique, requiring the input of material from two cells. Nevertheless, one can understand Schacht's arguments, for by 1855 a remarkable change had occurred in the interpretation of animal fecundation. In the 1840s and 1850s the view that sperm actually penetrated the egg was accepted.

THE NECESSITY OF SPERMATOZOA

During the 1840s the parasitic theory of sperm seems to have collapsed quite quickly. This collapse was due in part to the growing unpopularity of the theory of spontaneous generation and to the increasing popularity of the cell theory. The production of sperm inside the testis could no longer be ascribed to the spontaneous generation of parasites but rather had come to be considered the normal outcome of cell division or cell propagation. Once this view was accepted, the universal distribution of sperm in mature animals, as well as their absence from immature and old animals and from infertile hybrids, again suggested an essential role for them in fecundation. Then, in the early 1840s, two significant texts appeared. Albert von Kölliker and Rudolph Wagner argued that sperm were not parasites, but were essential components of the seminal fluid.

Both von Kölliker and Wagner were figures of some importance. The Swiss scientist Rudolph Albert von Kölliker had studied medicine at the University of Zurich. Then, in 1839, he had moved to Berlin to study in the laboratory of Johannes Müller. In 1841 he began a study of invertebrate sperm and received his doctorate for a thesis on this subject from the University of Zurich. In his thesis, he showed not only that motile sperm are usually present in most invertebrates but that they are formed from testicular cells only during times of sexual ripeness. Sperm, he further argued, ''have no animal organization and do not propagate,'' but rather consist of a transparent, homogeneous substance.[54] Because he viewed sperm as normal and essential components of the seminal fluid, it seemed absurd to von Kölliker that they could be thought of as parasitic organisms. "No other individual animal organisms could occur as normal or essential constituents in a living liquid of animal organisms," he wrote. "Will one believe that a fluid of an organism could be healthy and constituted in such a manner as to be capable of directing its functions if it were no longer submissive to the organism-as-a-whole and had to serve in the development and the life of an immense number of foreign organisms?" The sperm, or *Samenfaden,* he concluded, cannot be animals, but are "organized parts of the seminal fluid, elementary parts analogous to the blood corpuscles or eggs."[55]

Having argued that sperm are essential components of the seminal fluid, von Kölliker concluded his text with a theoretical discourse on their role in fecundation, a discussion that clearly shows the influence of Oken's *Naturphilosophie* lectures, which von Kölliker attended while in Zurich. Von Kölliker quoted the words of the *Naturphilosoph* Carl Gustav Carus: "In one or two individuals, two antagonistic substances are developed such that as soon as they touch each other immediately one of them is caused to grow into a third one different from both primitive substances." One of these primitive substances, von Kölliker argued, is the quiescent egg, which is endowed with latent life; the second, the sperm, is endowed with vitality and thus shows the highest form of development possible for part of an organism. The egg can never occur without sperm, he wrote, "for an egg is such a substance that can become stimulated only by another antagonistic in character in order to form a new individual from it."[56]

Somewhat reminiscent of the *Naturphilosophen,* von Kölliker argued that the motion of the sperm results from various organic polarities and that "the sperm, whose parts already exhibit various polar phenomena, excite also polarization in other parts with which they come in contact, similar in character to the magnet on iron."[57]

In 1837 and in his *Lehrbuch* (1842), Rudolph Wagner, professor of physiology at the University of Göttingen, basically agreed with von Kölliker that sperm are essential components of the semen, but he found it impossible to state anything conclusive about their organization. "One cannot say with complete certitude that they are true animals since it is not yet proven that they have an interior organization, a gut, etc. Nevertheless, their independent movement and manner of development suggest their animal nature."[58] Like von Kölliker, he, too, saw fecundation as merely a matter of contact, but being unsure of the nature of sperm, he assumed that the contact was between the egg and the semen as a whole.

Von Kölliker and Wagner continued to deny any *material* connection between egg and sperm, a claim which still seemed to them to be reminiscent of antiquated eighteenth-century animalculism. Von Kölliker admitted, thereby, his uncertainty as to how the characters of the male parent are passed to the offspring. Nevertheless, he did not consider this to be a problem of insurmountable difficulty. Johannes Müller, who also denied any material role for the semen or sperm, simply regarded the semen "as a kind of nutriment in which, no less than in the germ, are involved the specific form of the animal or vegetable, and all its individual peculiarities."[59] Von Kölliker dismissed the problem (somewhat mystically) as follows: "Just as the magnet on iron, an electrical body, communicates to a non-electrical body by mere contact its inherent activity and excites it in it, so also will the sperm excite in the egg the activity, the expression of the whole ideal impetus of male organisms in sexual relations which likewise must be implied from the same idea."[60] It appears that to these laboratory-based German physiologists, questions of inheritance were not of primary concern at this time.

Against this background Theodor Ludwig Bischoff (Fig. 2.7) began his lengthy studies on fecundation. After gaining his M.D. degree from Heidelberg in 1832 under the direction of Freidrich Tiedemann, and after occupying junior positions in physiology at the Universities of Bonn and Heidelberg, Bischoff turned to embryology, having been stimulated by the work of von Baer. In response to a problem posed by the Berlin Academy of Sciences, he investigated the development of the rabbit egg and in 1842 reported his results. He concluded that the essence of fecundation lay in the chemicallike interaction of egg and semen, "that fecundation is conditional," as he put it, "on a material reciprocal influence between semen and egg." He denied that this influence entailed the entry of sperm; that would be "absolutely impossible," he wrote. "I do not hesitate however in declaring absolutely the opposite view, that only the dissolved part of the semen enter the egg."[61]

Like others before him, Bischoff linked the views of Prevost and Dumas with older preexistence theories, but he did note that a remarkable similarity would exist between plants and animals if indeed sperm were found to penetrate the egg as the pollen tube did. It is clear that the botanical controversy influenced him, for unlike most investigators a few years previously, he seriously discussed the possibility of sperm penetration. Agreeing that mounting evidence illustrated the essential relationship of sperm to the fecundating ability of semen, Bischoff suggested, as others had earlier, that the active sperm kept the chemical semen in motion, and that this motion was essential for normal functioning. He argued that unlike the blood, the seminal fluid was not circulated, and that the sperm were needed to cause this motion.

By 1842, however, serious claims that the sperm did indeed penetrate the egg were appearing in the most unlikely places—notably, in Britain, where the level of microscopical work lagged disastrously behind that of Continental laboratories. In 1839, in a paper communicated to the Royal Society, Martin Barry had suggested that the sperm passed into the ovule while still in the ovary. The basis for this claim was his observation of sperm inside the ovary of a rabbit in two out of nineteen rabbits examined.[62] A year later, he had postulated that fecundation occurs "by the introduction of some substance into the germinal vesicle from the exterior of the ovary," and when he found a fissure in the egg, through which, presumably, the sperm entered, he had

FIGURE 2.7.   Theodor Bischoff. Reproduced by permission of the Museum of Comparative Zoology, Harvard University.

suggested that fecundation takes place upon the entry of the sperm. Barry had also pointed out the similarity of his views to the claim that the embryos of plants arise from cells within the pollen tube, although he did not suggest that the sperm contained the rudiments of the complete animal embryo.[63]

Martin Barry's claims needed to be taken seriously. Trained for a career in a Nova Scotian mercentile business, he had turned instead to medicine, gain-

ing an Edinburgh M.D. after studying also in London, Paris, Erlangen, and
Berlin. Like Bischoff, he, too, had spent time under Tiedemann at Heidelberg
and had become one of the few British microscopists who were conversant
with the German microscopical literature of the 1830s and 1840s. Moreover,
his embryological work had gained him the Gold Medal of the Royal Society
in 1839. Also like Bischoff, Barry was enormously influenced by the work of
von Baer; indeed, his first paper, which was published in 1837 under the title
"On the Unity of Structure in the Animal Kingdom," dealt at length with von
Baer's embryogenetic laws—laws which of course precluded any acceptance
of preexistence theories.[64]

In 1843, however, Bischoff and Barry became embroiled in a controversy
over the sperm question. In that year Barry reported that he *might* actually
have seen traces of the sperm within the egg cell lying inside a rabbit ovary.[65]
A year later, in a review of Bischoff's paper of 1842, his doubt turned to
certainty. He had, he claimed, actually seen the sperm in the interior of the
ovum.[66] By this time, however, Barry had undermined his case; at least, that
was the opinion of Bischoff, who described two of Barry's papers as "fantas-
tical." In the first of these papers, Barry argued that animal and plant fibers
were constituted of cells lying end to end; in the second, he claimed that
"every structure I have examined arises out of corpuscles having the same
appearance as corpuscles of the blood." Such structures included the sperm,
which Barry now described as "composed of a few coalesced discs." Loosing
faith in the worth of Barry's findings, Bischoff claimed: "I however never
could satisfy myself that one of these spermatozoa was contained in the
interior of the egg."[67]

Bischoff retained his chemical theory of fecundation in his second paper on
the subject, which was published in 1845. "The action of the semen appears
to be chemical," he wrote, "and the destiny of the sperm to preserve the
mixing of semen through their motion."[68] Still, he was far from satisfied with
this explanation of fecundation. Chemical mixing alone seemed insufficient,
for in nature a cell is produced; it could not be obtained by artificial means.
Thus the mystery was only partially explained by Bischoff's belief that the
mixing of semen serves as a condition for cell formation.

In the fall of 1843, however, Bischoff had moved to the University of
Giessen, where he would remain until 1854. There, under the influence of his
friend and colleague Justus von Liebig, Bischoff posed the essential question
of how fecundation could be explained totally in chemical terms. The cur-
rently held views that fecundation was brought about either by the mixing of
semen and egg contents, by the contact of semen and egg, or by the contact of
sperm and egg all seemed totally inadequate. None of these was a causal
explanation, he argued; they were merely statements of necessary conditions.
"Mysteries of this nature," he wrote, "persist as long as we do not succeed in
linking them to a class of phenomena already known." Such better-known
phenomena were at hand, however—for instance, von Liebig's theory of "the

excitation of chemical action through mere contact'': ''There is perhaps no more general law in nature than this, that a body in motion will impart its motion to another with which it meets, to the degree of resistance which the second body gives.''[69]

According to von Liebig, special kinds of chemical transformations exist in which the ''atoms'' of the transforming substance are merely rearranged to form a new compound. Such transformations reflect a disturbance in atomic equilibrium whereby atoms, having been set in motion, overcome their *vis inertiae* and rearrange themselves. Recent work on catalysts had suggested to Liebig that such a disturbance results not only from heat but also from contact with another compound, ''that a body in the act of combination or decomposition enables another body, with which it is in contact, to enter into the same state.''[70]

Fermentation provided the classic example of such a transformation, for in that process, ''a compound of nitrogen in the state of putrefaction and decay'' causes the molecules of sugar to rearrange themselves. Furthermore, should the transforming substance contain nitrogeneous compounds, or gluten, this gluten would be transformed into fresh quantities of yeast, while if gluten were absent, the yeast would disappear and the reaction would come to a halt.

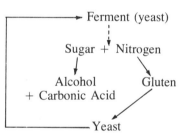

The elegance and popularity of this theory derived from the fact that it could be used to explain a host of other organic phenomena. It could explain, for example, the production of miasmatic and contagious diseases. The former occur when putrefying muscle or pus is placed upon a fresh wound, thereby causing disease and death. ''It is obvious,'' Liebig remarked, ''that these substances communicate their own state of putrefaction to the sound blood from which they were produced, exactly in the same manner as gluten in the state of decay.'' Contagious diseases occur when such poisons are generated in the body itself and are then propagated through the air:

> In small-pox, plague, and syphilis, substances of a peculiar nature are formed from the constituents of the blood. These matters are capable of inducing in the blood of a healthy individual a decomposition similar to that of which they themselves are the subjects; in other words, they produce the same disease. The morbid virus appears to reproduce itself just as seeds appear to reproduce seeds.[71]

Similar arguments could also explain the formation of, and the difference between, malignant and benign growths.

Liebig's contact theory provided the focus of Bischoff's famous paper on the theory of fecundation, which was published in 1847. To the numerous organic reactions that were explicable by Liebig's theory, Bischoff now added fecundation. "There appears to me no more doubt," he wrote, "that the fecundation of the egg is now to be brought under the law under consideration, that it must become acknowledged as one of the most striking examples and proofs of this law."[72] Then, for the first time, Bischoff admitted that the sperm played an essential role in fecundation and were not simply devices to ensure movement of the seminal fluid. The embryo, he argued, develops as a result of an inner motion of elements in the egg yolk, but such motion would quickly become disorganized and the egg would dissolve were not a "certain direction and intensity... given to it, and this is received from the spermatozoa." It was not, he added, the visible motion of the sperm that was imparted to the egg, but the invisible molecular motion within the sperm, of which the visible motion was merely a symptom. It was, in his view, this inner molecular motion that, "imparted to the atoms of the egg, stimulates that inner motion and their continued regulation, which constitutes fecundation." Thus, he concluded: "The semen operated by contact, by touching, through a catalytic power. That is to say it constitutes a material in a certain form of conversion and inner motion. This motion is imparted to another material, the egg, which opposes to it only very slight resistance."[73]

Adherence to the "contact theory" was widespread. Leuckart and Wagner, writing in Germany and also in the British *Cyclopaedia of Anatomy and Physiology*, and Jean Louis Quatrefages, writing in France, all referred to sperm as the "elementary constituents of the animal organization"; as the "direct agents of fecundation," not as parasites. All three believed as well that fecundation occurred after mere contact by the sperm with the egg and that no actual entry occurred.[74] It is clear, however, that Bischoff was less averse to the notion that the sperm actually penetrated the egg. In his 1847 paper he admitted that the sperm could actually penetrate the egg immediately after it escapes from the ovary, but he went on to say, "I do not wish to claim at all, that constituents of the semen penetrate into the interior of the egg and here only exert their influence." His opposition to penetration was based on his denial of any material role for the sperm. "The egg is a whole," he claimed, and has "all the constituent parts," and thus no part of the sperm can be converted into any part of the egg.[75]

What was implied in Bischoff's paper of 1847 and spelled out explicitly in a paper of 1852 was a mounting indifference to the question of sperm entry. Barry's claims could well be "eine Geburt der Phantasie" and fecundation a process of material contact only, but the basic inner motion could be imparted either by the sperm and seminal constituents' only coming into contact with the egg, or by their actual entry into the egg. "Since it requires material contact," he wrote, "so would it be perhaps possible that the dissolution of the sperm and the entry of its molecules through the yolk membrane could essentially promote the effect."[76]

At this point in the argument, the overriding importance of theory to a scientist's attitude toward debatable facts becomes clear. With a theory of chemical contact at his disposal, the question whether sperm did or did not actually penetrate the egg was a matter of little concern to Bischoff; his theory was applicable in either case. The only thing that Bischoff denied was that sperm contributed material to the next generation. Thus, he remarked in his 1852 paper, "If we consider the egg with its yolk, yolk membrane, etc., as a whole, which generally only needs a changing motion to become excited, then the difficult problem of the entry or non-entry of the sperm and its constituents is not at all essential."[77] Bischoff was therefore theoretically prepared to accept the mounting claims that indeed the sperm did penetrate the egg, claims that were now being made by British and German investigators.

In Britain the problem of fecundation was addressed by George Newport. Newport had received only a very elementary education, but by dint of tireless self-schooling, he had become a recognized expert on entomology. During the 1830s he became a surgeon's apprentice, eventually qualifying with a licentiate from the Society of Apothecaries and opening a practice in London in 1837. Because of his involvement in natural-history studies, however, his practice declined, and by the late 1840s he was engaged in research activities full time. In 1850 he presented the first in a series of papers on fecundation to the Royal Society, to which he now belonged. Accepting, as did everyone at that time, that sperm were not parasites, but were "merely elementary constituent parts of the male body," Newport correctly pointed out that the exact role of the semen and sperm was still in doubt. Bischoff, it should be noted, was still somewhat vague on the question whether the sperm alone or the total seminal constituents played the vital contact role in fecundation. The question to which Newport addressed himself was whether fecundation occurred "by any direct or palpable infiltration of seminal matter through the envelopes of the ovum." Having first proved that carmine granules had no deleterious effect on fecundation, he emersed forty-one frog ova in a carmine and seminal-fluid mixture and concluded that no carmine granules of a size comparable to a sperm passed into the egg. "The result of these experiments," he remarked, "was thus most unfavourable to the belief that the spermatozoa penetrate bodily through the membranes of the ovum." This conclusion seemed further strengthened by his repetition of Prevost and Dumas' experiments with coloring matter:

It seemed fair to conclude that the colour imbibed by the ova from ink, in MM Prevost and Dumas' experiment, was due to the admission of the chemically combined colours of the fluid, and not to an admission into the texture of the egg envelopes of solid particles held merely in suspension in the fluid. Consequently . . . there seemed less reason to believe that the spermatozoa bodies very much larger than the ink granules, could enter it.[78]

Newport then tested the effect of killing sperm with caustic potash at various intervals after contact had occurred between sperm and egg. Since segmentation of the egg occurred even after it had had only a few seconds' contact with the sperm, Newport concluded that "the act of impregnation . . . must take place or be commenced very rapidly; and, apparently almost at the instance of contact of the spermatozoon with the coverings of the ovum."[79] Having concluded, however, "that fecundation is commenced almost immediately [after] the fecundating body is in contact with the ovum," and having once again denied that any trace of sperm could be found within the egg, Newport tried to distinguish between two aspects of fecundation that had rarely been separated before. Fecundation, he argued, entails not only a process by which the egg is excited to develop, but also a process by which characters of the male parent are passed to the offspring. Simple contact may therefore be sufficient to induce egg division but not enough, he argued, "to determine the transmission of more or less of the material structural characteristics of the male parent to the offspring." Thus, he remarked, "possibly we may hereafter find that the first changes induced by contact of the impregnating body are completed by its diffluence and by the material constituents into which it is dissolved, being transferred to the yolk by endosmosis."[80] It was conceivable, in other words, that fecundation involved two steps, the first being actual contact between egg and sperm, after which the egg began to develop, and the second and later one being the physical entry of seminal material by which male characteristics were somehow passed into the egg.

The question of the inheritance of male characters had been of little interest to laboratory-trained investigators and had been glossed over or ignored by them. To a naturalist like Newport, however, such questions assumed a fundamental role.

Then, in 1852, in an experiment witnessed by Allen Thomson, professor of anatomy at Glasgow and one of the early pioneers in British embryology and microscopical anatomy, Henry Nelson reported that the spermatic particles in *Ascaris* (a nematode) actually passed into the substance of the ovule. Remarking that there could be no doubt whatsoever of this phenomenon since the completely transparent ovule consisted of a clear, gelatinous substance without a membrane. Nelson claimed with some justification that "the present investigation appears to be the first in which the fact of the penetration of the spermatozoa into the ovum has been distinctly seen and clearly established."[81]

Despite these findings, however, Nelson denied any material role for the sperm. Fecundation, he argued, is purely chemical in nature. The sperm are destroyed in the act, their particles acting to preserve the ovule from decay, to dissolve the vitelline granules and to transform the vitelline granules to embryonic granules. The division of the embryonic vesicle, he went on, is caused by "vitality inherent in it," the embryonic vesicle being a direct product of the germinal vesicle. Thus, "life does not originate at the period of fecundation . . . it is derived from the mother." Indeed, Nelson's view of the whole process had a curious affinity with earlier, ovist theories of preexistence:

A new life therefore is not generated during the development of the new being, by the happy combination of physical forces; but the same life bestowed by God at the creation, continues without intermission, transmitted from mother to offspring, pervading and redeveloping itself in each individual member of the species.[82]

Faced with such a report, the truth of which seemed incontrovertible, and realizing that their theoretical postulates could readily accommodate the fact of sperm entry, it is not surprising to find both Newport and Bischoff finally accepting the validity of Nelson's claim. In 1853 Newport reported that only those eggs of the frog in which sperm had been seen "sticking into the vitelline membrane" had developed. In a footnote to the paper, however, he finally reported having seen sperm actually within the vitelline membrane in contact with and penetrating the yolk itself.[83] A year later Bischoff confirmed these findings. "There can be no more doubt," he proclaimed, "that the sperm actually penetrate the frog egg," and neither could there be any more doubt that the sperm penetrate the eggs of mammals also.[84]

Nevertheless, it is clear that this discovery meant very little to Bischoff. Claiming that sperm entry was limited to those organisms whose sperm were mobile (that it was absent in those with immobile sperm), he asked: "Are we now further ahead in our insights into the nature of the fecundating process?" Admitting that this evidence of sperm penetration represented some advance in factual knowledge, he denied that it provided any proof of the "usual honest chemical process" as distinguished from the "catalytic or contact process": "Still the idea pleases me that the sperm is the exciter of that mobile and mixing phenomenon, with which the development of a new being out of a formless yolk-mass begins and that this excitation corresponds in type and manner to the so-called ferment."[85]

Bischoff's theory gained the support of Rudolf Leuckart, who in 1853 wrote a massive chapter on reproduction in Rudolph Wagner's *Handwörterbuch der Physiologie*. "The sperm operates by contact," he stated, "the same as a contagion or decaying body acts. Not through intimate relations with the egg, but in this way: that it imparts a certain motion to the molecules of the egg which transmitted from atom to atom produce new arrangements, new forms and new qualities."[86]

In spite of Leuckart's denial of sperm entry, it soon became apparent that the sperm cell did penetrate and dissolve within the egg cell. This view of the event received unexpected support from other areas, in addition to the pollen debate that was still raging among botanists. Curious events began to be observed in the algae, a group of plants assumed, since the demise of the Linnaean system, to be totally asexual. In 1836, for example, Jacob Agardh had ascribed reproduction in the algae to the production of sporules, each of which germinated into a plant similar to that which produced it.[87] In 1840, however, Gustav Thuret began his investigations of the algae by noting that *Chara* contained red spherical bodies that were analogous to antheridia in that they produced animalcules bearing "two tentacles of excessive fineness."[88]

The wealthy Thuret, a lawyer by profession and at one time attached to the French Embassy in Constantinople, had given up this career to devote himself to botanical research, remaining throughout his life unattached to any university. Being a wealthy amateur had profound advantages that were not shared by land-locked professionals until the opening of marine biological stations later in the century. His wealth enabled him to work on the seacoast of France examining algae in all their natural beauty. In 1844 he began his investigation of *Fucus*, the common rockweed of the Atlantic Coast. He reported that the blades of rockweed carried receptacles at their tips in which spore-forming conceptacles were buried and that the spores were shed into the sea. But he also noted the existence of other conceptacles. Similar in structure to the antheridia of several lower cryptogams, these conceptacles liberated small bodies that in structure resembled the spermatic animalcules of *Chara,* mosses, and liverworts.[89]

In 1847 Thuret shared the Académie des Sciences prize for his essay on algae reproduction, "Recherches sur les zoospores des algues," which was published in 1850 and 1851.[90] In this paper he noted the production of two spore types in the brown algae, a group which he was the first to classify, and reported that in *Enteromorpha* and *Vaucheria* two spores actually came together and united. In *Fucus* he again reported the existence of large, oval spores and "anthérides" containing antherozoids, each bearing two cilia. The oval spores, he maintained, were the true reproductive bodies of *Fucus* and readily germinated. The function of the antherozoids was a matter of conjecture; they never germinated and they bore a striking resemblance to typical zoospores of other algae. These bodies, he argued, were either sterile reproductive bodies or sexual products equivalent to those seen in other cryptogams. The latter theory seemed more likely, since it was more in conformity with a general theory of cryptogam sexuality. Thuret clearly believed at this time that the antherozoids found in *Fucus, Chara,* mosses, and liverworts were products of the male sex organs and that they influenced the other reproductive bodies in some unexplained way. Three years later, in 1853, after Thuret had moved to Cherbourg, he described the first artificially induced fecundation of "spores" by the antherozoids. This work, he wrote, "furnishes, if I am not mistaken, the first direct proof of the existence of true sexuality in the lower cryptogams. . . . One can doubt no more that these are fecundating organs, and that the antherozoids which they contain are the immediate agents of fecundation, although one has not been able still to observe directly the action of the latter on the female organ or archegonia.'"[91]

The large, oval spores of *Fucus,* he wrote in a paper of 1854, will not germinate unless put into contact with the antherozoids. But nothing authorized him to think, he added, that the antherozoids actually enter the spore as Newport, Nelson, and Bischoff had just claimed for the sperm. Thuret was simply not concerned with any causal analysis of fecundation; he was satisfied to report that fecundation occurred at the moment the antherozoids came into contact with the spore, thereby causing it to rotate.[92]

Across the Rhine, however, Nathaneal Pringsheim had come to different conclusions. Born in 1823, he was a member of that young brood of German botanists who, inspired by Schleiden, had turned their attention to the study of cell development and plant life-histories. Gaining his Ph.D. from the University of Berlin in 1848, he eventually took up a position as *Privatdozent* there in 1851 and began his work on the algae.

Pringsheim realized, quite correctly, that all the claims about the sexuality of the cryptogams, with the possible exception of Thuret's work on *Fucus,* were based on the structural similarity of antherozoids to sperm. "What is still lacking," he wrote, "for a clear and convincing proof, is the demonstration at least in a single case, of the entry of vegetable spermatozoids into the female organ and their action on it.'"[93] The proof came from his work on *Fucus* and *Vaucheria.*

Vaucheria, the widely distributed "water felt," not only produced large zoospores but also, as Pringsheim realized, contained two bodies, *Hornchen* and *Sporenfrucht.* The former, he observed, produced masses of dustlike spermatozoids that acted on the still-unorganized material of the *Sporenfrucht* to produce a membrane-bound embryonic cell. Thus, he concluded, "through the observation of the germination of this spore is the entire proof established, that this cell, formed as a result of the influence of the sperm, is the true reproductive cell of *Vaucheria* produced by the sexual act.'"[94] He also repeated Thuret's work on *Fucus* and again claimed that the antherozoids acted on a membraneless mass of "spore parts" to convert it to a membranous embryonic cell. What should be stressed here is Pringsheim's claim that fecundation did not involve the interaction of sperm with a ripe, membrane-bound cell; rather, the act of fecundation itself produced the cell. However, the new cell did contain within it the material of the penetrating sperm:

> The act of fecundation does not consist of the action of spermatozoids on a specific cell endowed with a membrane, an embryonic vesicle which becomes fecundated through its membrane, but rather that one or more spermatozoids penetrate a membraneless granular mass from which alone the membrane is formed enclosing one of the penetrated sperm and thus producing the embryonic plant cell with the capability of development.[95]

As has been mentioned, Hermann Schacht, who was still engaged in the controversy over the entry of the pollen tube into the embryo sac, had drawn inspiration from what he thought to be an analogous situation in animals—namely, the penetration of the egg by the sperm. In his paper of 1855 he confidently announced his *Befruchtungslehre der Pflanzen:* "that the origin of the germ *Anlage* inside the pollen tube as a certain truth no more may be doubted."[96] Later the same year, however, he expressed a marked change of opinion. Noting from the work of Pringsheim that the sperm could readily enter the female organ (what Schacht called a "membraneless protoplasmic mass"), and there make a direct contribution to the embryo, he also remarked that the physiological significance of fecundation was the same in plants as in animals. Surprisingly, however, he then described this physiological significance as a *Vermischung,* or "mixing," of male and female material.[97] The following year, 1856, Schacht finally admitted that his previous theory had been in error. "The first cell of the plant embryo," he now claimed, "does not arise in the pollen tube, as I believed until now, but rather causes, in an extremely characteristic manner, the formation of this first cell from a membraneless granular mass existing in the embryo-sac before fecundation." This was the view to which Schleiden finally subscribed, also in 1856.[98] In addition, Schacht placed himself in basic agreement with Hofmeister by denying that the pollen tube penetrated the embryo sac, claiming instead that it merely came into direct contact with it. It is not clear whether Schacht at this time denied any material contribution by the pollen or still assumed that pollen material actually passed across into the ovule, where a *Vermischung* took place.

After 1856, Schleiden's theory of pollen collapsed, and botanists explicitly denied that the pollen tube entered the embryo sac. They agreed with Hofmeister that the embryonic membranes remained intact and that a fluid passed by osmosis from the tube into the embryo sac. This issue remained unresolved until the end of the century. In 1882, for example, Julius von Sachs, who had long believed that "fertilization always consists in a union of the fertilising substance of the male cell with the protoplasm of the female cell," remarked: "It is not certainly established whether an extremely fine actual opening in the membrane of the pollen-tube facilitates the direct entrance of the fertilising protoplasm, or whether the membrane remains closed and the fertilising substance diffuses over as a true solution."[99]

The idea that the sperm and pollen material penetrated a naked mass of protoplasm rather than a membrane-bound cell clearly indicated that changes were taking place in the theory of cellular structure. The cell theory of organization, with its emphasis on the cell walls, was being replaced by a protoplasmic theory of life whose emphasis was on the cellular contents. The first clear statement that both plants and animals were constituted of a common basic substance, the protoplasm, rather than of cells, probably came from the pen of the German botanist Ferdinand Cohn. During the 1850s numerous other papers appeared on this subject and led to the gradual acceptance of two basic ideas: (1) that the contents of the cell were more essential than the cell walls; and (2) that cells could actually exist in a membraneless state. Pringsheim's views on the nature of the first embryonic mass were in keeping with the view that both the simplest organisms and the first stages in the development of higher organisms were merely "a small naked protoplasmic clump." In an address of 1856 to the Royal Microscopical Society, William Carpenter drew particular attention to Pringsheim's work on the algae:

> Instead of characterizing the simplest type of vegetable organization as a cell, having a distinct membranous envelope and liquid contents, we should more correctly describe it as a nucleated particle of protoplasm, that may either remain in that low grade of incipient organization of which a homogeneousness is the distinctive feature, or may make that first advance in organization which consists in the differentiation of its substance into the more solid envelope and the more liquid interior, the cell wall and the cell contents.[100]

Postulating a naked lump of protoplasm whose membrane formed only after the entry of the sperm made it somewhat easier to envisage how the sperm penetrated the egg. Nevertheless, others maintained that the sperm and antherozoids actually penetrated the vitelline membrane in order to reach the egg cell. In the case of the phanerogams, for example, it was generally accepted at this time that the pollen-tube membrane remained intact and the pollen material passed into the egg by fluid osmosis. Others, however, believed that the vitelline membrance was perforated by a micropyle through which the sperm passed. In 1855, for example, after examining the eggs of over one hundred insects, Leuckart concluded that a micropyle was a characteristic of insect eggs, and his view was adhered to by many others at that time.[101] As late as 1875, Austin Flint was still claiming that the fact of sperm penetration "almost by necessity pre-supposes the presence of orifices" in all eggs.[102]

It should be mentioned at this point that fecundation was generally thought to involve the entry of more than one sperm, but the actual number never seemed to be a problem of much concern; writers simply used the plural "spermatozoa" when referring to fecundation. The view that fecundation involved the entry of a variable number of sperm of course provided an explanation for the sex of the offspring: the greater the number of entering sperm, the greater the probability of a male offspring.

The theory that fecundation involved the entrance of a minute amount of male material into the egg either directly or through "endosmosis" remained unchallenged for the next twenty years, during which time very little further discussion took place. The reason for this seems clear: the explanation satisfied two fairly distinct groups of biologists who approached the problem of fecundation with very different questions in mind. It satisfied the physiologists, who were seeking a causal mechanism for the initiation of egg development, and it also satisfied field naturalists and plant breeders, who were concerned with the problem of paternal inheritance.

It had long seemed obvious to field naturalists and plant breeders that fecundation involved the interaction of male and female material, and by the middle of the nineteenth century, they had collected a mass of confusing data on the subject of plant hybridization. During the first half of the nineteenth century, while arguments over the nature and role of sperm and pollen were progressing, the Prussian, Dutch, and French academies of science established competitions for essays on the problems of hybridization. In every case the authors of the prize-winning essays—A. F. Wiegmann, Carl Fredrich von Gärtner, Charles Naudin, and others, including Gregor Mendel—noted that some hybrid characters assumed an intermediate condition, while others resembled either the male or female parent, and that whereas first-generation hybrids were intermediate between the parent forms, second-generation hybrids showed a more disorderly variation. It was clear, then, that the characters of the offspring involved the participation of both male and female material in some way or another, and that "the secret of reproduction depends on the fact that it is possible for two different cells to fuse together and become a unified whole."

> Once this process is accepted as a fact, we must recognize it as being natural and necessary for the union of the two cells, when they belong to differently constituted individuals, to produce a more or less intermediate form whose character does not change, no matter whether it was the ovule or the pollen cell that was taken from species a or b. Each of these two cells, the germinal vesicle or the pollen tube, bears within itself the type of the individual from which it was taken, and each of the two species supplies a numerically identical part, namely, a single cell, to the new individual. Even in reciprocal crosses, the union of both must give rise to one and the same intermediate formation which contains the same proportions of both species. . . . Since the sexual union of quite differently constituted individuals—i.e., hybridization—always yields a reproductive product which observes the mean between the types contained within the egg and pollen cell, we may regard this as a law that is valid for all cases of reproduction and therefore also applies to the production of varieties.[103]

The results of Gregor Mendel's experiments likewise convinced him that a union of the elements from the egg and pollen must take place. "If the influence of the egg cell upon the pollen cell were only external," he wrote in 1865, in obvious reference to Schleiden's theory, "if it fulfilled the role of a nurse only, then the result of each artificial fertilization could be no other than that the developed hybrid should exactly resemble the pollen parent, or at any rate do so very closely."[104]

To these plant breeders the entry of seminal or pollen material into the egg was regarded as the very essence of fecundation. Thus, explanations of fecundation which incorporated this interaction satisfied their particular requirements: the sperm and pollen carried paternal characters into the egg.

The second group of biologists, however, was much less sure whether a material mixing was necessary for fecundation to take place. Trained in German laboratories, skilled in microscopy, and physiological in their approach to the mysteries of fecundation, they remained unsure whether the mixing of male and female material played an essential or merely an accidental role in the process. Bischoff, of course, continued to believe that the essence of fecundation was a *Contactwirkung*, which would occur whether or not the sperm entered the egg. He was basically indifferent to the question. As mentioned earlier, he argued that the claim that sperm penetrated the egg might well represent a factual advance in science, but that this fact contributed nothing to the world's understanding of the basic process involved. The basic process for this second group of biologists was the physicochemical mechanism by which a sperm or pollen cell triggered an egg to develop. Mixing might well occur, the physiologists argued, but the essence of fecundation lay elsewhere.

Few biologists were able to separate these two problems and to conclude that two distinct operations were involved. Newport was one who could. Is the action of fecundation, he asked, to be explained as the diffusion of spermatic substances into the egg and an ensuing chemical fusion, or as an "expenditure on the yolk, of a force or power of vitality, which is inherent in the body of the spermatozoon?" Or as both of these?[105] Neither breaking down the sperm with a mortar and pestle, and then filtering and applying the liquid filtrate to the egg, nor applying the broken-down sperm directly to the egg caused fecundation to occur. Thus, he concluded, the substance of the sperm alone did not lead to fecundation, and fecundation "cannot be regarded as the result of a simple chemical combination of the substance of the spermatozoon with that of the egg, but essentially may be due to some dynamical influence in the body." To Newport, the sperm was the seat of a special kind of force by whose vitalizing influence the egg contents were stimulated to develop. Nevertheless, the fusion of sperm and egg substances was also necessary to ensure the transfer of paternal characters:

> Fecundation is not simply the result of a mere fusion or chemical mixture of the substance of the spermatozoon with that of the egg; although such fusion, probably, is necessary to the production of the organized body of the embryo, in determining its structural and psychical peculiarities, and its definite species, and more or less of which may, possibly, help to determine the sex, and the extent to which the structural and psychical peculiarities of the male parent are transmitted to the offspring.[106]

Georg Meissner argued in a similar way. He urged biologists to go beyond the contact theory. That theory could explain how the sperm acted as a kind of ferment to excite motion in the yolk, but the fact that mobile sperm and seminal particles penetrated the egg also provided a basis for the inheritance of male and female characters.[107] In 1859 Allen Thomson explained fecundation in terms of contact by a very small portion of the sperm, but he, too, stressed that the embryonic cell must in all probability "[take] its origin from the germ cell, or its nucleus, or from some part of it, in combination with a determinate portion of the sperm product.'' In other words, the new being must be formed from the interaction of the products of two cells, the germ cell and the sperm cell.[108]

This confusion over the role of sperm—whether they penetrated the egg to form part of the embryonic cell, or their penetration merely stimulated the female egg to develop—can be summarized by comparing two nineteenth-century physiological texts.

William Kirkes's *Handbook of Physiology,* which first appeared in 1848 and which by 1900 had gone through sixteen editions, reveals the strong impact of the contact theory and the basic indifference of its author to the problem of sperm entry. In the first edition, Kirkes wrote: ''Little that is certain can be said'' on the nature of the sperm and whether contact with the egg is essential, ''although it is probably so.'' By 1856, when the third edition was published, and by which time the findings of Newport, Barry, and Bischoff were known, Kirkes was still claiming that the necessity of contact was ''not quite determined,'' although again ''[it] probably is so.'' Only in 1873, in the eighth edition, did he admit the necessity of contact, but as late as 1881 he was still arguing that beyond actual contact nothing was known regarding the action.[109] Austin Flint's *The Physiology of Man,* on the other hand, stressed that fecundation involved a fusion of male and female elements.

Generally, however, the laboratory-based German physiologists adhered to Bischoff's contact theory and denied that sperm penetration was an *essential* mark of fecundation. Such a view remained virtually unchallenged until the 1870s. Wilhelm His, for example, the influential professor of anatomy at the University of Leipzig, argued in favor of ''die Theorie der übertragenen Bewegung'' in 1874 and denied that any material transfer was involved in fecundation.[110]

As Coleman has explained, the contact theory can be viewed as an example of the new physiological ideas that were in vogue in mid-nineteenth-century Germany. Physiology, it was argued, must strive to reduce all vital phenomena to physical and chemical principles.[111] Such an approach to the mystery of fecundation also revealed attitudes that were inherent in this new breed of professional, cloistered as he was in his laboratory. Indifferent, even hostile, to the world of the amateur naturalist, he remained generally oblivious to the problems of inheritance—problems that were of the utmost significance to naturalists and animal and plant breeders.

By attributing sexual reproduction merely to the activation of a special cell by contact with sperm cells, these physiologists clearly undermined any claim for the uniqueness and importance of sexual reproduction. The only difference between a spore or asexual germ cell and an egg cell seemed to them to lie in the stimulus needed for development: a spore required moisture, heat, and sunlight; an egg required sperm or pollen cells. As such, sexual reproduction appeared to be only a minor variant of asexual processes. "The only thing that we know with certainty," wrote Leuckart in 1855,

> is that the sperm, parts of which penetrate the vitellus and parts of which remain in the vicinity of it . . . are gradually dissolved. . . . What, however, becomes of the remains of these fecundating elements is at present completely unknown. It is extremely probable that the mass of the corpuscles becomes, after solution, mixed with the vitellus. We do not even know whether this mixture takes place only after the completion of fecundation . . . or whether it constitutes some essential element in the process of fecundation and development. Still less can we judge, in the second case, whether any remnants of the sperm directly participate in any way in the formation of the embryonic cells or in the construction of the embryo.[112]

Three years later, however, even that much certainty had disappeared.

THREE

The Decline of Sex

A series of intertwined events during the 1840s and 1850s added support to the conclusions that laboratory work on fecundation seemed to imply. Sexual reproduction did indeed appear to be fundamentally identical to asexual reproduction. And, adding insult to injury, asexual reproduction was found to be far more widespread than had ever been realized before. Not only was it found to be common among many invertebrate animals, but, even more surprisingly, the life cycle of a vast number of plants from mosses to angiosperms was discovered to include asexual stages. Sex no longer occupied center stage.

In the eighteenth century and the early years of the nineteenth, the type of development displayed by the vertebrates was accepted as the norm: once fertilized, the zygote developed directly into an immature adult. The only exception to this general rule appeared to be the insects, whose metamorphoses had been thoroughly examined during the eighteenth century. Studies of invertebrates during the first half of the nineteenth century, however, revealed that many planktonic animals were larval stages of well-known benthic organisms, which, like insects, underwent a remarkable metamorphosis during their life cycle.[1] Then, in 1842, Japetus Steenstrup described yet another mode of development: the alternation of generations, or *Generationswechsel*.

The term *Generationswechsel* had been used in 1819 by Adelbert von Chamisso, a student of the *Naturphilosoph* Lorenz Oken, but the Danish naturalist Japetus Steenstrup first drew the attention of the scientific world to it in 1842. Steenstrup had attended the University of Copenhagen during the 1830s, and from 1846 until his retirement he occupied the chair of zoology there. As a student he came into close association with the zoologist Johannes Reinhardt and the geologist Georg Forchhammer, both of whom worked in some capacity with Hans Christian Oersted, whose leaning toward *Natur-*

FIGURE 3.1. Japetus Steenstrup. Reproduced by permission of the Museum of Comparative Zoology, Harvard University.

philosophie is well known. Reinhardt belonged to a circle that included Oersted, Oehlenschlager, and Steffens, the latter having returned from Germany "a proclaimer of the new philosophical and poetic Gospels." In addition, none other than Johann Fichte himself had become part of the circle in 1807.[2] It is in this context, then, that we must understand the opening remark of Steenstrup's book, *On the Alternation of Generations,* which was translated into German in 1844 and published in English one year later. The book, Steenstrup wrote, would be of interest to those concerned with "the harmonies of development in Nature."

Referring to the life histories of polyps and acalephs (later combined as the Coelenterata), salps, and trematodes, Steenstrup pointed out that their eggs produced neither an adult animal nor a larval stage but an offspring "which at no time resembles its parent, but which, on the other hand, itself brings forth a progeny, which returns in its form and nature to the parent animal." He noted also that the return to the maternal form could occur directly in the offspring of this first progeny or after the appearance of numerous other generations. As a result, "the maternal animal does not meet with its resemblance in its own brood, but in its descendents of the second, third, or the fourth degree or generation," the number of such generations being specific for each animal.[3] Given the fact that these "preparative generations" did not change directly into the adult or another developmental stage, but rather eliminated offspring "from the germs contained within them," Steenstrup termed them *ammenden Generationen,* or "nursing generations." He visualized them as female generations in which the sex organs had been aborted, similar in some respects to the workers of insect colonies, which likewise had lost their reproductive organs in order to take on the role of feeding, fostering, or nursing the young.[4] These *Ammen* were widely distributed in the animal kingdom and included the hydroid and scyphozoan polyps, solitary salps, the sporocysts and rediae of trematodes, and the scolex of cestodes (tapeworms).

Steenstrup's work illustrates the widespread belief, held also by the *Naturphilosophen,* that the whole of organic nature, not just individual ontogenies, exhibits goal-directed development. Not only were the life cycles of the forms that showed *Generationswechsel* directed toward the production of the "perfect," sexually mature form, but by comparing such life cycles one with another, it should be possible to discover "how [each] begins and advances, so that at last we discover to what it tends."[5]

At the simplest level—in the Campanularian polyps for example—the generations were very similar and were fused together into a set of polyps, or "an outward unity." Thereafter, Steenstrup argued, there was a tendency toward "becoming free and accomplishing a perfect growth," with the sexually mature, perfect medusae generations being set free from the polyp, often at an early age.[6] Similarly, while in tapeworms the perfect generations (the proglottids) remained connected to the nursing generation (the scolex), in trematodes the different nursing generations were separated from the perfect adult worm and from each other. In aphids, in which a series of nursing individuals arose from the unfertilized ova, "all the individuals are free and enjoy the power of free locomotion and undergo a metamorphosis." In the words of Steenstrup, development was now exhibited "in the external, more free, and nobler form."[7]

Finally, as in the colonial insects, the nursing activity was provided voluntarily by an "internal, obscure and irresistible impulse," in opposition to the purely material physical nursing of the lower forms.[8] The physical nurses of the lower forms were thought to lack an ovary but to possess the rudiments of the oviducts and uteri in which the germs were incubated. These nurses were thus born "with germs in the organs in which the embryos are afterwards nourished." Such germs were, in other words, "committed to their fostering care." In the voluntary feeders of the insect colony, on the other hand, the oviducts and uteri were lacking, physical nursing was therefore impossible, and so "the propagative instinct in a physical and corporeal sense passes into the will for the propagation of the species." In this case, therefore, *Generationswechsel* was no longer present, and development took place through several broods of one generation, not through several generations.[9]

Although naturalists were enormously stimulated by Steenstrup's ideas, it is clear they found the term "alternation of generations" to be extremely obscure. This obscurity arose from their repeated attempts to define the term in morphological rather than developmental terms. The word "generation" is a genetic term that refers to a group of individuals having a common origin. However, because such different generations must, by definition, contain a different set of individuals, many naturalists attempted to define *Generationswechsel* in terms of an alternation of individuals. They thus stressed the morphological criteria by which individuals are recognized rather than the genetic criteria by which generations are formed. In the case of the vertebrates, such an approach would have been quite adequate. Vertebrate individuals are intuitively obvious; moreover, individuals so recognized on morphological terms would be identical to generations defined in developmental terms. But in plants and many invertebrates, no such harmony exists. In the first place the nature of their individuals is notoriously difficult to define. Are plant shoots, individual polyps on a compound stalk, medusae, tapeworm proglottids, and trematode stages within the snail true individuals or are they merely parts of a compound individual? And how does one define an individual so that such questions can be answered? Not surprisingly, no common answers to these riddles were forthcoming, and while both botanists and zoologists thought they were discussing the alternation of generations, they were in fact dealing with the alternation of individuals and became embroiled in semantic arguments over the nature of individuality. However, and this point seems to have escaped them entirely, the mere alternation of individuals does not necessarily imply also an alternation of generations.

If the life cycle of any organism involves the alternation of two or more sets of individuals, such individuals represent different generations only if they arise from different parents or if they arise from the same parent at different times. In other words they must differ from each other in the place or time of their origin. Two sets of individuals arising from different parents at different times clearly belong to different generations, and even if they arise from the same parent, these individuals are considered to represent individuals

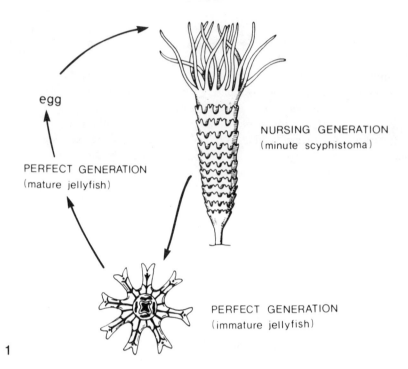

egg

NURSING GENERATION
(minute scyphistoma)

PERFECT GENERATION
(mature jellyfish)

PERFECT GENERATION
(immature jellyfish)

1

NURSING GENERATION (polyps)

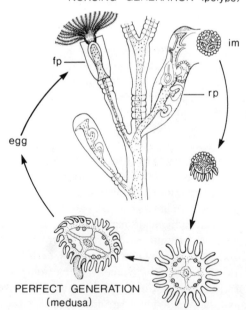

im

fp

rp

egg

PERFECT GENERATION
(medusa)

2

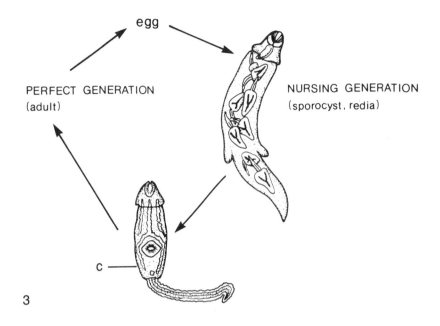

egg

PERFECT GENERATION
(adult)

NURSING GENERATION
(sporocyst, redia)

c

3

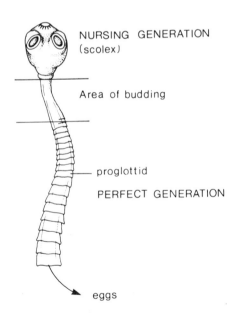

NURSING GENERATION
(scolex)

Area of budding

proglottid

PERFECT GENERATION

4   eggs

FIGURE 3.2. Examples of Steenstrup's alternation of generations in animals: (*1*) acalephs; (*2*) polyps (*fp*, feeding polyp; *rp*, reproductive polyp; *im*, immature medusa); (*3*) trematodes (*c*, cercaria): (*4*) tapeworm. After Allen Thomson, "Ovum," in *The Cyclopaedia of Anatomy and Physiology*, ed. Robert Todd, vol. 5 (1859).

of different generations if they arise at sufficiently wide time intervals. Thus, to give an obvious example, a brother and sister born within a few years of each other will be considered as belonging to the same generation. If this time difference extends beyond some arbitrary, fixed time-scale, they will be said to represent individuals of different generations. But the structure or form of these individuals is totally irrelevant to the issue.

Can Steenstrup's alternating individuals be considered a true alternation of generations? They clearly cannot on the basis of their place of origin. The sexually mature, perfect individuals give rise to the nursing individuals, but the latter do not produce the perfect individual again; they merely act as foster parents to eggs laid by the perfect females. Thus, all the individuals in the life cycle, whether *Ammen* or perfect forms, have a common origin; both forms are derived from the same supply of eggs and on the basis of this criterion belong to the same generation. However, they can be construed as belonging to different generations because, although they arose from the same source, they did so at different times.

This, however, was not the criterion used by Steenstrup. To him, the nursing and perfect individuals represented an alternation of generations because they were individuals that differed morphologically in "form and nature"—criteria that I consider to have been basically flawed and a source of much unnecessary confusion.

By presenting morphological criteria for his "generations," Steenstrup suggested that a parallel existed between *Generationswechsel* in animals and metamorphosis in plants. He remarked at the conclusion of his book that the nursing system "reminds us of the propagation and vital cycle of plants," a subject close to the hearts of German plant morphologists:

> It is certainly the great triumph of Morphology, that it is able to show how the plant or tree (that colony of individuals . . .) unfolds itself through a frequently long succession of generations, into individuals, becoming constantly more and more perfect, until, after the immediately precedent generation, it appears as *Calyx* and *Corolla* with perfect male and female individuals; stamens and pistils.[10]

The alternation of generations in animals was thus equivalent to so-called plant metamorphosis—to the alternation of vegetative and flowering shoots on a tree, for example. A maple (Fig. 3.3) begins growing by putting forth a series of vegetative shoots, but eventually produces terminal buds that form the flower. Thus, the life cycle of the maple can be construed as an alternation of vegetative shoots with flowering shoots equivalent to Steenstrup's nursing and perfect generations respectively.

This theory of plant metamorphosis was derived from the idea that all plant organs have a basic identity. Goethe believed that they were modifications of an idealized leaf. Although the name of Goethe is traditionally associated with this concept of "leaf metamorphosis," it was, in fact, Linnaeus who first

FIGURE 3.3. The morphological alternation of generations in plants. A branch bears two lateral vegetative buds (generation I) and a terminal flowering bud (generation II: the perfect individual). Courtesy of M. Primrose and M. J. Harvey.

conceived of the morphological identity of leaves and flower parts and who used the term "metamorphosis" to describe the *actual* (not idealized) change from a leaf phase to a flowering phase, drawing a conscious analogy to insect metamorphosis. Plants, like insects, he argued, must transform before they can generate. "They proceed from the herb, in which they had remained concealed, like the silk-worm in its caterpillar, and exhibit their true form naked in the flower, whose only business likewise is to propagate its species. The evolution of flowers is exactly similar to the exit of insects from their caterpillars." In other words, "Vegetables, like insects, are subject to a metamorphosis."[11] Nevertheless, Goethe was mainly responsible for popularizing the concept. No longer using it in a purely ontological sense, he expressed it in terms of an idealized morphology. Unfortunately, however, he chose to retain the word "metamorphosis."

Plant morphologists saw the plant not as a unity but as a plurality, the individual elements of which were either leaves or shoots. Many idealistic morphologists also visualized animals in this way. As Guédès has remarked, "In both animals and plants, one could find a basic unity, an ideal individual, which by associations and modifications, realized the multiplicity of the plan of organization."[12] When put into a developmental framework, this beautiful and poetic concept provided that each elementary individual, in order to reach its full potential, had to pass through a series of changes, either discontinuously, as in insects and plants, or continuously, as in vertebrates. Thus, some animal and plant morphologists argued that animal metamorphosis, plant metamorphosis, and Steenstrup's *Generationswechsel* were manifestations of a single biological phenomenon: all involved a succession of distinct morphological individuals.

An extraordinarily lengthy and complex discussion followed on the relationship between Steenstrup's alternating generations and animal and plant metamorphosis, all of which centered on the problem of individuality. Are a tree and a hydrozoan polyp individuals or a colony of individuals? If a plant is a colony, what is the nature of the individuals of which it is constructed? Is the plant individual the cell or the shoot? Are larval stages of insects true individuals and thus equivalent to Steenstrup's nurses and plant shoots?

Fortunately, however, only morphologically inclined naturalists became embroiled in the argument over individuality. Having little sympathy for such scholastic niceties, the new breed of developmentalists came to define *Generationswechsel* in very simple and direct terms. To them *Generationswechsel* meant an alternation of sexually and asexually produced individuals within a single cycle of development. Their criteria were developmental, not morphological; they recognized these individuals as members of different generations because they originated from different sources at different times; and they conveniently ignored Steenstrup's claim that his nurses were merely foster parents.

By using these developmental criteria, naturalists could argue that the alternation of generations was widely based. Allen Thomson could, in 1859, include in its domain the life cycles of polyps, acalephs, echinoderms, mollusks, salps, cestodes, trematodes, annelids, and aphids.[13] Asexual reproduction was no longer limited to a few insignificant lower animals. Also, by using these criteria, naturalists were able to distinguish between the alternation of generations and animal metamorphosis. The latter involved no asexual reproduction, but merely the development of one larval stage directly into another.

This simple solution to the problem was short-lived, however. Even the developmentalists believed or implied that such individuals must also be morphologically different to be termed a true alternation of generations. Moreover, by the end of the 1850s, the distinction between asexual and sexual reproduction had become so clouded that morphological criteria of individuality were again dominant. Developmental criteria of origin were once again replaced by morphological criteria of form and structure, and with this reversal to morphological arguments the relationship between animal metamor-

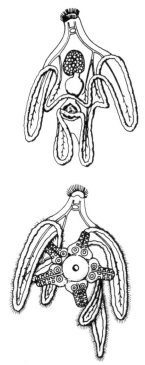

FIGURE 3.4.   A starfish larva (*above*) budding off of a single young starfish (*below*) (after Müller).

phosis and the alternation of generations again became the focal point of controversy. The controversy became particularly evident after Johannes Müller published his reports on the development of echinoderms. While some echinoderms went through a typical metamorphosis from a larval stage to an immature adult stage, others seemed to exhibit a kind of alternation of generations. In the latter cases, as in some starfish, the larva is not converted directly into an adult, as is the case in animal metamorphosis, but seems to bud off a *single* young adult (Fig. 3.4). On this basis Müller concluded that indeed animal metamorphosis and the alternation of generations were closely related to each other. The starfish seemed intermediate between the two. On one hand the larval stage clearly did undergo a form of asexual reproduction, but on the other only one adult was formed in the process and the adult so formed retained the larval gut.[14]

PLANT LIFE-CYCLES

The extension of asexual reproduction into many animal groups previously believed to reproduce by entirely sexual means was as nothing compared to events that took place in botany. Here, a new developmental and cellular approach, initiated primarily by Matthias Schleiden, led to the realization that earlier assumptions about the distribution of sexual and asexual reproduction in plants were incorrect. The realm of asexual reproduction was dramatically increased by the amazing discovery that all plants—cryptogams and phanerogams—reproduced by both sexual and asexual means. Asexual reproduction was no longer limited to the lowly cryptogams.

I have already discussed the controversy generated by Schleiden's claim that the pollen tube contained the embryo that was to be implanted in the embryo sac. However, Schleiden's argument that spores and pollen grains were similar enabled him for the first time to unify the cryptogams and phanerogams. Before that time, partly as a result of the use of the phanerogams as the basic plant model with which all others had to be compared, cryptogams had been defined in essentially negative terms. Thus, to the Linnaeans, cryptogams had been plants in which phanerogamlike sex organs were invisible or hidden, while to the French School and other adherents of the natural method, they were essentially plants without phanerogamlike sex organs, or even without sex organs at all. The emphasis, however, had always been on a comparison of the fully matured lower plant with the mature phanerogamic plant.

Schleiden's perspective was entirely at odds with these traditional approaches. The basis of Schleiden's new botany rested on developmental history, not the structure of mature forms. As he wrote in his *Principles of Scientific Botany:*

The mode in which one cell forms many, and how these, dependent on the influence of the former, assume their proper figure and arrangement, is exactly the point upon which the whole knowledge of plants turns; and whosoever does not propose this question to himself, or does not reply to it, can never connect a clear scientific idea with plants and their life. From the total neglect of this point it is no wonder that most of the notions in botany are enveloped in a dark and formless mysticism.[15]

In addition, he wrote in his famous methodological introduction to his text, "I tried to avoid one basic fault from the start, namely, to begin with the Phanerogamia and from them logically to derive the Cryptogamia. [Such an approach] is contrary to all reason. To proceed from the simple to the compound is the most common procedural method; to explain the compound by way of the parts, rather than vice versa, is the least-disputed hermeneutic rule."[16]

Schleiden's methodology, together with his views on the relationship between spores and pollen, led him to devise a new plant classification scheme. It naturally collapsed as soon as his ideas on the nature of pollen collapsed, but between its inception in 1842 and its demise in 1856, it was one of two that successfully linked together cryptogams and phanerogams in a meaningful way. The other scheme, as we shall see, was devised by Wilhelm Hofmeister.

Schleiden's classification scheme, which is outlined in Fig. 3.5, was clearly based on the presence of the ubiquitous spore and on the primacy of cell development. Unlike his pollen theory, it received little if any support from other botanists, but it does represent the state of botanical knowledge of the lower plants immediately before the 1840s, a decade when profound changes would take place, culminating in the publication of Hofmeister's radically different ideas in 1851. The unifying role of the classification scheme was expressed very clearly by Schleiden. "It removes," he wrote, "the enigmatical separation between *Cryptogamia* and *Phanerogamia,* and at

ANGIOSPORAE —plants with spores enclosed within a parent cell or sporan-
(covered spores)   gium.
                   Plantae aquaticae: plants growing in water (algae)
                   Plantae aereae: plants growing in air (lichens, fungi)

GYMNOSPORAE—plants with isolated, independent spores, contained free in the
(naked spores)   cavity of cellular tissue (sporocarp, anther); plants with axis,
                 leaves, and vascular bundles.
                 Plantae agamicae: spore forms new asexual plant directly
                 (liverworts, mosses, Lycopodiaceae, ferns, Equisetaceae)
                 Plantae gamicae: development of spore (i.e., pollen) requires
                 that it be brought under the influence of certain cells of the
                 parent plant (rhizocarps, phanerogams)

FIGURE 3.5.   Matthias Schleiden's classification of plants

a point where a higher unity is first to be expected, and explains the most widely different facts from one natural law instead of from two. This simplification of the grounds of explanation is, however, one of the most important methodic claims of a sound natural philosophy."[17]

As far as the *plantae agamicae* were concerned, Schleiden was naturally adamant that the so-called antheridia, carried on the tips of the moss plant, were not sexual organs and bore no resemblance to the so-called sex organs of the higher plants. Antheridia, he reported, "neither in the history of their development, nor in their structure, nor in their physiological relation, . . . show the most remote analogy with the anthers of the Phanerogamia." Those who attributed a sexual function to these organs were guilty of "fine-spun fancies," Schleiden wrote, even though he himself never hazarded a guess as to what their functions might be.

To Schleiden, the *plantae agamicae* exhibited three stages of development. At the lower end of the scale were the liverworts, whose spores wholly developed into a new plant. In the mosses, ferns, and fern allies, however, the spores immediately formed a "tubular utricle," the extremity of which formed a proembryo. The antheridia formed in the tissues of the moss plant represented not sex organs but the early stage of the sporocarp, in which the spores would eventually form (Fig. 3.6).

A similar life cycle was observed in the ferns (Fig. 2.6). There, again, the spore developed into a tube, which in turn gradually grew into a flat, bilobed proembryo, the prothallus. The prothallus then germinated into the fern frond, the leaves of which carried the spore-bearing sporocarps. Again, Schleiden heaped scorn on those who believed that the fern bore sex organs. The following passage illustrates the polemical character of his writing: "The *formal* mania for discovering antheridia in the *Cryptogamia* for a long time failed to find support in the class of the Ferns. . . . Fortunately for those who delight in sporting with words without affixing any definite ideas to them, a few glandular hairs . . . were found near the capsules in some species of Ferns; they were pronounced to be anthers, and the discoverers rejoiced in the self-satisfaction of having followed the course of science." Using this type of rhetoric, he continued: "I am surprised that no one has as yet insisted upon the presence of the organs of sense, as eyes and ears, in plants, since they are possessed by animals; such an assumption would not be a bit more absurd than the mania of insisting upon having anthers in the *Cryptogamia,* simply because they are found in the *Phanerogamia.*"[18]

Schleiden recognized two groups of sexual gymnospores, the rhizocarps and the phanerogams. As explained in the previous chapter, he argued that in the phanerogams the spores or pollen grains lying in the cavity of the sporocarp or anther needed to be implanted into a second organ, the ovule (which bore the multinucleate embryo sac), in order for development to take place. It should be noted that the ovule was considered to have no counterpart in the lower plants.

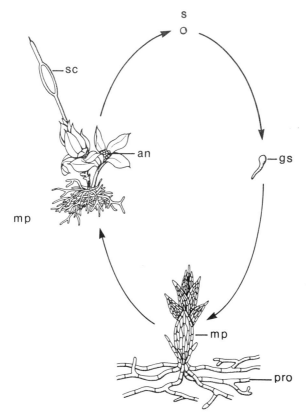

FIGURE 3.6. The life cycle of mosses according to Schleiden: *an*, antheridia; *gs*, germinating spore; *mp*, moss plant; *pro*, proembryo; *s*, spore; *sc*, sporocarp.

In the rhizocarps, on the other hand, Schleiden argued, both the pollen *and* the ovule were released from the parent plants as independent bodies, and the pollen subsequently penetrated the ovule to implant the embryonic cell. (Botanists will realize that the so-called pollen and ovule are, in fact, the micro- and macrosporangia.) The so-called ovule, Schleiden noted, was made up of a very large starch-, mucilage-, and oil-containing embryo sac surrounded by a membrane with a papilla at one end, the whole being enclosed in a cellular sac. Eventually the cells of the papilla broke through the enclosing membrane, and it was with this projection, Schleiden maintained, that the pollen tube made contact, penetrated between the cells, and thus entered the embryo sac.[19]

This, then, was Schleiden's unifying system, a system based on the universal presence of spore cells. Schleiden's *Principles of Scientific Botany*, in which this scheme was detailed, established, in the words of Ferdinand Cohn, "two maxims as supreme watchwords of science: a study of developmental history as the key to all morphology, and the study of the cell's structure and life as the key to plant physiology."[20] "It has been my object, in the present work," Schleiden concluded, "to indicate the proper path, and to open an entrance into it according to the best of my ability. May better men continue the work!"[21]

One of the "better men" to be profoundly influenced by the new developmental and cellular methodology was Wilhelm Hofmeister. Born in Leipzig in 1824, the son of a music publisher, he entered the Realschule there and was thus debarred from entry into a university. By 1841 he had entered his father's business, but had also begun to study botany, influenced no doubt by his father's interest in the subject. During this early period of his life Hofmeister read Schleiden's *Principles* and developed an interest in developmental history and microscopy, a specialty for which he was eminently suitable, being acutely short-sighted. By 1847 he had begun to publish, and by 1851 the University of Rostock had awarded him an honorary doctorate, observing that he was one "who, by observations instituted with the greatest exactness, developed with the finest perception and expounded with the nicest fitness, has admirably illuminated, increased and strengthened the physiology of plants." Indeed, so magnificent was his work that, in 1863, despite his lack of any formal training, he was appointed to the chair of botany at Heidelberg, replacing Theodor Bischoff, who had died a few years before. Although a disciple of Schleiden's new botanical methodology, it is a reflection of Hofmeister's greatness that he quickly came to disagree with Schleiden in four major areas: the method of cell formation, the relationship between spores and pollen grains, the nature of fecundation, and the relationship between cryptogams and phanerogams.

Hofmeister's disagreement with Schleiden over the last three of the above-named areas clearly stemmed from Hofmeister's early work on pollen, which I have already discussed. Once Schleiden's pollen theory had been attacked, it followed that Hofmeister would disagree with Schleiden over the nature of fecundation and the relationship between cryptogams and phanerogams. Hofmeister went far beyond this, however. He not only attacked Schleiden's complete scheme but also put forward his own great unifying system. This system formed the basis of his 1851 text, *Vergleichende Untersuchungen der Keimung, Entfaltung, und Fruchtbildung höherer Kryptogamen, und der Samenbildung der Coniferen,* a book which surely ranks among the world's greatest scientific masterpieces—in content if not in style.

The key to Hofmeister's profoundly original interpretation of plant life-cycles seems to have been his views on the nature of cell formation. Schleiden, in his *Beiträge zur Phytogenesis* of 1838, and Theodor Schwann, in his *Mikroscopische Untersuchungen* of 1839, both had expressed essentially similar views. Not only were plants and animals constituted of an "aggregate of fully formed individualized, independent, separate beings,"

FIGURE 3.7. Wilhelm Hofmeister. Reproduced from Karl Maegdefrau, *Geschichte der Botanik* (Stuttgart: Gustav Fischer Verlag, 1973), by permission of the publisher.

the cells, but these cells had a common mode of formation. Schleiden argued that the nucleus, or cytoblast, of new cells arose from mucus granules within parent cells. A delicate, transparent vesicle then appeared on the surface of the cytoblast and expanded to form the new cell. This process—endogeny, as it was called—was denied by Schwann, who argued instead in favor of an *exogenous* formation of new cells. Schwann ascribed the origin of most cells to a process that occurred in *extracellular* material, the cytoblastema, a process that was very similar to endogeny in that it occurred "in a fluid, or in a structureless substance." The corpuscles that first appeared in this cytoblastema were considered by Schwann to be either nonnucleated cells or, in most animals, the rudiments of the cell nucleus. The remainder of the nucleus

was then deposited around these rudiments, and thereafter the cell membrane
was formed by the deposition of new molecules:

> The following admits of universal application to the formation of cells; there is,
> in the first instance, a *structureless* substance present, which is some times quite
> fluid, at others more or less gelatinous. This substance possesses within itself, in
> a greater or lesser measure . . . a capacity to occasion the production of cells.
> When this takes place the nucleus usually appears to be formed first, and then
> the cell around it.[22]

One of the most important theoretical implications of both of these theories
of cell formation—collectively called theories of free cell-formation—was in
the new meaning they gave to the nature of animal and plant individuality. If
the cell never divided but always propagated by endogenous or exogenous
free cell-formation, the cell itself was the indivisible unity, the true individual. Thus the plant body came to be viewed as a composite, multicellular generation, or as Carl von Mercklin remarked, a "Zellencomplex der
Zusammengesetzen Pflanze." Von Mercklin, clearly adhering to Schleiden's
view of the ubiquity of spores and to the analogy between the pollen grain and
the spore, regarded the single-celled spore thrown off from the multicellular
plant as the plant *Anlage,* the plant individual in its simplest form. From this
spore, he wrote, "begins a sequence of greater or lesser metamorphoses, in
which cycle the production of spores marks each time a new epoch or beginning." In the fern, he argued, there are two important stages of metamorphosis: the first when the spore metamorphoses into the minute prothallus
stage, and the second when the prothallus metamorphoses into the fern frond.
The second stage of the metamorphosis, he noted, begins only after "spiral
threads," arising in special organs on the underside of the prothallus, enter
other special organs on the prothallus, there to dispose a cell to develop into a
frond. But this activity, he further argued, was not one of fecundation, "for
the product of this process is not, as in the Phanerogams, the origin of a new
individual but only the continuing development of the prothallium and the
appearance of a new metamorphosis of the fern plant. The frond, as this
metamorphosis is called, is the partial continuing development of a cell of the
prothallium and designates only a stage in the developmental history of
ferns."[23]

In 1850, in a review of von Mercklin's text, Hofmeister took exception to
this interpretation of the fern life-cycle and used the word *Generationswechsel* in a way that it had never been used before by botanists. Ferns, he
argued, display an alternation of generations, for their life cycle involves the
alternation of a gamete-forming multicellular generation (the prothallus) with
an asexual, spore-forming multicellular generation (the frond).[24] Hofmeister
had come to this conclusion because, by 1850, botanists no longer believed
that cells arose only by free cell-formation. Schleiden's and Schwann's
theories had been subjected to mounting criticisms, and it was now being

argued that two methods of cell formation occurred in plants: cells propagated by free cell-formation or they divided. This distinction had theoretical implications, for a cell that divided could no longer be visualized as an indivisible plant individual.

Such views of cell formation had been articulated by two influential botanists during the 1840s: Hugo von Mohl and Karl von Nägeli. New cells, argued von Mohl, arise either by division or "through the formation of secondary cells lying free in the cavity of a cell." The latter process, he continued, is preceded either by the division of the original nuclei into as many daughter nuclei as there are daughter cells or by the appearance of new nuclei each time. Von Mohl's most significant point, however, was his claim that the process of free cell-formation involving the *de novo* production of new nuclei and new cells occurred *only in the embryo sac of phanerogams and during the production of spores*. In all other instances cells divided.[25] An essentially similar conclusion was reached by von Nägeli.

According to von Nägeli, cells formed by one of two methods. The first—parietal cell-formation—occurred when "the whole contents of the parent cell became divided into two or more portions." The second—free cell-formation—took place when "a nucleus originates in the contents of the parent-cell. This accumulates on its surface, by attraction, a greater or smaller quantity of the contents of the parent-cell, which, at least at the periphery, consist of homogeneous mucilage. This portion of the contents becomes coated by a membrane over its entire surface."[26] The occurrence of both methods was clearly outlined by von Nägeli:

> Parietal and free cell-formation would, according to my views, extend through the vegetable kingdom within the following bounds: *Parietal: the vegetative cell-formation of all plants, further, the reproductive cell-formation of many Algae and Fungi. Free: the reproductive cell-formation of most* (not all) *plants,* namely, the *germ-cell formation of many Fungi, many Algae, and of Lichens; the formation of the spores inside the special parent-cells in the four-spored Cryptogamia (?), the formation of the pollen-cells inside the special parent-cells in the Phanerogamia (?), and the formation of the endosperm-cells in the Phanerogamia.*[27]

Hofmeister also held to this belief in his early work on the phanerogams. In 1847, for example, in his description of fecundation in the Oenothereens, he reported that the immature embryo sac contained nothing but numerous granules floating in a viscous, mucilaginous liquid. Then, at the micropyle end of the embryo sac, a granular agglomeration took place and from two to four free-floating nuclei arose *de novo*; around one a new cell formed, "the true egg of the plant." Once this mother cell was formed, Hofmeister continued, its nucleus divided into two, such that "one half of the content of the cell collects around each daughter nucleus." However—and this I think was to prove very significant—Hofmeister reported that the mother cell did not appear as a result of the division of a preexisting cell or nucleus; rather, it was the product of true free cell-formation.[28] A year later Hofmeister reported that the pollen mother cell also was a "frei gewordenen Mutterzelle."[29] Such views were reiterated in 1849 in his first text on the subject, *Die Entstehung des Embryo der Phanerogamen:* "At some time before fecundation a number of free cell-nuclei are formed in the liquid contents of the embryo sac—sometimes before, sometimes during, and sometimes after the dissolution of its primary nucleus."[30]

The significance of the view that both pollen and egg mother cells arose by free cell-formation, but that all subsequent cells arose by division, lay in the definitions of "generation" and "individual" that could be deduced from it. All cells arising by division from the original pollen and egg mother cells could be seen as representing a single generation of cells, or in other words, a multicellular individual. But since the actual mother cells did not arise from the division of a preexisting cell, they no longer represented part of the preceding generation of cells, but rather represented the initial stage of a new cell generation or a new multicellular individual. The stage at which free cell-formation took place in the life cycle of a plant thus represented the transition from one cell generation to another, from one multicellular individual to the next. If, however, a plant life-cycle involved two stages of free cell-formation (spore formation and gamete formation), then two transitions occurred and an alternation of cell generations took place. This, in fact, was the basis of Hofmeister's criticism of Mercklin's view of the fern life-cycle (Fig. 3.8).

When in 1849 Hofmeister had turned his attention to the mysterious cryptogams, they had been receiving particular attention for about a decade. As Schleiden's work had illustrated at the beginning of this period, those who proclaimed the universality of sex in all plants could claim only marginal

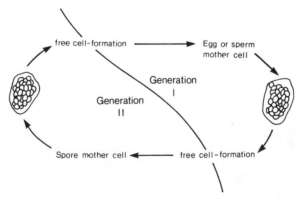

FIGURE 3.8. Free cell-formation and the concept of alternating cell generations according to Hofmeister.

corroboration from the cryptogams. There was little support for the view that the so-called antheridia on the tips of the moss plant were male sex organs and that the spores were seeds. More and more botanists had come to realize that the reproductive bodies of these plants were spores, not seeds; containing only a formless liquid, spores lacked any of the plant rudiments that were visible in seeds. Thus, most classificatory schemes of that period, Schleiden's included, were based on the belief that all or most cryptogams were asexual.

Those who believed that the antheridia of mosses were male sex organs received a further setback in 1844 when von Nägeli discovered antheridialike organs, similar in structure to those found on mosses, on the microscopic prothallus of the fern and not on the fern frond, where supporters of plant sexuality had quite rationally assumed them to be. Von Nägeli reported that the prothallus of the fern bore glandular organs filled with opaque granules. From such organs, he reported, actively moving small cells arose bearing spiral filaments identical to the *Samenfaden* of mosses and liverworts. "On two grounds," von Nägeli argued, "one must come to the conclusion that the antheridia of cryptogams and the spiral-filament organ of the fern prothallus are identical: 1. because they are identical in all essential characteristics. 2. because they are so different from all other known plant organs, by virtue of their characteristics, that they cannot be included among them." They must, von Nägeli continued, be considered male sex organs; "one cannot doubt for a moment their identity." But he had to admit that serious objections could be raised against such an assumption, the most important of which was their curious location.[31]

Naturally, to Schleiden and others, the location of antheridia on such an unlikely spot as the fern prothallus proved conclusively that they had no reproductive role to play either in ferns or in mosses and liverworts. It seemed inconceivable that these spiral filaments could fertilize a cell to produce a seed when the latter (if indeed they were seeds) arose very much later and, above all, after the prothallus had disappeared. This very understandable confusion arose, of course, from a basic misconception. Those who argued for plant sexuality were assuming that the antheridia were male sex organs and that the spores were seeds; those who argued for the asexual nature of these plants were assuming that the spores were not seeds and that sex organs were entirely lacking. Both sides in the controversy therefore argued that the plants must be entirely sexual or entirely asexual—a not unwarranted assumption.

There was considerable confusion also concerning those allies of the fern which produce two types of spores (known today as the heterosporous forms). In the homosporous ferns, only one type of spore is liberated from the underside of the leaf, and these spores germinate to produce the minute prothallus. In the heterosporous fern allies, however, two types of spores are produced: large megaspores that germinate to produce a minute prothallus, and smaller microspores that develop into a few cells while still retained in the microspore wall. They had been described earlier by the Liège professor of physiology

Antoine Spring, who had termed the organ that produces the large spores an
"oophoridium" and the one that produces the small spores an "antheridium," a choice of terms which naturally suggested a sexual role for
both.

In 1847 and 1848 Karl Müller described the development of two such
heterosporous forms, *Lycopodium* and *Isoetes*. He related how spores from
the "oophoridia" germinated to form a blunt, rounded *Keimkorper* that was
analogous to the prothallus of the fern. The "antheridia," on the other hand,
produced spores by free cell-formation, but, Müller claimed, they failed to
develop. The function of these small spores remained a mystery to Müller, but
he regarded the oophoridia as ovule producers, the ovules being capable of
development without fecundation. Thus, Müller argued, these plants formed a
link between the true asexual plants and the rhizocarps. Schleiden had described how the rhizocarps also produced two types of spores, but he had
noted that the smaller spore developed only after implantation into the larger
spore. The most essential characteristic of *Lycopodium* and *Isoetes,* Müller
concluded, was the fact that the plant "formed in the interior of the ovule
without the occurrence of impregnation."[32]

In 1846, von Nägeli added further confusion to the debate by claiming that
the "oophoridia" and "antheridia" of the rhizocarps were indeed sex organs.
Contrary to Schleiden, however, he argued that the latter never formed a
pollen tube, but instead released "spiral filaments" similar to those liberated
in ferns, mosses, and liverworts. He described these "spiral filaments" as
true spermatozoa, not infusorians as others believed.[33]

The belief that the "antheridia" of the ferns and their heterosporous allies
were indeed spermatozoa-producing organs received additional support from
Le Comte Leszczyc-Suminski and Gustav Thuret. In a paper of 1849, the
former argued that in addition to bearing antheridia, as suggested by von
Nägeli, the fern prothallus also carried organs that contained at their base a
small cell which he termed an ovule. Furthermore, Suminski reported, the
spiral filaments of the antheridia actually entered the canal-shaped neck of this
organ and penetrated the basal cell (Fig. 3.9). "As soon as the small head of
the spiral filament reaches the middle of the embryo sac," he noted, "it is
separated from the filament and forms in the sac a closed globule, the embryonic vesicle." Thus, like Schleiden, Suminski believed that the entering
filament carried the embryo that was to be implanted in the embryo sac. As he
concluded:

> In ferns, the proembryo appears as an intermediate formation between the spore
> and the plant; it has to be regarded as an animated complete individual. For,
> separated from the mother plant, it takes root in the soil, is nourished, and
> develops by itself. With fecundation, it becomes the nourishing receptacle of the
> flower; with the production of the embryo it becomes the fruit; with germination
> it becomes the embryo and thus furnishes the young germ with its first nourish-
> ment; it replaces the albumen of phanerogams.[34]

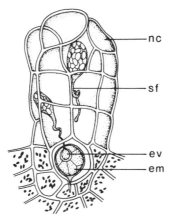

FIGURE 3.9.   Fecundation in the fern (after Suminski). A spiral filament, released
from the fern's antheridia, enters the embryo sac: *em,* embryo sac; *ev,* embryonic
vesicle; *nc,* neck cell; *sf,* spiral filament.

This was the first clear statement that the life cycle of ferns involved both
spore and gamete formation.

At this stage of the debate, Hofmeister entered the picture. In a paper that
clearly illustrates the direction his thinking was taking, he pointed out that the
spores of ferns and their heterosporous allies all originated in the same way as
the egg mother-cell in phanerogams—by free cell-formation. Initially, he
noted, the spores were filled with starch, oil droplets, and albuminous matter;
they contained not "the slightest trace of a cellular body." Then the cell
appeared *de novo.* In the case of the large spores of the rhizocarps, for
example, this newly formed cell within the large spore divided rapidly to
produce a nuclear papilla as described by Schleiden. "I see no reason,"
Hofmeister wrote, "to apply any other name than that of pro-embryo to this
body." The small spores, on the other hand, produced only moving spiral
filaments, which, Hofmeister postulated, penetrated through the tip of the
proembryo into a large, central cell, the ovule of the plant. "Scrupulous
investigations place me in a position perfectly to confirm what, in my opinion,
is the most important point of Suminski's exposition," Hofmeister wrote,
"the regular origin of the young plant . . . in the interior of one of the organs
which he calls ovules."[35]

At this point Hofmeister began his comparison of the liverworts, mosses,
ferns, and fern allies. He noted the striking similarity between the antheridia
and archegonia of mosses and those of ferns, but pointed out their markedly
different localities—the tips of the moss plant, but the prothallus of the fern.
Given these facts, he argued, any analogy between the fern prothallus and the
moss plant (i.e., the sites of the antheridia and archegonia) is prevented by the
common assumption that the moss plant and its "fruit"—namely, the part

consisting of the stalk and the capsule—represent a single, multicellular plant generation equivalent to the fern frond. "This assumption," he wrote, "is fundamentally false." Only the moss "fruit" is equivalent to the fern frond, since both arise from the direct division of a single, free-formed cell located at the base of the archegonia. It is the moss plant and the fern prothallus that are equivalent, in that both bear the archegonia. Thus, in neither mosses nor ferns is the life cycle established by the continuous, repeated division of the spore cell; rather, in each case the spore forms a distinct multicellular individual— the fern prothallus and the moss plant—which in turn produces gametes by free cell-formation.

> In both the development is subjected, if I may so express it, to an inversion, since in a cell, enclosed by an organ of essentially similar structure in both great groups of plants, is developed an independent cellular body, unconnected morphologically [i.e., arising by a process of free cell-formation] with the parent cell, from which body results in the Mosses simply the development of the fruit, in the Filices also far the greater part of the vegetative growth.[36]

With these distinctions, Hofmeister not only established "the theory of sexuality of mosses and Filices," but he drew attention to the curious fact that they reproduced alternately by sexual and asexual means, displaying what he would later call an alternation of generations (Figs. 3.10 and 3.11).

Hofmeister's views were supported a year later by George Mettenius. Describing the development of a broad range of heterosporous forms, Mettenius stressed the basic similarity between their development and that of ferns. In both cases a spore formed a proembryo furnished with sex organs, and from the ovule of this proembryo the young, leafy plant arose. Like Hofmeister, Mettenius believed that the ovule developed only after excitation by the spiral filaments released from the antheridia of the ferns and from the small spores of the heterosporous forms. Like Hofmeister also, he pointed out the error of comparing the moss plant and its fruit to the fern frond, concluding that "the fern proembryo corresponds to a living portion that the moss reaches only after the production of the leafy stem, [and] that in this respect the proembryos of fern and mosses cannot be compared with each other."[37]

But botanists like Mercklin and Braun continued to interpret these new findings in morphological terms. Braun, for example, viewed the prothallus-to-frond transition as a shoot metamorphosis. As he described it, "Many ferns would appear as shootless plants, in the strictest sense of the word, were not the entire fern stem itself a second generation—that is, a shoot produced from the thalluslike proembryo."[38] Mosses and liverworts appeared to be less of a problem to these morphologists since, like the phanerogams, they were leafy plants, some of whose shoots produced sex cells from which the seed-bearing "fruit" arose. Thus, Braun was able to report, "the phenomenon of the essential and necessary succession of shoots long known in the vegetable kingdom agrees completely with that occurring in the animal kingdom, the

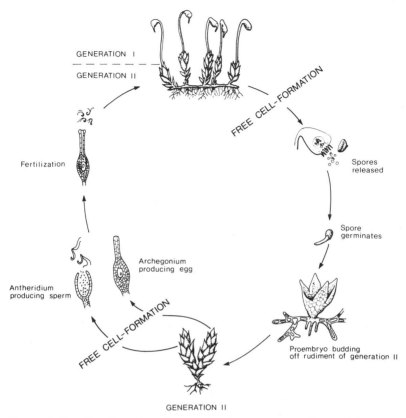

GENERATION I

GENERATION II

FREE CELL-FORMATION

Spores
released

Spore
germinates

Proembryo budding
off rudiment of generation II

GENERATION II

Fertilization

Antheridium
producing sperm

Archegonium
producing egg

FREE CELL-FORMATION

FIGURE 3.10.   The alternation of generations in mosses according to Hofmeister.

so-called alternation of generations brought into its true position by Sars and Steenstrup.''[39]

Hofmeister, however, argued from a developmental, not a morphological, framework. He noted that ''the most striking similarity existed in the processes by which the young plant of ferns, Selaginella and Rhizocarps, *arose* from the prothallium on one hand and the fruit of mosses and liverworts *arose* on the other'' (italics added). These plants, he argued, were clearly separate individuals from those formed by the spores, and thus, he concluded, ''there takes place throughout the higher crytogams a *Generationswechsel*.'' It was not a *Generationswechsel* in the sense of a plant metamorphosis, however. Rather, he described it as one in which

the plants which arise from the spore—the prothallus of ferns, the moss plant in the usual sense—reach the goal of their existence with the production of sperm and archegonia. . . . The destiny of the new individual that arises inside the

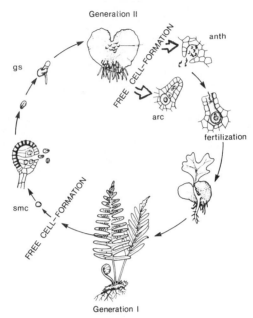

FIGURE 3.11.   The alternation of generations in ferns according to Hofmeister: *anth,* antheridium; *arc,* archegonium; *gs,* germinating spore; *smc,* spore mother cell.

archegonia following fertilization is the production of great quantities of specifically constituted propagatory cells, the spores.[40]

Arthur Henfrey in England agreed with Hofmeister. Admitting that one could interpret the situation in mosses and liverworts in one of two ways—either as an alternation of generations in Hofmeister's sense, or as a simple analogy with phanerogams in which the spore-forming capsule is merely a fruit—he argued that in the higher cryptogams the plant arising from the ovule had "a distinct individual form and existence." "We seem to have a reʳ alternation of generations," he concluded.[41]

What was lacking in Hofmeister's use of the concept was some understand-ing of the relationship between cryptogams and phanerogams. Did the latter display a developmental, as well as the well-known morphological, alterna-tion of generations? The answer came in 1851 with the publication of Hof-meister's monumental *Vergleichende Untersuchungen der Keimung, Entfal-tung, und Fruchtbildung höherer Kryptogamen und der Samenbildung der Coniferen,* a book which one reviewer described as the beginning of "a new era" in botany. It was a book "which illuminated the hitherto dark relation-ships of these plant families to each other and to those of the higher classes, and [it] demonstrated the thread that runs through the diversity of phenomena."[42]

The thread is apparent, however, only in a short review chapter, for the bulk of Hofmeister's text followed what the same reviewer described as "everywhere the inductive method, leading the reader directly to the microscope." Beginning the first sentence of the book with a fact, the author paid no further heed to stylistic niceties. Instead, we have a monument to that brand of nineteenth-century science which relished painstaking, exacting detail and *hypotheses non fingo*. Even the crucial sentence of the book, which appears as late as page 139, conveyed only half the truth. "Mosses and Ferns," Hofmeister wrote, "show one of the most remarkable examples of a regular alternation of two generations that differ much in their organization."[43] He failed to mention one of the most important aspects of his work, the extension of the alternation-of-generations concept to the conifers.

In his short review chapter, Hofmeister pointed out that in both mosses and ferns the structure of the archegonia is the same and that in both a "frei entstehende Zelle" is produced. After fertilization this cell divides repeatedly to form the moss "fruit" and the fern frond respectively. In the mosses, he noted, the vegetative life is confined to the gamete-forming generation, while the spore-forming generation draws its nourishment from the first generation. In ferns this process is reversed, since the spore-forming generation is the vegetative one. Hofmeister then concluded his book with a reference to the conifers. "In more than one respect," he wrote, "the course of development of the embryo of the conifers stands intermediately between those of the higher cryptogams and the phanerogams." He argued that the so-called endosperm within the embryo sac of conifers was equivalent to the prothallus produced by the large spores of the higher cryptogams, and that the "corpuscula" they bore were equivalent to the archegonia of the higher cryptogams. "The filling up of the embryo sac with the albumen [in conifers] may be compared with the origin of the prothallium in the Rhizocarpeae and Selaginellae," which differ only in their mode of fecundation: the former by a pollen tube, the latter by free-swimming spermatic filaments. Thus, "the embryo sac of the conifers may be regarded as a spore that remains enclosed in its sporangium, for the prothallus that it produces never comes to light. The fertilizing matter must make its way through the tissues of the sporangium to reach the archegonia of this prothallus."[44]

During the 1850s, Hofmeister continued his work on fecundation in both cryptogams and phanerogams, paying particular attention to the conifers. His numerous publications from this decade were finally drawn together and incorporated into a much extended version of his 1851 text, which, surprisingly, appeared only in English. This text, entitled *On the Germination, Development and Fructification of the Higher Cryptogamia, and on the Fructification of the Coniferae,* was published by the Ray Society in 1862, three years after the *Origin of Species* appeared.

In the English text, Hofmeister added a substantial section to the conclusion of his review chapter, the evolutionary significance of which soon be-

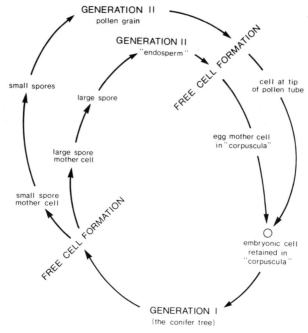

FIGURE 3.12.   The life cycle of the conifer according to Hofmeister

came apparent to botanists. ''The phanerogams,'' he noted, ''form the upper terminal link of a series, the members of which are the Coniferae and Cycadeae, the vascular cryptogams, the Muscineae, and the Characeae. These members exhibit a continually more extensive and more independent vegetative existence in proportion to the gradually descending rank of the generation preceding impregnation.'' In other words, in *Chara* and the mosses, the gamete-forming generation represents the dominant vegetative stage of the plant, while in the ferns and their allies ''this state of things is reversed.'' In the ferns and their allies, the spore-producing individual represents the dominant vegetative phase, while initially the gamete-producing generation is limited to the prothallus body. In most ferns the prothallus bears both archegonia and antheridia, but in a few of them these sex organs occur on different individuals, there to form male and female prothalli. In the heterosporous forms (*Selaginella*, etc.), on the other hand, the male and female prothalli exhibit remarkable differences in their mode of development, size, and form. The minute, male prothallus develops from a microspore and forms motile antherozoids, or sperm, while the larger, female prothallus develops from a megaspore and bears the archegonia. Finally, in the conifers, the male prothallus is reduced to the pollen grain and the female prothallus to the ''endosperm,'' which is now retained in the spore-producing generation and not shed as in the other forms. Basically, this retained female prothallus constitutes the seed. In the flowering plants the female prothallus disappears almost entirely. The principal difference between conifers and flowering plants, Hofmeister noted,

> is the development of the endosperm and of the corpuscula, a process exactly analogous to the formation of the prothallia and archegonia of the vascular cryptogams, and which is entirely wanting in the phanerogams. The whole series of developmental processes which occur in the Coniferae between the filling of the embryo-sac with the cellular tissue of the endosperm and the production of the germinal vesicles in the corpuscula, is entirely passed over in the phanerogams. Here the germinal vesicles are formed immediately in the embryo-sac.[45]

Thus, in phanerogams, "there is no vital phenomenon analogous to the development of the prothallia and of the endosperm of gymnosperms." The true endosperm of the flowering plants is produced only after fertilization and thus is in no way analogous to the "endosperm" of the gymnosperms or to the

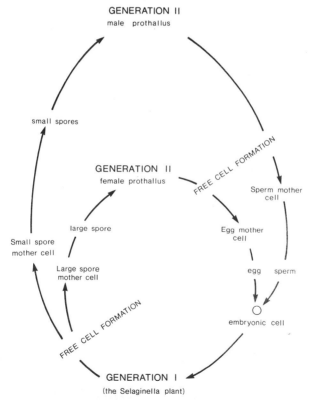

FIGURE 3.13.   The life cycle of a heterosporous fern ally according to Hofmeister

prothallus of the vascular cryptogams. According to Hofmeister, the gamete-producing generation has disappeared in the phanerogams, and thus they no longer display an alternation of generations.

Both editions of Hofmeister's text were obviously explosive documents. On the one hand, Hofmeister had expanded the range of sexual reproduction by showing that it occurred in the cryptogams, a group which many botanists had regarded as entirely asexual. On the other hand, he had extended the range and importance of asexual reproduction enormously, not only by extending its occurrence to the higher plants, but more significantly by showing that in them the asexual generation is predominant. It was profoundly disturbing to realize that conifers were basically asexual spore producers and that the sexual generation had been reduced to a minute pollen grain and a parasitic "endosperm." Shortly thereafter it became apparent that Hofmeister's scheme applied also to the flowering plants, in which the sexual generation had been even further reduced to a very minute pollen grain and an equally minute embryo sac.

Even modern-day botanists have failed to come to terms with Hofmeister's work. By retaining an essentially Linnaean terminology, they have managed to confuse generations of botany students. The male sex organs of lower plants are still called "antheridia," meaning antherlike; but this is a misnomer. They are not antherlike; anthers produce spores, while antheridia produce sperm. Similarly, botanists still insist on using Linnaean terminology in the naming of conifer cones and flower parts. Cones are referred to as male and female, but in reality they produce only asexual spores. An even more confusing distinction is the labeling of stamens and anthers as "male-reproducing organs" and pistils as "female-producing organs." As became apparent over a century ago, however, both stamens and pistils produce spores; they are not sex organs. It is these spores that produce a very reduced sexual generation, which in turn produces sexual nuclei.

One implication of Hofmeister's work soon became apparent: it seemed to illustrate quite magnificently a progressive evolution from the mosses to the seed plants. Another implication, however, was never articulated. Vascular plants reproduce primarily by releasing asexual spores, yet nevertheless retain sexual reproduction. This suggests that sexual reproduction exists for reasons other than reproduction, a view that was totally at odds with the biological and social theories of sexuality in the Victorian era.

PARTHENOGENESIS

The undermining of the importance of sexual reproduction that had taken place with the discovery of asexual stages in the life cycles of so many organisms was carried even further by work on animals that were known to propagate *sine concibutu*. The fact that aphids could propagate "without sexual influence'' had been established by Bonnet in the eighteenth century and had formed an important argument for those who favored the ovist theory of preexistence. By the mid nineteenth century, continued interest in these strange beings had revealed that eggs from an impregnated female overwintered and hatched in the spring. These eggs produced wingless females that were viviparous and produced young without any male influence. After numerous generations of such wingless viviparous females, males and oviparous females were eventually produced late in the summer. These then copulated to produce the winter eggs.

In 1843 Richard Owen proposed a mechanism to account for these strange processes and, not surprisingly, the model he used was sexual. Believing that the sperm endowed the egg with a ''spermatic virtue,'' a ''fecundating principle,'' or a ''spermatic force,'' Owen argued: ''The vitelline cells retain their share of the fecundating principle in so potent a degree, that a certain growth and nutritive vigour in the insect suffice to set on foot, in the ovarian cells, a repetition of the fissiparous and assimulative processes by which they transform themselves in their turn to productive insects, and the fecundating force is not exhausted . . . until a seventh, ninth, or eleventh generation.''[46]

With the appearance of Steenstrup's book, Owen believed that all the cases mentioned by Steenstrup could be explained in exactly the same manner as he had explained vivipary in aphids—by a process he termed ''parthenogenesis.'' To his credit, Owen had little time for the protracted discussions of individuality that were then in vogue. To argue whether or not two beings produced from the division of a parent were new individuals or only part of the original individual was as pointless, he wrote, as those arguments by the wits of ancient Greece who ''debated the question of the identity of the Old Ship of Theseus which had undergone so much repair that no part of the original wood remained.'' What was required, instead, was an explanation of the phenomena.[47]

Owen argued that after receiving the sperm, the germ divides to form ''secondary or derivative impregnated germ-cells.'' Certain of these germ cells form the body of the organism, but others remain unchanged and may themselves form another individual or individuals—similar or dissimilar to the parent—to the extent that they ''retain their individuality and spermatic power.'' For example, in *Hydra,* in which a great number of such unchanged cells exist, almost any part of the animal can form a new polyp. On the other hand, these cells may form ova instead, in which case the animal would reproduce sexually. In the same way, marine hydroid polyps may bud off new individuals but retain them to form a compound body. These may then produce ova, or their derivative germ cells may instead form ''a generative individual of a particular form'' which develops, becomes detached, and ''carries the contained ova to a distance from the composite and fixed group of nutritive individuals.''[48]

This parthenogenesis, ''or power of propagation by the virgin larval polyp,'' depends, Owen argued, on the sperm's procreative force not being exhausted. By the means of this same force, he continued, the parasitic trematode can assume its many forms, with "one impregnated ovum developing many individuals of the second form, each of which finally passes into a third form."[49] The production of such individuals, as in the viviparous aphids, will continue, he further argued, until "the spermatic virtue of the ancestral coitus has been exhausted," and "the condition which renders this seemingly strange and mysterious generation of an embryo without precedent coitus possible, is the retention of a portion of the germ-mass unchanged."

> In proportion to the number of generations of germ cells, with the concomitant dilution of the spermatic force, and in the ratio of the degree and extent of the conversion of these cells into the tissues and organs of the animal is the perfection of the individual, and the diminution of its power of propagating without the reception of fresh spermatic force. In the vertebrate animal the whole of this force . . . is exhausted in the development of the tissues and organs of the individual. . . . The completion of an embryonic or larval form by the development of an ovarian germ-cell, or germ mass, as in the Aphis, without the immediate reception of fresh spermatic force, has never been known to occur in any vertebrate animal.[50]

Similarly in plants, the series of generations preceding the flower are exactly equivalent to the succession of viviparous aphids, the planula and polyps leading to the mature medusa, and the larval forms leading to the trematode. In the plant the parent individual buds off a generation of vegetative shoots each year without the male influence until "finally, we have the energies of the successive generations of the compound plant exhausted in the perfection of the female individual, the pistillum."[51]

According to Owen, therefore, that which causes an individual to reproduce asexually "is the same in kind though not in degree" as that which causes the fecundated egg to develop—namely, the spermatic force. And since "the force is exhausted in proportion to the complexity and living powers of the organism," it is more strongly manifested in plants than in invertebrates. "An impregnated germ cell," he concluded, "imparts its spermatic power to its cell-offspring; but when these perish, or when the power is exhausted by a long descent, it must be renewed by fresh impregnation. But nature is economical; and so long as sufficient power is retained by the progeny of the primary impregnated vesicle, individuals are developed from that progeny without the recurrence of the impregnating act." To Owen, then, sexual and asexual reproduction differed only in nonessential details; both fundamentally depended on the existence of the spermatic force; both were essentially sexual.[52]

In 1851, in response no doubt to those who regarded the alternation of generations as equivalent to animal metamorphosis, Owen coined the term "metagenesis." He defined it as the cycle of change caused by the pollen force or sperm force that is "carried through a succession of individuals and

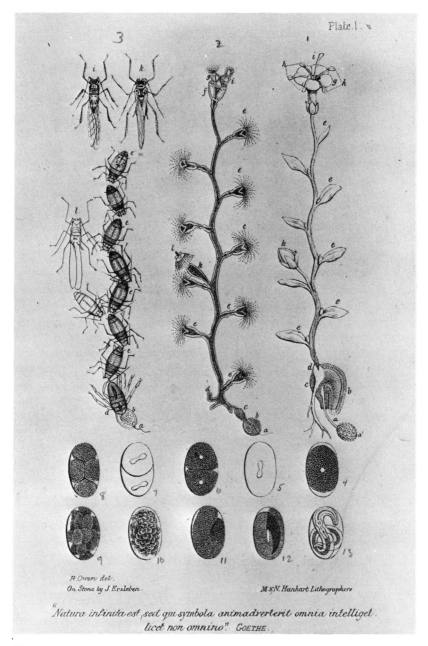

FIGURE 3.14.   The alternation of generations in aphids, polyps, and flowering plants. From Richard Owen, *On Parthenogenesis; Or, the Successful Production of Procreating Individuals from a Single Ovum* (London, 1849). Reproduced by permission of the Museum of Comparative Zoology, Harvard University.

not completed in a single life-time."[53] In so doing he thereby subsumed the alternation of generations and plant metamorphosis under the phenomenon of metagenesis and limited animal metamorphosis to events in insects, crustaceans, etc. Identical views were also presented in the second edition of his *Lectures on the Comparative Anatomy and Physiology of the Invertebrate Animals,* which was published in 1854.

Given the present meaning of the word "parthenogenesis," it is curious that Owen's use of the term was meant to imply a central role for sexual means of reproduction in the life cycle of plants and animals. A totally different perspective arose in 1856 with the publication of Theodor von Siebold's *Wahre Parthenogenesis.* Von Siebold completely undermined the significance of sexual reproduction by pointing out the widespread occurrence of parthenogenesis, which he defined as "reproduction by actual females . . . furnished with perfectly developed virgin female organs, which produce eggs capable of development without previous copulation in an unfecundated condition."[54]

Von Siebold, by this time professor of anatomy at Munich and one of the most prestigious figures in German science, had had a varied career. Graduating in medicine from the University of Berlin in 1828, he spent many years in a vain attempt to gain a university position. Finally, in 1841, he replaced Rudolph Wagner at Erlangen, but moved to Freiberg four years later. There, with Albert von Kölliker, he coedited the prestigious *Zeitschift für wissenschaftliche Zoologie,* which was first published in 1848. The political turmoil of those years, however, led him to Breslau, where he succeeded Jan Purkinje, and eventually, in 1852, he settled in Munich, where he remained until his retirement in 1883.

During these years many papers had appeared in support of the argument that some insects, especially bees and some lepidopterans, reproduced without fecundation.[55] These had been ignored, as being contrary to one of nature's laws, or had been explained away by reference to the seminal receptacle—a storage organ for sperm that enabled them to fecundate many generations of eggs, often in the absence of copulation.

In 1845 the Silesian minister and beekeeper Johannes Dzierzon had published, in a very obscure beekeeping journal, an account of the life of bees in which he suggested that drones developed from unfertilized eggs. One of Dzierzon's supporters, Baron von Berlepsch, had further substantiated this claim by showing that when the seminal receptacles of queen bees were removed, or when the sperm within them were immobilized by a thirty-six-hour exposure to freezing temperatures, the queen bees produced only drones.

In 1855 both von Siebold and Leuckart visited von Berlepsch's apiary in order to test this hypothesis. By then Leuckart had discovered the micropyle in the insect egg through which he assumed the sperm traveled. Both men realized that it might be possible to observe the sperm in this region and thus to ascertain whether only queen and worker eggs actually contained them.

Initially both men failed to establish any legitimate proof, but then von Siebold hit upon the idea of gently crushing the eggs so that they would rupture at the end opposite the micropyle. When he did this, the egg contents flowed out through the ruptured opening, leaving an empty space at the micropyle end. Examining this open space, he found spermatic filaments in thirty of fifty-two worker eggs examined, but none in thirty-seven drone eggs examined. Thus he concluded that eggs, "when they are laid without coming into contact with the male semen, develop into male bees, but, on the contrary, when they are fertilized by male semen, produce female bees."[56]

On the basis of these and similar findings in *Psyche*, *Solenobia*, and *Bombyx*, von Siebold declared that "the hitherto generally admitted proposition of the fecundation theory, that the development of the eggs can only take place under the influence of the male semen, has suffered an unexpected blow by parthenogenesis."[57] A year later Ludwig Radlkofer argued that parthenogenesis also occurred in some plants—notably, in *Coelobogyne*, *Cannabis*, *Mercurialis*, and *Bryonia*. He claimed that his findings represented "the upsetting of a physiological law which is supposed to have been positively established by the most recent researches."[58]

At this time, however, von Siebold believed that aphids displayed an alternation of generations, not parthenogenesis, because the viviparous "nurses" were not identical to the oviparous females. The former, he argued, lacked a seminal receptacle and bore *Keimstock* organs rather that true ovaries. Because they were so different, von Siebold could not accept that they reproduced parthenogenetically: "Under the term parthenogenesis I do not understand reproduction by asexual nurse-like larval creatures, but a reproduction by actual females, that is to say, by individuals furnished with perfectly developed virgin female organs, which produce eggs capable of development without previous copulation and in an unfecundated condition."[59]

The discovery of parthenogenesis and the realization of its widespread occurrence presented microscopists with a vexing question. If eggs could develop without fertilization, they then appeared to behave like "germs," buds, and spores. Was there, then, any fundamental difference between them, and if not, what was the significance of sexual reproduction? Such questions had been implied, of course, in the theory that the sperm acted merely to stimulate the egg, but now even that activity appeared to be somewhat incidental.

In 1857 Thomas Henry Huxley addressed this problem. Unlike von Siebold, Huxley argued that histologically "the agamic offspring of *Aphis* is developed from a body of precisely the same character as that which gives rise to the true egg."[60] To emphasize this similarity, Huxley coined the term "pseudova" to distinguish the products of von Siebold's *Keimstock* organ from the products of the ovary. Despite these histological findings, however, Huxley believed in the essential primacy and uniqueness of sexual reproduction; the ova and pseudova might be histologically identical, but physiologically they were distinct. "It cannot be said," he wrote, "that the sole difference between them is, that one requires fecundation and the other not."[61]

However, Huxley was obviously aware of the pitfalls that lay ahead. "Time was," he remarked, "when the difficulty of the physiologist lay in understanding reproduction without the sexual process. At the present day, it seems to me that the problem is reversed, and that the question before us is, why is sexual union necessary? Far from seeking for an explanation of the phenomena of germination in the transmitted influence of the spermatozoon, the philosopher acquainted with the existing state of science will seek, in the laws which govern germination, for an explanation of the spermatic influence."[62] This passage, it should be noted, was read to the Linnaean Society two years before the appearance of Darwin's *Origin of Species*.

John Lubbock, a friend of Charles Darwin's, carried Huxley's findings to their logical conclusion. In a paper to the Royal Society, which Darwin read for him, he reported on the peculiar pattern of reproduction in *Daphnia*, noting that two types of eggs were produced: agamic eggs, which developed without fertilization; and ephippial eggs, which presumably needed to be fertilized (and whose name is derived from the special boxlike modification of the carapace, or ephippium, in which these eggs are carried). Assuming that the agamic eggs were actually buds, and yet finding no structural or developmental differences between the two types, Lubbock concluded: "I believe that the development of the ephippial and agamic eggs of *Daphnia* . . . is sufficient evidence for the fundamental identity of eggs and buds."[63] Two years later, having examined numerous insects, he declared that a gradation existed between buds at one extreme and true eggs—which required fertilization—at the other. Fundamentally, however, they were all identical; no differences of any real significance existed between buds, parthenogenetic eggs, and true eggs.[64]

These English naturalists were emphasizing the implications of the discoveries and theories on fecundation that had been put forward by the German microscopists a few years earlier. The contact theory had implied that eggs and asexual germ cells differed only in the type of initiating stimulus required. With the discovery of parthenogenesis these earlier implications had become very explicit. In the words of Frederick Churchill, biology had reached a stage of "sex in crisis." "Physiology has become deprived of an apparent law," Leuckart wrote in 1858; it is now "only a principle of experience." As for the necessity of fecundation, he added: "Scientifically this necessity has never been recognized, nor proven in particular cases."[65]

Not everyone could accept that physiology had been deprived of its sexual law. In a *Nachtrag* appended to his *Handwörterbuch der Physiologie*, Rudolph Wagner took issue with Bischoff's and Leuckart's theory of fertilization through contact. Such a theory, and terms like "molecular motion" and "catalytic force," he argued, were meaningless words used by some to hide their ignorance. Fecundation, he continued, was a phenomenon of a different sort, for it involved the passing of male characteristics to the offspring. Wagner's argument implied that sperm were absolutely essential for fecundation and that they needed to penetrate the egg so as to participate materially in the development of the next generation. The discovery that indeed the sperm did penetrate and dissolve in the egg provided evidence for the sort of mechanism for male inheritance in which he had long believed.[66]

It is significant that Wagner occupied a chair at the University of Göttingen at a time when that institution was the center of an antireductionist school of physiology. The materialistic and mechanical approach to physiology practiced by the Berlin School was considered by those at Göttingen to be totally inapplicable to the biological world. Instead, the antireductionists took a more organismic approach, arguing that each organism had to be seen as a functional whole. In the words of Wagner, "Life as embodied in the organism could not be reduced to matter and motion,"[67] and neither could fecundation. Wagner reacted strongly to the ideas about sexual reproduction that had been generated by the studies on parthenogenesis:

> I must unfortunately say that one of the most unpleasant of facts, [parthenogenesis] has been introduced into physiology, which for the hope of so-called general laws of animal life-phenomena is most distasteful. It is impossible, considering the glorification of our highly vaunted progress in the theoretical understanding of the life processes, for it to be welcomed or particularly encouraged; and sincerely speaking, I can be as little pleased about it as a physicist would be if suddenly one or more exceptions to the law of gravitation were discovered.[68]

Clearly, a profound change had taken place since the eighteenth century. Sex had lost status; it was no longer universal or the model for all reproduction. Many animals reproduced asexually, and plants seemed to display a steady progression in favor of asexual modes of reproduction. Even in sexual reproduction, however, the histological similarity between buds, spores, and ova and the very limited role played by the sperm seemed to suggest that no fundamental difference existed between sexual and asexual reproduction. It seemed that a spore and an egg differed only in the stimulus that was needed to initiate development, and with the widespread occurrence of parthenogenesis, even that difference disappeared. Biologically the mid nineteenth century had become a sexless age.

DARWIN AND SEX

These changing ideas had a considerable impact on Charles Darwin. The appearance of the cell theory, the debates over the role and behavior of sperm and pollen, the work of Wilhelm Hofmeister, and the increasingly asexual interpretation of the sexual process all took place while Darwin was struggling to understand the origin of his much-needed variations. In his first notebook, which he began in July 1837 before the onset of the cell theory and at a time when sperm were usually considered to be parasites, Darwin presented the standard argument of plant and animal breeders that asexual, or "coeval," generation was quite different from sexual reproduction. Whereas asexual reproduction produced individuals that were all alike, sexual reproduction "seems a means to vary or adaptation." But, he asked, "with this tendency to vary by [sexual] generation, why are species all constant over the whole country?" The answer lay in the traditional view of blending inheritance: the interbreeding of individuals carrying opposite variations destroyed these variations. Thus Darwin initially believed that variations arising during the production of eggs and semen were normally kept in check by interbreeding.[69] As he stated in his third notebook, which was written in 1838, "One of the final causes of [the] sexes [is] to obliterate differences.... if animals became adapted to every minute change [in the environment], they would not be fitted to the slow great changes really in progress." Darwin also viewed fecundation in the traditional way: "Woman makes bud, man puts primordial vivifying principle."[70]

Sexual reproduction thus resulted in the *gradual* accumulation of blended characters, which was necessary if species were to adapt to a very slowly changing environment. "Without sexual crossing," Darwin noted in his fourth notebook, of 1839, "there would be endless changes, and hence no feature would be deeply impressed on it [the species], and hence there could not be *improvement*."[71] Thus, at this stage of Darwin's thinking, sexual reproduction was not only unique but also necessary as an evolutionary mechanism. With asexual reproduction no changes would occur; with egg production alone, widely divergent individuals would arise; only through crossing would the requisite slow changes ensue. But Darwin also felt that "the absolute necessity that every organic being should cross with another [is] the weakest part of my theory."[72] Indeed it was the weakest part, for plants commonly propagate asexually, and in the 1830s English botanists like Hooker believed that cryptogams were totally asexual.

By mid-century, when there seemed to be no essential difference between buds and ova and when asexual reproduction had assumed a more dominant role, Darwin downplayed the importance of sexual reproduction. "We do not know," he wrote in 1861 (in contrast to his views in the 1830s), "why nature should thus strive after the intercrossing of distinct individuals. We do not in the least know the final cause of sexuality."[73] In the *Origin of Species* itself he also downplayed the importance of crossing. "Sports," he noted there, arise from buds, "but it is the opinion of most physiologists that there is no essential difference between a bud and an ovule." As a result, "variability may be largely attributed to the ovules or pollen, or to both, having been affected by the treatment of the parent prior to the act of conception. These cases anyhow show that variation is not necessarily connected . . . with the act of generation.''[74] When the final edition of the *Origin* appeared in 1872, Darwin claimed again that ''the importance of crossing has been much exaggerated,''[75] and in 1876 he noted that ''the mere act of crossing by itself does no good.''[76]

Variations now arose because, during several generations, individuals were subject to slightly different conditions. These conditions then caused the reproductive organ—whether ovaries, testes, or sporangia—to ''depart from its normal function of like producing like'' and instead to produce variation: ''The chief cause of variation did not supervene during pregnancy, or the formation of the seed, or during the act of impregnation; but in the action of the life of the parents, on the separate male and female elements of reproduction.''[77]

What was the significance of sexual reproduction to Darwin now that it had lost its evolutionary implications? Crossing, he had noted in 1858, is a ''subordinate Law of Reproduction''; through it individuals ''have their vigour and fertility increased.''[78] Darwin's theory of evolution reflected mid-nineteenth-century views on sexual reproduction. Sex was essentially sexless, a process of little significance to the overall scheme of things, a process of rejuvenation that played no unique role in the evolutionary story. It was, moreover, a process that involved a totally asexual mechanism—the gemmules:

> All forms of reproduction graduate into each other and agree in their product, for it is impossible to distinguish between organisms produced from buds, from self division, or from fertilized germs; such organisms are liable to variations of the same nature and to reversion of character, and as we now see that all the forms of reproduction depend on the aggregation of gemmules derived from the whole body, we can understand this general agreement. Sexual and asexual generation are fundamentally the same.[79]

An evolutionary theory could not be based on any special sexual mechanism.

FOUR

The Sexless Age

Despite all the discoveries described in the last three chapters, assumptions about the role of eggs, pollen, and seminal fluid changed very little between the time of Linnaeus and that of Darwin. The dominant view in the eighteenth century—that in the egg alone lay the material of the future generation—was basically shared by biologists in the mid nineteenth century.

Of course, in the nineteenth century, unlike the eighteenth, material from the relatively minute sperm and pollen was assumed to enter and dissolve in the egg or embryo sac, thereby providing a tangible vehicle for the passage of male characteristics to the offspring. However, the widespread occurrence of parthenogenesis and the assumed similarity of buds, spores, and eggs had undermined the significance of this mixing. In both centuries the essence of sexual reproduction was generally assumed to be the stimulating effect of seminal fluid, sperm, or pollen material on the egg. Basically, therefore, in both centuries, reproduction was a uniquely female occupation in which the role of the male was very limited or even entirely unnecessary.

Despite these obvious similarities, however, biologists in the nineteenth century interpreted sexual reproduction very differently than their eighteenth-century predecessors had. In the eighteenth century, sexual reproduction was thought to be virtually universal and quite distinct from the much less important asexual methods of reproduction. By the mid nineteenth century, however, asexual reproduction had risen in stature. Not only had its boundaries been extended to encompass many of the lower animals as well as the entire assemblage of vascular plants, but in addition, eggs were now assumed to be virtually identical to asexual cells. As a result, sexual reproduction had lost status. It was no longer considered universal, unique, and the model of all reproduction, as it had been in the eighteenth century; rather, it was merely a minor variant of the more fundamental, asexual process. As Darwin pointed out in 1868, the distinction between asexual and sexual modes of reproduction "is not really so great as it first appears,"[1] and "it is satisfactory to find that sexual and asexual generations are fundamentally the same." Parthenogenesis, he noted, "is no longer wonderful; in fact, the wonder is that it should not oftener occur." Darwin's satisfaction rested on his belief that all forms of reproduction graduate into each other: "It is impossible to distinguish between organisms produced from buds, from self division or from fertilized germs."[2]

In a curious way, nineteenth-century biological views on sexual reproduction mirrored nineteenth-century social views on sex. Socially and biologically, sex was denied. Indeed, one wonders whether biological theories of sex were at least partially determined by the social views these naturalists shared. In other words, nineteenth-century biological theories of sexual reproduction can be thoroughly understood only within the social context of the nineteenth century. Indeed, the differences between eighteenth- and nineteenth-century theories of sexual reproduction reflect, in part, those centuries' differing social values.

The clearest example of this social and scientific interaction was the fact that, biologically and socially, sexual reproduction was thought to represent a particularly clear example of the division of labor. This concept had first been introduced by Bernard Mandeville in his *Fable of the Bees* and was further elaborated by Adam Smith. "Man," Mandeville wrote,

> naturally loves to imitate what he sees others do, which is the reason that savage people all do the same thing. This hinders them from meliorating their condition. For if one will wholly apply himself to the making of Bows and Arrows, whilst another provides Food, a third builds Huts, a fourth makes Garments, and a fifth Utensils, they not only become useful to one another, but the Callings and Employments themselves will in the same Number of Years receive much greater Improvements, than if all had been promiscuously follow'd by every one the Five.[3]

By the nineteenth century, Mandeville's concept had been extended from man's society to the biological world. Rudolf Leuckart in Germany and Henri Milne-Edwards in France continually referred to it and included sexual differentiation in its compass. With the appearance of sex organs, Leuckart pointed out in 1851, a division of the tasks depending on these organs took place, a single individual being viewed as incapable of fulfilling both roles with equal facility. In terrestrial forms, where internal fertilization is required and where the care, protection, feeding, and rearing of the young is so important, it is not surprising to find distinct male and female individuals appearing: the former to seek out the female; the latter to care for the young. Naturally, each sex, having a different role, requires distinct equipment and organization. In some cases, Leuckart noted, the male and female come together only to reproduce, but in others the male has additional roles to play. He protects society and receives weapons, courage, and the endurance to do it. However, such a division of labor is not restricted to dioecious forms but is also present in insect societies, in polymorphism, in colonial plants, in polyps, and in Steenstrup's alternation of generations (which he described as "a polymorphism that is brought about by a division of labor in the area of ontogeny [*Entwicklungslebens*]").[4] A similar and more extensive treatment of these phenomena appeared in Leuckart's massive article "Zeugung" in 1853.

Similarly, Henri Milne-Edwards, in his *Eléments de zoologie* of 1834, attributed the different modes of reproduction found in animals to the "principe de la division du travail." In simple organisms, he explained, any individual has the capacity to produce new individuals by fission, budding, or asexual spores. In the more advanced forms, however, the reproductive germs are produced in a special organ, the ovary. Initially, these forms were hermaphroditic, but further extension of the division of labor led to the dioecious condition.[5] In sexual reproduction, he argued in 1863, we see "la loi de perfectionnement par la division du travail et la spécialité des instruments."[6]

From the perspective of the division of labor, sexual reproduction was very simply *the means of reproduction employed by those species in which a distinctive reproductive individual, the female, had been formed.* In social terms, this division of labor into males and females allowed the male to expend his energies in more noble and civilized pursuits while placing the entire burden of procreation and childbearing on the female. Friedrich Engels expressed this facet of sex most succinctly when he wrote: "The first division of labor is that of man and wife in breeding children. . . . It develops the welfare and advancement of one by the woe and submission of the other."[7]

The place of women in Victorian society was further fixed by the belief that any society that ignored such obvious biological differences was in danger of collapse. "Any system which ignores this division of labour," wrote Lynn Linton, "and confounds these separate functions, is of necessity imperfect and wrong."[8] Such biological laws naturally became a constant source of frustration to those who demanded sexual equality and women's rights, for built into them was the assumption that females were not only different but inferior. This concept of the division of labor also provided a biological rationale for stereotyping female and male character traits and for fixing the roles the two sexes played in society. A woman, wrote James Weir in 1895, "[is] by function a mother, by virtue of her surroundings a housewife." To quote his particular stereotyping, "Woman is a creature of the emotions, of impulses, of sentiment, and of feeling; in her the logical faculty is subordinate."[9] The female of the species existed only to procreate; she and sex existed for no other purpose. "The Almighty, in creating the female sex, had taken the uterus and built up a woman around it."[10]

One of the most vivid examples of the way biological and social theories of sex became intertwined can be seen in the work of William Brooks of Johns Hopkins University. Brooks was a student of Louis Agassiz's and a teacher of Thomas Hunt Morgan's. In the 1880s, when questions of inheritance were beginning to preoccupy biologists and when the women's rights movement was growing in intensity in the United States, Brooks published his *Law of Heredity*. In it, he made very clear the commonly held distinction between inheritance and variation. Heredity, or "a power to produce a definite adult animal, with all its characteristics, even down to the slightest accidental peculiarity of its parents," resides, he argued, in the egg. In a quite distinct process, however, variations arise which soon become a part of this heredity. Brooks believed that laws of heredity were needed to explain how such variations arose. Only in sexual reproduction, he argued, did variations arise which were not inherited—that is to say, not part of the egg's makeup.[11]

After discussing at length the relative merits of "evolution" and "epigenesis" as explanations for the hereditary characteristics of the egg, Brooks concluded that the ovum "really contains, in some form or other, actually or potentially, the future organism with all its hereditary characters"—in brief, that the egg must be a complex of material particles corresponding to "each of the hereditary characteristics."[12] How, then, did variations arise? Brooks's answer, which came by way of Darwin's theory of gemmules, was that each cell, being an individual, can throw off minute germs when the conditions of life change. These germs, or gemmules, are gathered up and stored in the male gametes, so that at fecundation each gemmule conjugates with "that particle of the ovum which is destined to give rise in the offspring to the cell which corresponds to the one which produced the germ." Thus the male gemmules were the source of variation, and by conjugating with a specific particle within the egg, they transmitted to descendants a tendency to vary in the part that was initially affected by the change in environmental conditions. "The male element," Brooks wrote, "is the organizing and the female the perpetuating factor; the ovum is conservative, the male cell progressive."[13]

As evidence for this theory, Brooks reprinted in his book an essay entitled "The Condition of Woman from a Zoological Point of View," in which he had put forward the social argument that a fundamental difference exists between the sexes. Not surprisingly we read:

> Among the lower animals and most plants both sexes are united in the same individual, but the law of physiological division of labor, the principle that an organ or organism, like a machine, can do some one thing better and with less expenditure of force when it is specially adapted to this one thing than when it is generally adapted for several functions, would lead to the preservation by natural selection of any variations in the direction of a separation of the sexes, and we should therefore expect to find among the higher animals what we actually do find: the restriction of the male function to certain individuals, and the restriction of the female function to others. From this time forward the male is an organism specialized for the production of the variable element in the reproductive process, and the female an organism specialized for the production of the conservative element.[14]

From this biological perspective, Brooks then characterized the two sexes in the usual way:

> If this is so, and if the female organism is the conservative organism, to which is intrusted the keeping of all that has been gained during the past history of the race, it must follow that the female mind is a storehouse filled with the instincts, habits, intuitions, and laws of conduct which have been gained by past experience. The male organism, on the contrary, being the variable organism, the originating element in the process of evolution, the male mind must have the power of extending experience over new fields, and, by comparison and generalization, of discovering new laws of nature, which are in their turn to become rules of action, and to be added on to the series of past experiences.

Not surprisingly, therefore, Brooks went on to argue that the occupations and therefore the education befitting females differed fundamentally from those suited to the male. Men "pursue original trains of abstract thought"; they become scientists, poets, and artists, and at lower levels of society, competitive tradesmen. "Women, on the other hand, would seem to be better fitted for those occupations where ready tact and versatility are of more importance than the narrow technical skill which comes from apprenticeship or training, and where success does not involve competition with rivals."[15] He admitted that such scientific views on the subject were identical to what he called the old-fashioned male view of women. However, he concluded, "the positions which women already occupy in society and the duties which they perform are, in the main, what they should be if our view is correct; and any attempt to improve the condition of women by ignoring or obliterating the intellectual differences between them and men must result in disaster to the race."[16] There was nothing unique in Brooks's view of the sexes, for the idea of a conservative female and a variable male was widespread. Brooks's writing clearly showed however, how biological theories were fitted into the prevailing social attitudes.

Moreover, the parallels between social and biological theories went deeper. Sex as a division of labor implied that sexual reproduction was a special kind of asexual reproduction. In some protophytes, Herbert Spencer explained, the entire organisms coalesce, while in slightly higher forms, only the contents of two cells come together. Then, as we proceed higher in the series, the reproductive elements become distinct, arising from distinct organs. Spencer argued, however, that these germ cells were in no way special: "They have not been made by some unusual elaboration, fundamentally different from all other cells... they are not so peculiar as we are apt to assume." They arise, he theorized, like all other cells, by an asexual division from organs that have no special structures. Thus, he concluded, "there is no warrant for the assumption that sperm-cells and germ cells possess powers fundamentally unlike those of other cells."[17] In other words, the nineteenth-century female, whether plant, animal, or human being, not only existed to procreate but did so by essentially asexual means. Eggs were simply cells produced in the same way as all other cells. In biological terms, the egg differed little from an asexual spore; in social terms, the mid-Victorian female was urged to behave as if indeed she did procreate asexually.

The Victorian ideal of love provides a clear example of this asexual approach. The ability to love was central to the Victorian ideal of womanhood; it was a "woman's all—her wealth, her power, her very being." According to Sarah Ellis, however, this love was strictly platonic in nature:

> In woman's love is mingled the trusting dependence of a child, for she ever looks up to man as her protector, and her guide; the frankness, the social feeling, and the tenderness of a sister—for is not man her friend? The solitude, the anxiety, the careful watching of the mother—for would she not suffer to preserve him from harm?[18]

The relationship between a mother and her child was exactly akin to the relationship between husband and wife; a woman was a mother to both. E. J. Tilt expressed those ideas in *Elements of Health and Principles of Female Hygiene* when he described the mother as a "*nutrix,* the chief nourisher and supporter of mankind, whether to an infant seeking milk at her breast, or to suffering humanity requiring love's watchful tenderness to restore it to health."[19]

In her famous book on household management, Isabella Beeton probably best describes the Victorian ideal of a woman and her household in a fairly affluent middle-class setting. A woman's role was to serve others—her children and her husband. As the author urged on her reader: "A good wife is Heaven's last best gift to man, —his angel and minister of graces innumerable, —his gem of many virtues, —his casket of jewels —her voice is sweet music —her smiles his brightest day; her kiss, the guardian of his innocence; her arms, the pale of his safety, the balm of his health, the balsam of his life; —her industry, his surest wealth; —her economy, his safest steward; —her lips, his faithful counsellors; —her bosom, the softest pillow of his cares; and her prayers, the ablest advocates of Heaven's blessings on his head."[20] To attain this high ideal, a woman had to manage her household properly; a rigorous set of rules had to be obeyed. Isabella Beeton described all of this in exacting detail in the first volumes of her book. She began with a quote from *The Vicar of Wakefield,* which, although written in the eighteenth century, conveys the middle-class Victorian ideal of womanhood:

> The modest virgin, the prudent wife, the careful matron are much more serviceable in life than petticoated philosophers, blustering heroines, or virago queens. She who makes her husband and her children happy, who reclaims the one from vice and trains up the other to virtue, is a much greater character than ladies described in romances, whose occupation is to murder mankind with shafts from their quiver or their eyes.

Lynn Linton, author of a series of essays, constantly urged women to fit into the Victorian ideal. They do not lead armies, she wrote, but "they make the characters of the men who lead and are led."[21] Such a woman must, she warned, be generous, capable, modest, dignified, "a tender mother, an industrious housekeeper, a judicious mistress." The girl of the period who dyes her hair and paints her face pays "fearful moral penalties." Men may amuse themselves with her, but "when they go into their mother's drawing rooms, with their sisters and their sister's friends, they want something of quite a different flavour." The women of England who "gather around them love, homage and chivalrous devotion" do not seek self-gratification. Rather, their lot in life is to suffer and serve, being "most at liberty to devote [themselves] to the general mood of the whole, by cultivating cheerful conversation, and adapting [themselves] to the prevailing tone of feeling, and leading those who are least happy, to think and speak of what will make them more so."[22] They were, above all, sexless, childbearing, child-rearing, man-serving goddesses who had been placed on a pedestal of virginal white purity.

Any physical sexual relationship bespoiled this image of purity. As R. E. Sencourt observed, "It was an age when nothing was thought so hideous as nakedness, and nothing so wicked as impassioned love."[23] Under the leadership of the Wesleyans, a series of reforming societies were established in the early Victorian era whose express purpose was to regulate all forms of enjoyment. According to some historians, this move toward prudery and away from the sensuality of the previous century was the device of a civilization which felt itself to be only a step away from barbarism. The middle classes—the center of the sexless cult—could certainly see around them, in city slums, the activities of a seemingly depraved people. To stop themselves from slipping back into this state, as well as to set themselves apart from the sexually promiscuous aristocrats, they needed to enforce rigorous rules of behavior. Societies for the Reformation of Manners, for the Suppression of Vice, and for the Prevention of Prostitution, the Association for Securing a Better Observance of Sunday, the Religious Tract Society, and the Proclamation Society, whose goal was to suppress "licentious publications," bore down relentlessly on the middle classes. The attitude toward sex which they encouraged seemed to mirror the attitude toward sexual reproduction displayed by nineteenth-century biologists: women, like organisms, reproduced in an essentially asexual manner.

Duncan Crow, whose description of sexual intercourse significantly paralleled the biological outlook of the period, was I assume, unaware, of the validity of the parallel to which he alluded. "Ideally," he wrote, "women would produce children by parthenogenesis; failing that, male impregnation should take place in a dark bedroom into which the husband would creep to create his offspring in silence, while the wife endured the connection in a sort of coma, thereby precluding any stigma of depravity which would have been incurred by showing signs of life. Silence was important."[24] Ideally, too, animals would reproduce parthenogenetically, the male sperm being an accessory device of no real significance to reproduction.

However, sexual intercourse was not just unnecessary; it was considered a bestial act. As such, it became the great unmentionable. Just how the young Victorian middle-class female changed from a modest virgin into a prudent wife and mother was never discussed. She was expected to produce children in abundance, but how this came about was, in the words of Duncan Crow, as hidden as "an Eleusinian mystery." How could she know? All reference to sex was banned in correct society; it was something barbaric that needed taming. In her autobiography, Annie Besant described the dilemma: "So I married in the winter of 1867, with no more idea of the marriage relation than if I had been four years old instead of twenty. I had been guarded from all pain, shielded from anxiety, kept innocent on all questions of sex." No wonder marriage came as a "terrible shock to a young girl's sensitive modesty and pride."[25]

In his study of Victorian sexuality, Fraser Harrison, pointed out that the entire sexual function "remained either a non existent concept or a nebulous target for unresolved uninformed speculation."[26] E. J. Tilt totally ignored the bestial issue, remarking only that a woman is "the matrix in which humanity is cast."[27] How, where, when, and by whom? one might ask.

Consider Thomas Hardy's description of Alec d'Urberville's act, whereby poor Tess became "a maiden no more":

Darkness and silence ruled everywhere around. Above them rose the primeval yews and oaks of The Chase, in which were poised gentle roosting birds in their last nap; and about them stole the hopping rabbits and hares. But, might some say, where was Tess's guardian angel? where was the providence of her simple faith? Perhaps, like that other god of whom the ironical Tishbite spoke, he was talking, or he was pursuing, or he was in a journey, or he was sleeping and not to be awaked.

Why it was that upon this beautiful feminine tissue, sensitive as gossamer, and practically blank as snow as yet, there should have been traced such a coarse pattern as it was doomed to receive; why so often the coarse appropriates the finer thus, the wrong man the woman, the wrong woman the man, many thousand years of analytical philosophy have failed to explain to our sense of order. One may, indeed, admit the possibility of a retribution lurking in the present catastrophe. Doubtless some of Tess d'Urberville's mailed ancestors rollicking home from a fray had dealt the same measure even more ruthlessly towards peasant girls of their time. But though to visit the sins of the fathers upon the children may be a morality good enough for divinities, it is scorned by average human nature; and it therefore does not mend the matter.[28]

This treatment of sex, which by the very absence of explicit detail created a scene of intense sensuousness, contrasts sharply with the more explicit and cruder treatment rendered by many eighteenth-century writers. The exploits of Tom Jones provide a case in point. Whereas Victorians typically tried to distinguish "love" from bestial animal "passion," Henry Fielding expressed the common eighteenth-century attitude that love without passion was unnatural:

That which is commonly called love, namely, the desire of satisfying a vora-
cious appetite with a certain quantity of delicate white human flesh, is by no
means that passion for which I here contend: this is indeed more properly
hunger; and as no glutton is ashamed to apply the word love to his appetite, and
so to say he loves such and such dishes; so may the lover of this kind, with equal
propriety, say he hungers after such and such woman. I will grant... that this
love for which I am an advocate, though it satisfies itself in a much more
delicate manner, does nevertheless seek its own satisfaction as much as the
grossest of all our appetites: and lastly, that this love, when it operates towards
one of a different sex, is very apt towards its complete gratification, to call in the
aid of that hunger which I have mentioned above.[29]

When Tom caught philosopher Square in bed with Molly—a somewhat delicate situation, since Molly was carrying Tom's child—Square reasoned thus:
"Fitness is governed by the nature of things, and not by customs, forms, or
municipal laws; nothing is indeed unfit, which is not unnatural." "Right!"
cried Jones, "what can be more innocent than the indulgence of a natural
appetite; or what more laudable than the propagation of our species?"[30] It
seemed "very strange and absurd" to Fielding "to deny the existence of a
passion of which we often see manifest instances," yet that is precisely what
transpired in the nineteenth century.

Moreover, according to Linnaeus, such sexual delights were shared by the
plants. While a young student at Uppsala, Linnaeus penned these immortal
lines:

The petal of the flower in itself contributes nothing to generation, but only
serves as the bridal bed, which the Great Creator arranged so beautifully, and
garnished with such precious bed-curtains, and perfumed with so many deli-
cious scents, in order that the bridegroom with his bride may therein celebrate
their nuptials with so much greater solemnity. When the bed has been so
prepared, is the time for the bridegroom to embrace his darling bride, and loose
himself in her.[31]

His teachers, Celsius and Rudbeck, we are told, were delighted with Linnaeus's paper. It is hard to imagine a youthful Linnaeus of the nineteenth
century describing the flowers of plants in quite these terms, and it is even
harder to imagine his professors at that time doing anything but censoring it.
Beauty in one century had become sinful and bestial by the next.

Even in his serious published work, Linnaeus openly discussed the "nuptials of plants." He discussed, too, his observation that the genitals of animals
in rutting time have "a ranke strong smell" compared with the very agreeable
scents of flowers. Such language would have been unthinkable in the
nineteenth century, when even reference to "the body" was considered inde-
licate. Overtly sexual words became completely taboo, and were to be clothed
with coy euphemisms. Indeed, the whole area between the neck and the knees
in human anatomy was often simply referred to as "the liver." Linnaeus, to

be sure, would have found such social conventions a bore of the first magnitude.

Neither can one imagine some of the etchings by James Gillray, a major eighteenth-century English cartoonist, being accepted for publication in reputable nineteenth-century magazines. The two reproduced here (Fig. 4.1) are sufficiently explicit and require no comment.

Paintings, however, became a popular art form in the nineteenth century, and classical art in particular came to be seen as a "civilising influence, offering as it did an escape from vulgar reality into a sort of cultural dreamland." Much of classical art had emphasized the beauty of the naked body, and in the nineteenth century the necessary respectability was ensured by presenting naked females in a Grecian setting and by using models from classical mythology. Within such settings, paintings such as Alexandre Cabanel's *Birth of Venus* and Lord Leighton's *The Bath of Psyche* were generally assumed to lie within the bounds of Victorian propriety. Indeed, the latter painting, which shows a maiden posed at the side of a pool gazing at her nude reflection in the water, was described by one art critic as "a suave conception informed with a spirit of perfect chastity, through its inimitable purity of feeling and classic refinement." Such paintings explicitly characterized the human female as a classical, sexless goddess, and their popularity, of course, lay in the opportunity they presented for Victorian gentlemen to view pictures of nude females behind the façade of "high art."[32]

A similar, although far less marked, contrast can be seen in French novels of the eighteenth and nineteenth centuries. Two distinct attitudes toward sex, love, and marriage can be seen in the eighteenth-century novel. The first, which was sentimental and romantic, accepted the Christian ideal of respect for celibacy and rejection of sexual activity. The second, "le roman érotique," rejected this ideal, projecting instead "an anti-Christian view of love by arguing that it was purely physical, and by separating it from marriage."

> Heroines in the erotic novel typically took the lead in love, and they frequently seduced youths. The enterprising heroine often took her lover to her apartment or room, which contained furniture with paintings designed to stimulate and arouse. . . . Love in the classical novel was sometimes Platonic, sublime, and *spirituel*. . . . in the erotic novel it was but a meeting of two skins. Consequently a variety of objects such as furniture, mirrors, paintings, were used in order to maximize physical contact.[33]

Similarly, a new genre of art appeared in eighteenth-century France. Like the erotic novel, this "bedroom art" depicted love as completely physical, an act in which only pleasure counted. "In both the "*roman érotique* and bedroom art the boudoir was furnished with sumptuous beds, couches, erotic statuary, and paintings designed to arouse the passions."[34] The close association between the erotic novel and "bedroom art" is shown in Nicolas Lav-

Lady Godina's Rout. _ or _Peeping-Tom spying out Pope-Joan. Vide Fashionable Modesty.

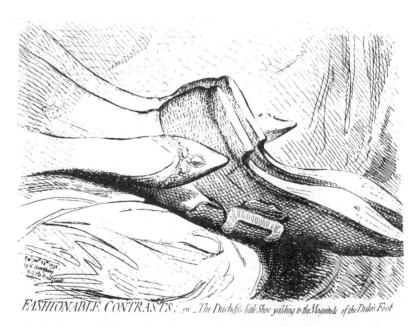

FASHIONABLE CONTRASTS; _ or _The Dutchess's little Shoe yeilding to the Magnitude of the Duke's Foot

FIGURE 4.1. Two etchings by James Gillray published in the 1790s. Reproduced by permission of the Trustees of the British Museum.

FIGURE 4.2. Lord Leighton's *The Bath of Psyche*. From Peter Webb, *The Erotic Arts* (New York: Graphic Society, 1975). Reproduced by permission of The Tate Gallery, London.

FIGURE 4.3.   Nicolas Lavreince's *Le Roman dangereux*. Reproduced from Warren
Roberts, *Morality and Social Class in Eighteenth-Century French Literature and
Painting* (Toronto: University of Toronto Press, 1974).

reince's *Le Roman dangereux* (Fig. 4.3). Here, a scantily clad lady lies on her
bed with an erotic novel open on the floor beside her. A man creeps up to take
advantage of the effect created by the book.

In the nineteenth century such novels and paintings were, like prostitutes and mistresses, hidden behind the public façade of married life, but marriage guides stressed the desirability of satisfying sexual desires; prudery was far less marked in nineteenth-century France than in England. Nevertheless, French society was more moralistic in the nineteenth century than it had been in the eighteenth. The Catholic clergy, in particular, constantly stressed the sanctity of marriage, the need for a strong family life, and the belief that the purpose of sex was procreation, not pleasure. It recommended the suppression of sexual desire and continence in marriage. Thus the church sought to maintain Christian morality by preaching self-control; the sexless life was to be cherished.[35]

In January 1857 Gustave Flaubert was put on trial for publishing the infamous novel *Madame Bovary*. It was, said prosecutor Ernest Picard, an offense against ''la morale publique'' and against ''la morale religieuse.'' The public morality, he argued, was offended by lewd passages, among which the forest scene involving Emma Bovary and Rudolphe Boulanger was especially stressed. ''Les amants arrivent jusqu'aux limites extrêmes de la volupté,'' Picard remarked.[36] The scene, as painted by Flaubert, was certainly sensuous, but unlike the erotic novel of the eighteenth century, it carried no explicit reference to the sexual act: ''The cloth of her robe caught against the velvet of his coat, she threw back her white neck, swelling with a sigh and swooning with tears with a long shudder and hiding her face, she surrendered herself to him.''[37] The major concern of the prosecution, however, was not so much the scene's language as its seeming glorification of adultery. ''That, sirs,'' Picard stressed, ''was for me more dangerous and more immoral than the downfall itself.''[38] It would be more accurate, Picard argued, to title the book ''Histoire des adultères d'une femme de province.'' The book was, he concluded ''un crime pour la famille.''[39]

Although Flaubert was acquitted, the author of a book like *Madame Bovary* would not have been subject to such a trial in the eighteenth century. Social attitudes toward sex had changed in France, as they had in Britain, though perhaps not to the same degree.

An openness to sex can be seen in the popular medical literature of eighteenth-century England. The ''Aristotle series'' was filled with the sort of sexual information and guidance that was totally lacking in the following century. As the reader of *Aristotle's Master Piece* learned, virginity, that most noble of nineteenth-century virtues, grows useless if kept, ''a stale virgin, (if such a thing there be), being looked upon like an old almanack out of date.'' The series' readers were enjoined to enjoy ''the act of coition or carnal copulation'' within the confines of marriage. The happy pair should ''survey the lovely beauties of each other,'' while the bridegroom might heighten the joy by whispering an ''amorous rapture'':

I will enjoy thee, now, my fairest come,
And fly with me to Love's Elysium,
My rudder with thy bold hand, like a try'd

And Skilful pilot, thy shalt steer, and guide,
My bark in Love's dark channel, where it shall
Dance, as the bounding waves do rise and fall.
Whilst my tall Pinnace in the Cyprian strait,
Rides safe at anchor and unlades the freight.[40]

In sharp contrast to views commonly expressed in the nineteenth century, eighteenth-century women were said to actually enjoy sexual intercourse and to share with man the "voluptuous itch, which begets in them desire to the action." "Women are never better pleased than when they are often satisfied this way," the reader learns from *Aristotle's Compleat and Experienc'd Midwife*, "which argues the pleasure and delight they take therein, which pleasure and delight they say, is double in women, to what it is in man."[41]

Similarly in sharp contrast to nineteenth-century views, "carnal copulation" was seen by the authors of the "Aristotle series" to be a healthy pastime. "It eases and lightens the body, clears the mind, comforts the head and senses, and expels melancholy," whereas, again in contrast to nineteenth-century thinking, "[with] the omission of this act, dimness of sight doth ensue, and giddiness." Holding to a spermist doctrine of generation, the authors of the "Aristotle series" argued that "the seed of man retained above its due time, is converted into some infectious humor."[42]

But if nineteenth-century females were considered sexless beings, nineteenth-century males shared a similar fate. In contrast to much eighteenth-century advice, nineteenth-century males were warned against excessive seminal emissions. One medical pamphlet even went so far as to suggest that a healthy man "does not and cannot have sexual emissions. Health does not absolutely require that there should ever be an emission of semen from puberty to death, though the individual live a hundred years." For husbands, as for their wives, love must be platonic, or, as this idea was expressed in 1870, "Only lust creates semen, pure love, never any." Once more we are reminded that sexual intercourse was considered to be a lustful and bestial act even within wedlock. There were even good medical reasons why the semen should be preserved within; it was, after all, the source of an energizing principle or, if Dumas was to be believed, the source of the whole nervous system of the offspring. In an era much impressed by the law of energy conservation, therefore, to lose it was to deprive the ardent male of much-needed and irreplaceable nervous energy, and to lose it in excess would lead to drastic complications—the disease spermatorrhoea.

Victorian males seem to have gone to extraordinary lengths to conserve their precious semen. For example, gadgets could be purchased that were designed to prevent "nocturnal leakage,"—night being the most dangerous time for the poor male. Among these were leather penis-rings with inward-turned steel points that would awaken the sleeper when the penis erected; knotted towels that prevented the sleeper from lying on his back; "cork cushions'' that kept the thighs apart; and an alarm system that was triggered by an erect penis closing the electric circuit![43]

The French physician François Lallemand, professor of medicine at Montpelier, was a recognized authority on spermatorrhoea, a disease which "degrades man, poisons the happiness of his best days and ravages society,"[44] and which was considered to be ultimately responsible for quite terrible consequences. Excessive seminal losses, effected by "venereal excesses," masturbation, "nocturnal leakage," inflammation of the genitals from such activities as sleeping on the belly (heating and titillating the genitals), lascivious ideas, and horsebackriding, if allowed to continue unchecked, were considered to be the causes of a continuum of deleterious malfunctions that culminated in the disintegration of the nervous system. As an article in *Lancet* described it, "At length, epilepsy, catalepsy, mania, or some other disease of the nervous system, makes its appearance, and the patient is relieved from his horrid state of existence by a premature death."[45]

Lallemand described the case of a man who, because of "unaccustomed exercise" of the genital organs following his marriage, lost one kidney, suffered brain damage, and eventually died. From such a series of symptoms, Lallemand concluded, "the ever-increasing influence of spermatorrhoea over the whole economy, but especially over the cerebro-spinal system, can be appreciated."[46]

Lallemand is an especially important figure, for he was one of the very few who supported Dumas' thesis of fecundation (that the sperm contribute the nervous system to the offspring). In 1841 he examined human semen using the new achromatic microscopes of Oberhoeuser and Irecourt and noted that the sperm, agitated "like tadpoles crowded together in a pool of stagnant water," were more active and enjoyed a greater longevity after "l'acte venerien" than after unnatural "pollutions diurnes et nocturnes."[47] He went on to argue, on medical grounds, that sperm could not be parasites, for if they were, loss of them would presumably lead to an improvement in health, not disease. On the contrary, he argued, "the prolonged sojourn of zoosperms augments the physical force and moral energy" of the male, and it is those "exhausted by involuntary seminal losses" who "fall into an even more deplorable state.''

> But if they were parasites, why would their expulsion have such deplorable effects? If one recalls that a single drop of this liquid in a normal state contains thousands of living beings, one will understand that their production has other consequences than that of mucus; that their composition, energy, vivacity etc. are modified by the passions, by images and lascivious ideas, by copulation etc. These would be impossible to explain if one regarded the zoosperms as parasites.[48]

Following the vitalistic teachings of the Montpelier School, Lallemand conceived of sperm as "living tissue" and as "an organized living part" having the elements necessary to form the cerebral spinal system. "In effect," he wrote, "the male, more ardent than the female in all species, furnishes the zoosperm . . . which becomes the first element of the cerebrospinal system, that is to say of all exterior life."[49] In Victorian society, the male was the center of force and vitality. Since this force would be dissipated by the loss of sperm, it followed, on medical grounds, that the sperm must contribute this force to the next generation. It made medical sense, therefore, to claim that the sperm actually contributed the material of the cerebrospinal system, just as Dumas had argued.

Nevertheless, the production of children had to go on, and thus the male was faced with a curious dilemma. Why be a father if the effects of releasing sperm in sexual intercourse were as deplorable as those of releasing them during masturbation? The answer was provided in William Acton's widely read *The Functions and Disorders of the Reproductive Organs*, which was first published in England in 1857. Before the age of twenty-five, *continence* must be practiced; after the age of twenty-five and the onset of marriage, it could be replaced by *moderation*. Before this magic age, Acton urged, total abstinence is required of the growing male, in thought, word, and deed, and he must be constantly reminded of the dangers of spermatorrhea and masturbation. But then suddenly "the whole being of the male cries out, at this period of his life, for, not the indiscriminate indulgence, but the regulated *use* of his matured sexual powers. And at this time, therefore, but not before, the medical man will recommend marriage."[50] (The poor woman, meanwhile, remains totally unaware of this crying out, particularly since Acton's book treats only of the male reproductive organs.) At last, with marriage, the male could "moderately" gratify his sexual passions. Still, the ever-present threat of spermatorrhoea required that the male partake of intercourse at the most only every seven to ten days, a prescription which fell short of that given to eighteenth-century males, who were often advised to "enjoy" it every day. Even Acton was forced to admit that certain pleasurable sensations were involved, but he added, hastily: "This pleasurable sensation, however, is of momentary duration; like a battery, it exhausts itself in a shock."[51]

Acton provided his readers with biological reasons for the caution and moderation he advocated. Not only were the sperm and semen bestowed with a considerable amount of energy, but, he added with some reservations, the semen that were not emitted would be reabsorbed into the male's body, there to augment "in an astonishing degree the corporeal and mental forces."[52] When carried to the brain by the blood, this energy would be "coined into new thoughts—perhaps new inventions—grand conceptions of the true, the beautiful, the useful, or into fresh emotions of joy and impulses of kindness."[53]

Even pregnancy itself and child rearing were seen by Acton as devices by which the female could stifle the male's sexual desires and thereby augment this highly desirable reabsorption of the semen:

If the married female conceives every second year, during the nine months that follow conception she experiences no great sexual excitement. The consequence is that sexual desire in the male is somewhat diminished, and the act of coition takes place but rarely. And, again, while women are suckling there is usually such a call on the vital force made by the organs secreting milk that sexual desire is almost annihilated. Now, as all that we have read and heard tends to prove that a reciprocity of desire is, to a great extent, necessary to excite the male, we must not be surprised if we learn that excesses in fertile married life are comparatively rare, and that sensual feelings in the man become gradually sobered down.[54]

As astonishing as it may seem to us, Acton went on to reiterate the Victorian notion that "the best mothers, wives and managers of households know little or nothing of sexual indulgence." "Love of home, children and domestic duties are the only passions they feel," he argued. Nevertheless, he deplored the fact that young married women displayed such ignorance of sex.[55] But how could it be otherwise, when society made a point of denying women access to any sexual information?

Nineteenth-century scientists noted other reasons, too, for limiting a male's sexual activity. To be parsimonious in seminal emissions was considered a sign of civilization, as an observer of nature would see. The higher forms of life, Herbert Spencer noted in 1852, propagated less frequently than did the lower forms; a long-lived oak tree "does not develop as many acorns as a fungus does spores in a single night."[56] According to Spencer, the ability to propagate varied inversely with the longevity of an individual life. Civilization, he further noted, was correlated both with man's longevity and with his cranial capacity. The average cranial capacity of an Englishman, he reported, was 96 cubic inches, compared with only 82 cubic inches for the African and 75 cubic inches for the Australian aborigine. Believing that African and Australian natives were sexually promiscuous and short-lived, he argued that fertility decreased with increased civilization, and that thus, spermatic energy was channeled into cerebral activity, thereby increasing the cranial capacity and enhancing longevity. Self-discipline, therefore, that most cherished of middle-class virtues, was the key to English civilization:

The contrast between a Pacific Islander, all whose wants are supplied by nature, and an Englishman, who, generation after generation, has had to bring to the satisfaction of his wants ever-increasing knowledge and skill, illustrates at once the need for, and the effects of, such discipline. And this being admitted, it cannot be denied that a further continuance of such discipline, possibly under a yet more intense form, must produce a further progress in the same direction—a further enlargement of the nervous centres and a further decline of fertility.[57]

Utopia, Spencer noted, would be reached with zero population growth!

In passing, one might note that many Victorian middle-class males did in reality enjoy numerous sexual adventures both before and after the age of twenty-five. Not only was prostitution rampant in nineteenth-century society, but domestic servants and lower-class girls in general (who were certainly more worldly than their middle-class female peers) provided an ever-available market for learning, practice, and perfection. The class hierarchy of English society made it absolutely impossible for a girl to reject a young man's advances, despite the fact that, if caught, the repercussions would fall only on the female.[58]

I am not arguing that biological theories of sexual reproduction in any way determined social attitudes or vice versa. However, as many historians of science now acknowledge, scientists and their theories cannot be divorced from their social contexts. The central role of sex in theories of reproduction in the eighteenth century, compared with its low status in the nineteenth, reflected changing social attitudes toward sex as well as a change in the social rank of scientists. Eighteenth-century biology was dominated by the wealthy, often aristocratic amateur; nineteenth-century biology, more and more by members of the sexually more prudent middle classes. The biological theories to which the nineteenth-century scientists subscribed—an almost sexless egg-laying female and a reproductively insignificant energizing male—were as much a reflection of these middle-class values as they were the result of the biologists' scientific discoveries.

FIVE

Asexual Progression

In 1919, the English botanist and historian Robert Harvey-Gibson pointed out that the *Origin of Species* had appeared at a very opportune time, "for it supplied the key required to elucidate the wonderful discoveries that had been made by Hofmeister."[1] In one sense, Gibson was correct. With the collapse of Schleiden's system of classification in 1856 and the appearance of the *Origin of Species* three years later, evolutionary botanists in Germany began to discuss Hofmeister's work at great length. In another sense, however, Gibson was incorrect. A truly *Darwinian* interpretation of the alternation of generations did not appear until 1890, and then not in Germany but in Britain. It came from the great Glasgow botanist Frederick Orpen Bower. He, and only he, introduced the key concept of adaptation into the discussion.

German evolutionary botanists were basically indifferent to the concept of adaptation, and with their strong morphological heritage, approached Hofmeister's work from a very different perspective. Their goal was limited to discovering algal and flowering-plant homologues to the two alternating stages that were known to exist in the higher cryptogams and conifers, and to *describing* the alternating generations throughout the plant kingdom. They offered little or no explanation for the existence of this peculiar phenomenon or for the reduction of the gamete-producing generation in the ferns and conifers.

To a morphologist, the key to understanding living organisms lay not in Darwinian adaptation but in discovering the underlying structural unity by which Nature had formed all beings according to one or a few basic body plans. The construction of idealized body plans, or archetypes, which contained potentially all the organic forms pertaining to that archetype was one of their prime concerns, as was the description of homologous organs. Homologous organs were, in the words of Richard Owen, "the same organ in different animals under every variety of form and function."[2] Two organs were said to be homologous if both could be derived from the same organ of the same idealized archetype. Thus, as was mentioned in previous chapters, plant leaves, sepals, petals, stamens, and carpels all were considered to be homologous since all could be derived from an idealized leaf of a flowering plant archetype (Fig. 5.1).

Morphologists occupied a very important, although often underrated, place in the history of nineteenth-century biology, both before and after Darwin. They not only dominated much of German biology but had representatives in France and Britain too. In France, for example, Etienne Goeffroy Saint-Hilaire not only claimed that the organization of vertebrates could be reduced to one uniform type, but he even suggested that the articulates (the segmented annelids and arthropods) also could be reduced to this vertebrate type. Likewise, the English morphologist Richard Owen believed the vertebrate skeleton was ideally made up of a series of identical vertebrae, and from this ideal he constructed the archetypal vertebrate.

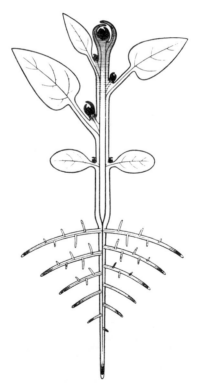

FIGURE 5.1.   The archetypal plant (after von Sachs). Reproduced from Wilhelm Troll, *Allgemeine Botanik* (Stuttgart: Ferdinand Enke, 1948), by permission of the publisher.

Although the morphologists were a diverse group, they all believed that organic diversity could not be ascribed to the physical conditions in which organisms existed. Such physical conditions, Louis Agassiz argued in his *Essay on Classification*, modify features that are of only secondary importance, such as the thickness of mollusk shells, hair color, and body weight. "Neither the plan of their structure nor the various complications of that structure are ever affected by such influences." As a result, he continued, "nothing is more striking throughout the animal and vegetable kingdoms than the unity of plan in the structure of the most diversified types." It was a system that could have been called into existence only by "One Supreme Intelligence as the Author of all things."[3]

Many morphologists easily accepted the new evolutionary framework without altering their scientific methodology. The search for homologies and archetypes went on as before. The reasons for the lack of change are abundantly clear. The existence of uniform body plans, wrote Darwin in his *Origin of Species*, can be explained by "the theory of natural selection of successive slight modifications," in which "there will be little or no tendency to modify the original pattern, or to transpose parts." Thus, he wrote, "If we suppose that the ancient progenitor, the archetype as it may be called, of all mammals, had its limbs constructed on the existing general pattern, for whatever purpose they served, we can at once perceive the plain signification of the homologous construction of the limbs of the whole class." Natural selection, according to Darwin, provided an explanation for the question, "Why should the sepals, petals, stamens, and pistils in any individual flower, though fitted for such widely different purposes, be all constructed on the same pattern?" They appear that way, he wrote, because "the unknown progenitor of flowering plants [consists of] many spiral whorls of leaves."[4] Thus Darwin and the evolutionary morphologists made the error of concluding that the idealized archetype represented an actual ancestor and that homology could be defined by reference to common ancestry. The theory of evolution therefore had the effect of buttressing the discipline of morphology by providing a new justification for what they were already doing. After Darwin, to describe an archetype was not to deal with hypothetical constructions of the mind but to reconstruct real ancestors. Post-Darwinian biology was dominated by those who looked for this basic unity of plan, sought to reconstruct common ancestors, watched for homologies, and attempted to build up phylogenetic trees.

William Coleman has described how Darwin's theory gave new legitimacy to the study of animal morphology, a methodology which had seriously declined by the middle of the nineteenth century. The theory of descent provided "a new conceptual structure which posed problems that only comparative anatomy seemed competent to resolve." In the words of one of the foremost animal morphologists, Karl Gegenbauer, Darwin's theory "allowed what previously had been designated as *Bauplan* or *Typus* to appear as the sum of the structural elements of animal organization which are propagated by means of inheritance.''[5]

A similar rejuvenation of the morphological approach took place in botany. Julius von Sachs, a post-Darwinian botanist, integrated the evolutionary theory into his morphological writings. After gaining his Ph.D. degree in 1856 from the University of Prague, where he had studied under Jan Purkinje, von Sachs met Hofmeister, and in 1860 the two began editing *Das Handbuch der physiologischen Botanik*. Eight years later, after holding teaching posts in Poppelsdorf and Freiburg, von Sachs was appointed professor of botany at Würzburg, where he remained for the rest of his life.

Initially, at least, von Sachs was an avid Darwinist, accepting both the theory of descent and the mechanism of natural selection. "In the struggle for existence," he wrote in the final chapter of his immensely influential *Lehrbuch der Botanik*, "only those varieties survive and reproduce their kind which are better adapted, through some property they possess, to endure the struggle. . . . By the undesigned reciprocal influences of plants and their living and physical environment, specialities of organisation finally arise which could scarcely be better adapted for the preservation of the plant under its special local conditions, and which give the impression of being the result of the greatest ingenuity and foresight.''[6] It should be pointed out, however, that von Sachs's discussion of evolution and adaptation was restricted to the last chapter of his book. In this chapter he accepted that the diversity of plants corresponded to variety in the conditions of life, but in the rest of the text he retained a morphological outlook, stressing continually that "there are only three or four morphologically distinct forms of structure, axis, leaves, roots, and trichomes.''[7]

In an earlier chapter entitled "Morphology of the External Conformation of Plants,'' von Sachs argued that plants may be studied from two *distinct* points of view: the physiological and the morphological. The former, he wrote, is concerned with the way in which plant parts adapt to perform physiological work; the latter ignores function and deals with the development of parts and the relationship of these parts in space and time. Thus, "from a morphological point of view, stems, leaves, hairs, roots, thallus-branches, are simply members of the plant form; but a *particular* leaf, a *particular* portion of the stem, etc., may be an organ for this or that function, which it is the province of physiology to investigate.'' Metamorphosis, therefore, "is the varied development of members of the same morphological significance.''[8] In von Sachs's view, all organs of vascular cryptogams and phanerogams were metamorphosed axes, roots, leaves, or trichomes (outgrowths of epidermal cells), whereas in the lower cryptogams, which lacked true roots, stems, or leaves, the organs were metamorphosed thallomes.

Moreover, both the existence of these morphological members and their subsequent metamorphoses could now be explained by Darwin's theory of descent. According to von Sachs, metamorphosis resulted from the adaptation of different forms to definite functions, while "what we call the common law of growth of a class, or in other words its *Type,* is the result of all the plants of this class being descended from one ancestral form or Archetype, as Darwin terms it."[9] Thus, while the morphologists accepted Darwin's theory of descent, they were not interested in explaining the causes of such changes. As far as they were concerned, their task was not to *explain* the existence of the alternation of generations by reference to adaptation, but rather to describe its existence in as wide a variety of plants as possible. The primary task of botanists, von Sachs argued, should be "to demonstrate and compare the homologies in the alternation of generations in different classes."[10] Which organs in flowering plants, algae, and fungi were homologous to the fern prothallus and fern frond? This was the basic question posed by most botanists in the post-Darwinian period. The theory of descent, not the concept of adaptation, provided the key to the puzzle.

To answer a question of this nature, it was once again necessary to come to an agreement over the criteria by which the alternation of generations was to be defined. Only then could the homologies be recognized. Given the morphological emphasis in German botany, both old and new, and the absence of any clear distinction between sexual and asexual reproduction, it is not surprising that botanists again focused on morphological criteria. Alternating generations differed from each other not in their mode of propagation, as Hofmeister had argued, but in their morphological form: one generation was thalluslike; the other bore stems and leaves. Having taken this stand, however, botanists were then again faced with the problem of distinguishing this phenomenon from that other morphological concept, plant metamorphosis.

The great German morphologist Ernst Haeckel was one of the first evolutionists to attempt to make this distinction on purely morphological grounds. In 1866, in his *Generelle Morphologie,* he explored again the relationship between Hofmeister's alternation of generations, plant metamorphosis, animal metamorphosis, and Steenstrup's alternation of generations. His interpretation differed from the views discussed in Chapter 3 of the present volume in its evolutionary perspective, in particular in its definition of ontogeny as a short recapitulation of phylogeny—the biogenetic law.

The belief that a parallel existed between the stages of ontogeny of higher organisms and the sequence of adult organisms below them in the chain of life had long been applied to the animal kingdom. It became a central theme in the speculations of early nineteenth-century *Naturphilosophie,* whose adherents postulated that the laws that regulated the historical progression of species also regulated ontogenies. As J. F. Meckel expressed it, "The development of the individual organism obeys the same laws as the development of the whole animal species: that is to say, the higher animal, in its gradual development, essentially passes through the permanent organic stages that lie below it."[11] After Darwin, many authors, in particular Ernst Haeckel, placed the ancient doctrine of recapitulation in a transformist framework: the living world, as

well as individual eggs, developed such that the stages of an individual ontogeny recapitulated the adult stages of the organism's phylogeny.

Applying a morphological criterion of individuality, Haeckel sharply distinguished the phenomenon of alternation of generations (both the Steenstrup
and the Hofmeister version) from animal and plant metamorphosis. In Haeckel's opinion, as in the opinion of many other morphologists, there could be
no absolute definition of individuality. Rather, a hierarchy of individuals
existed in the organic world. Each organic entity was made up of a multitude
of individuals of a lower category, and each was also a constituent of an
individual of a higher category. Six orders of individuality were recognized in
the plant and animal kingdoms, and the individuals of each grade were constituted of a complex of individuals of the immediately lower grade. Each
individual in this hierarchy was termed by Haeckel a "morphological individual" and was represented by

1. *Plastiden:* the cell or elementary organism
2. *Organe:* the organ systems in plants and animals
3. *Antimeren:* "Das Gegenstucke," (each half of a bilaterally symmetri-
   cal animal, or the two cotyledons)
4. *Metameren:* the segments of animals and the internodes of higher
   plants
5. *Personen:* individual higher animals, plant shoots, and individual
   coelenterate polyps
6. *Cormen:* the colonial plant stock and polyp colony

Thus, for example, the *Cormen* of the higher plant was constituted of a
hierarchy of subordinate morphological individuals running from the shoot
down to the individual cells.[12]

There were two important aspects of this concept of individual hierarchy.
First, any one of these morphological individuals could also occur as a
"physiological individual," or *Biont,* defined by Haeckel as an individual
"which [is] able to carry on an individual existence for a shorter or longer
time completely independently; an existence which manifests itself in all cases
in the activity of the most general organic function, in self-preservation."[13]
Thus, the *Cormen* of the higher plant represented a physiological individual,
while the *Personen* could be either a morphological individual of a higher
plant or coelenterate, or a physiological individual of a vertebrate. Second,
according to the biogenetic law, all the subordinate categories of individuality
were passed through during the ontogeny of any physiological individual.
Thus Haeckel was also able to define a physiological individual as "that
completely developed organic individual which has reached the highest grade
of morphological individuality which belongs to it as the mature, fully grown
representative of the species."[14]

On the basis of this distinction between morphological and physiological individuals, Haeckel classified sexually reproducing organisms into two groups: those in which the egg-to-egg cycle involved only one physiological individual, or *Biont,* and those in which two or more *Bionten* were produced.

The one-*Biont* cycle, or *hypogenesis,* occurred in organisms in which the sexually mature form was produced after an unbroken succession of stages. The individuals passed through during this ontogeny were only morphological in character and were lower in quality (*Rang*) than those of the fully de-veloped sexual form. Hypogenesis occurred in such organisms as vertebrates, the phanerogams, and insects. In the latter case, called *hypogenesis metamorpha,* the lower-grade morphological individuals possessed "pro-visional organs which are lost during the transformational processes."[15]

Metagenesis, on the other hand, occurred when the life cycle "is repre-sented by two or more different physiological individuals (*Bionten*). [The] series of forms [that] arise out of every fertilized egg . . . dissociate them-selves into at least two physiological individuals and thus become interrupted at least once before they conclude with sexual maturity. Thus, within the life cycle of the species there is always a combination of sexual and asexual reproduction."[16] As Haeckel made clear on numerous occasions, however, the alternation of sexual and asexual reproduction within a single life cycle was not the major distinguishing feature of metagenesis. Metagenesis de-pended on the establishment of two or more *Bionten.*

Following Steenstrup, Haeckel termed the *Biont* produced from the fecun-dated egg the *Amme,* or "nursing generation." In the animal kingdom it occurred in some arthropods, annelids, trematodes, cestodes, tunicates, bryo-zoans, echinoderms, and coelenterates, while in the plants, it was restricted to the vascular cryptogams: mosses, ferns, and the heterosporous fern allies. Steenstrup's alternation of generations in animals was thus equivalent to Hofmeister's alternation of generations in the cryptogams.

According to Haeckel, however, no such metagenesis occurred in the phanerogams. Here, shoots, morphological individuals of the fifth order, propagated other shoots to eventually establish the stock, or cormus. "The cormus," he wrote, "is, however, a single morphological individual of the sixth and highest order, and as such, at the same time, the physiological individual which represents the species as a concrete living unity." Thus, the alternation of shoot individuals represented only an alternation of morphological individuals and differed fundamentally from metagenesis. The cormus, like the vertebrate, he argued, developed by passing through a series of subordinate morphological individuals that finally produced the sexual, flowering shoot. Vertebrates differed from phanerogams in that the fifth stage of their morphological individuality represented their final stage, while phanerogams went one step further to the sixth stage. "The cormus is the specific form of the mature *Biont* in phanerogams, as the Person is in verte-brates and arthropods."

We believe that we are not mistaken when we differentiate all these asexual reproductive stages, which proceed to a single sexually produced and sexually mature *Biont,* from true alternation of generations, which always results in two or more *Bionten.* We propose to designate the former with the name of *Generationsfolge* or *Strophogenesis.* We can term, therefore, the apparent alternation of generations in phanerogams as strophogenesis of *Cormi,* the individual development of vertebrates and arthropods as strophogenesis of *Personen.*[17]

What was so peculiar in Haeckel's deliberations was his failure to recognize that Hofmeister's scheme applied also to the conifers. Haeckel simply restricted Hofmeister's concept to the vascular cryptogams! It is hard to believe that Haeckel was unaware of the extension of Hofmeister's concept to the conifers; it is more likely that he chose to ignore that aspect of Hofmeister's work because it was incompatible with his own scheme. Haeckel would have had difficulty defining the sexual generation of conifers as a *Biont,* since it was so reduced as to be virtually incapable of "self-preservation." By ignoring this extension of Hofmeister's work, however, Haeckel failed to appreciate its evolutionary implications.

Other morphologists had even greater difficulty appreciating the significance of Hofmeister's scheme. They were unable to accept the criteria by which Hofmeister distinguished the alternating generations and thus they could not accept the equivalence of the fern prothallus and the leafy moss plant. As mentioned in Chapter 3, Hofmeister ignored the morphology of fully formed plants, stressing instead their life cycles and the activity of their cells. He viewed the leafy moss plant and fern prothallus as equivalent because of their similar cellular origins and because both produced sexual gametes. To many morphologists, however, such equivalence made no sense. To them, equivalence had to be based on morphological features. A fern prothallus and a leafy moss plant could not be equivalent: the former was an undifferentiated thallus form; the latter, a fully differentiated body. The moss plant, on the other hand, was morphologically equivalent to the fern fronds; the fact that the former produced gametes and the latter produced spores was of little significance. There was no way to resolve this conflict, of course; it illustrated the fundamental division in botany between the Schleiden School of developmentalists and the older morphological school. Furthermore, the post-Darwinian era witnessed the strengthening of the morphological outlook. It was a time when that program of research was buttressed by the search for morphological archetypes. In addition, Hofmeister's fundamental division between gamete- and spore-producers lost its credibility when the distinction between sexual and asexual modes of reproduction became so unclear.

A good example of the confusion can be seen in the writings of Alfred Kirchoff, Hermann Schacht's first student. Contrary to Haeckel, Kirchoff realized that Hofmeister's scheme applied also to the conifers. By referring to the "endosperm" (that is, the remnant of the sexual generation) of conifers as the equivalent of the fern prothallus, Hofmeister had, according to Kirchoff, "uncovered the transition of leaf-forming cryptogams through rhizocarps and *Selaginella* to the phanerogamic plants." In doing so he had "broken the inhibiting dividing wall between both great halves of the higher plant cycles" and had established a truly monistic conception of the plant kingdom.[18] But Kirchoff used morphological criteria to define the alternation of generations. For him, the transition between an undifferentiated thallus body and a differentiated cormophyte body was the basis of this alternation. Thus, he wrote, "Cormophytes, or axis-leaf plants, are thallus plants in the first generation and true leaf-building plants only in the second generation." As a consequence, Kirchoff believed, the fern prothallus was equivalent not to the leafy moss plant but to its proembryo, and fecundation could not be the transition point, for "it has nothing which is essential to it."[19] Moreover, Kirchoff did not accept the theory that the proembryo and leafy moss plant constituted a single generation; rather, he believed that the life cycle of mosses and ferns went through an alternation of generations involving three *Bionten:* the prothallus, or proembryo; the leafy moss plant and fern frond; and the *Sporenfrucht.*

Kirchoff's classification scheme was based on this view of the alternation of generations. The cormophytes, those plants in which development involved a transition from a thallus plant to an axis-leaf plant, were divided into two groups: those in which the spores were undifferentiated, and those which had macro- and microspores. In the former group, Kirchoff argued, fecundation could occur either at the conclusion of the first generation, as in ferns, or in the course of the second generation, as in mosses. Equivalence, therefore, was based on form, *not* on the site of fecundation.[20]

Today one can fully appreciate the difficulty nineteenth-century morphologists had with Hofmeister's scheme. To understand it one needed to take a dynamic developmental approach to the subject rather than a static morphological one. As the Bohemian botanist Ladislav Celakovsky remarked in 1868:

> Kirchoff's concept stands and falls with his assertion that the distinction between thallus and leaf-bearing plant is the deepest there is in the plant kingdom. . . . Above all, it is to be emphasized that the contrast between stem and leaf is certainly not so fundamental as those with a morphological and systematic bias assume it to be.[21]

Celakovsky's views were a complete antithesis of Kirchoff's. Celakovsky, like von Sachs, had been a pupil of Jan Purkinje's at the University of Prague. An early convert to Darwin's theory, Celakovsky realized that the theory of descent provided the key to understanding Hofmeister's scheme. "The comparative investigations of Hofmeister, especially the remarkable relationships

HETEROSPOROUS FERN ALLIES : Antiphyte dominant , producing two
types of spores

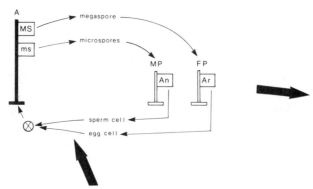

FERN : Antiphyte dominant ; protophyte small , free living .

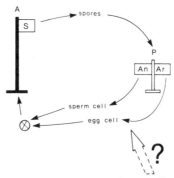

MOSS : Protophyte dominant , antiphyte ''parasitic'' on protophyte .

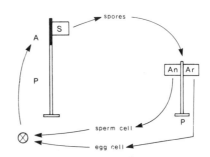

a : anther
⊥A : Antiphyte generation
An : Antheridium
Ar : Archegonium
c : Cone

en : Egg nucleus
□FP : Female protophyte
generation
□FPe : Female protophyte
generation or embryo-
sac

GYMNOSPERMS : Antiphyte dominant. Male protophyte reduced to air-borne pollen grain; female protophyte to "parasitic" body retained in ovule.

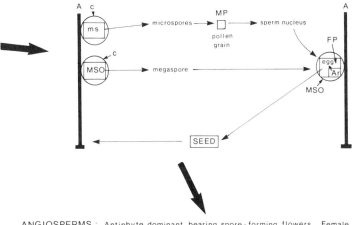

ANGIOSPERMS : Antiphyte dominant, bearing spore-forming flowers. Female protophyte further reduced to minute embryo sac inside ovule.

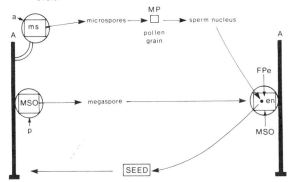

The Botanical Progression:
from mosses to angiosperms

MP: Male protophyte generation
ms: Microsporangium
MS: Megasporangium

MSO: Megasporangium or ovule
p: pistil
P: Protophyte generation
S: Sporangium

FIGURE 5.2.   The botanical progression from mosses to angiosperms

of *Generationswechsel,* have yielded a sure base from which a genuine monistic conception of the plant kingdom as a unitary developmental whole has become possible,'' he wrote.[22]

Celakovsky was the first botanist to see the complete evolutionary implications of Hofmeister's work. From a phylogenetic perspective he argued that the sexual generation was more primitive than the asexual one and thus he termed it the *protophyte,* as opposed to the secondary, asexual generation, or *antiphyte.* The alternation of generations was, therefore, an alternation between the primitive, protophyte generation and the secondary, antiphyte generation. As such it differed fundamentally from the phenomenon of shoot metamorphosis, with which it had been long confused. According to Celakovsky, the phenomenon of alternation of generations in plants could be understood only from this phylogenetic perspective, and it was marked by four essentially phylogenetic characteristics.

First, he argued, both *Bionten* carry out different physiological functions: the protophyte, that of fecundation, the antiphyte, that of true propagation. Second, if one generation is morphologically differentiated, the other has a subordinate, less-developed form. Third, in the overall development of the plant kingdom, the protophyte reaches its highest level of development in the mosses; thereafter it slowly regresses and finally disappears in the flowering plants. The antiphyte, on the other hand, progresses from a single cell in the algae to an organ in mosses to a more highly developed generation in the ferns and phanerogams. Finally, the protophyte never possesses vessels and bundles, while the antiphyte can show a high degree of differentiation.

From this phylogenetic perspective, Celakovsky criticized any attempt to divide the plant kingdom into morphological groupings such as thallophytes, cormophytes, etc. Instead, he argued that

> the true distinction between these divisions is not only in their respective formation of both generations in the life cycle but also still more in the direction of development which these generations follow in both divisions. In the cellular plants the protophyte perfects itself through progressive differentiation and also the antiphyte, beginning with zero, develops in a parallel direction but no further than an externally undifferentiated body, a mere organ (the spore capsule of the moss). This developmental direction is completely terminated in the highest leafy mosses. The construction of the plant kingdom does not start with these, but directly from the lower, leafless liverworts. Here a new developmental direction branches out which consists of a completely opposite formation of protophyte and antiphyte, and terminates in the complete atrophy of the protophyte.[23]

Therefore, since Hofmeister's alternation of generations represented an alternation between a phylogenetically primitive generation, the protophyte, and a more advanced generation, the antiphyte, it bore no resemblance to the numerous other phenomena with which it had long been confused.

In 1868 the first edition of Julius von Sachs's *Lehrbuch der Botanik* appeared. At this time, von Sachs defined the alternation of generations in morphological terms, differentiating the generations on the basis of different laws of growth:

> In every plant a turning point occurs at least once in the course of development, where the initial growth laws are suddenly lost and henceforth growth follows an essentially different law. Sooner or later this law is lost and the first growth law is valid once again. Every such turning point begins with an individual definite cell. In this periodically changing occurrence of different growth laws in the developmental cycle of plants lies the essence of the alternation of generations.[24]

On this basis he argued not only that the moss plant contained three generations, the proembryo, leafy stem, and "fruit"—because each was morphologically distinct and followed different growth laws—but also that the endosperm of phanerogams was equivalent to the fern prothallus because "the endosperm, like the prothallus of higher cryptogams, is a thallus form."

Von Sachs also argued that shoot metamorphosis was not equivalent to the alternation of generations. "[The] flower shoot of a plant obeys the same growth law as the nonflowering leaf shoot; both are not morphologically but physiologically different."[25] It seems strange, perhaps, that von Sachs should argue that leafy and flowering shoots obey the same law. From a morphological perspective, however, all plant parts may be referred back to a few original archetypal forms such as the axis, leaf, and trichome. In the development of each plant these forms metamorphose into particular organs. Because flowering buds and nonflowering buds are both derived from leaves, both follow identical growth laws and differ only in their final physiological function. This metamorphosis, von Sachs argued, should not be termed an alternation of generations. "It seems more suitable," he concluded, "to designate alternation of generations exclusively as the morphological difference of generations that always arise from individual cells in a developmental cycle."[26]

By the time the third edition of his textbook appeared in 1873, von Sachs had modified his earlier definition of the alternation of generations. His emphasis had shifted from growth laws to the origin of the generations. A generation, he wrote, comprises those plant structures which arise from similar reproductive cells that "are separated from the organic structure of the plant." Thus, an alternation of generations occurred when "the generations which proceed from one another are dissimilar and produce dissimilar kinds of reproductive cells."[27] Von Sachs was now clearly moving toward an emphasis on differing reproductive cells, for an alternation of generations could still occur even if the only difference between the generations was the production of asexual spores and sexual gametes respectively. Even here, however, his distinction was still basically morphological: his claim was not so much that spores and gametes were actually produced but that "the morphological

distinctions of the alternate generations [were] only observable in the preparation for production."[28] The generations still differed morphologically, but the
difference was now restricted to the structure of their reproductive organs: one
generation bore antheridia and archegonia; the other bore sporangia.

Von Sachs's textbook was important because it stressed the overall importance of alternating generations to a full understanding of the relationship
between different plant groups: "It is one of the most important problems of
morphology and systematic botany, not only to demonstrate the alternation of
generations in different classes of plants, but also to compare the process
according to definite principles."[29] Thus we have von Sachs's emphasis on
the homologies that exist between the different generations in different plant
groups.

This approach became apparent in von Sachs's systematic treatment of the
plant groups. Not only were mosses, liverworts, and vascular cryptogams
marked by a distinctive alternation of generations, but another pattern also
was observed. In proceeding from the ferns and Equisetaceae to the rhizocarps
and Lycopodiaceae, "the development of the prothallium becomes continually simpler and its morphological differentiation less pronounced." Von
Sachs argued that the prothallus, beginning in ferns as an independent thallus
body, becomes differentiated into male and female prothalli. Then, in the
rhizocarps, the female prothallus is reduced to a small appendage of the
macrospore, while in the Lycopodiaceae it is reduced still further to form a
mass of tissue entirely within the spore. These two groups, he went on to
argue, form a bridge to the phanerogams.[30]

In the phanerogams, von Sachs pointed out, the alternation of generations
is concealed in the formation of the seed, and the changes seen in the vascular
cryptogams are carried one step further. In the cycads and conifers, "the prothallium, which is now known as the Endosperm, remains during its whole
existence enclosed in the macrospore or Embryo-sac; it produces before
fertilization archegonium-like structures, the 'Corpuscula', in which the Germinal or Embryonic Vesicles arise."[31]

Von Sachs also realized that the so-called endosperm of the conifers was
not equivalent to the endosperm of flowering plants, which arises only after
fertilization. Thus, he wrote, the prothallium of the higher cryptogams "does
not appear to have anything to correspond to it in Angiosperms" unless the
antipodal cells represent the "last occasional occurrence of the rudiment of
the true prothallium." The pollen grains of phanerogams were strictly comparable to the microspores of the higher cryptogams. Both contained "the
male fertilizing principle, which, passing into the oosphere, causes it to form
the embryo." But whereas in the cryptogams this principle was carried by a
spermatozoid or an antherozoid, in the phanerogams it was carried by a pollen
grain. Thereafter the pollen grain germinated like a spore, and the pollen tubes
thus produced "transmit[ted] by diffusion the amorphous soluble fertilising

substance into the embryonic vesicle." "The general result of these observations," von Sachs concluded,

> is that the Phanerogam, with its pollen-grains and its embryo-sac, is equivalent to the spore-producing (asexual) generation of the heterosporous Vascular Cryptogams. But as in Vascular Cryptogams the sexual differentiation first makes its appearance (in Ferns and Equisetaceae) on the prothallium only, and next (in Rhizocarpeae and Lycopodiaceae) on the spores themselves, so, in Phanerogams, this process is carried back a step further, the sexual differentiation arises still earlier, being manifested not only in the formation of embryo-sac and pollen-grains, but also in the difference between ovule and pollen sac, and even earlier in the distinction between male and female flowers, and last of all in the dioecious condition of the plants themselves.[32]

According to von Sachs, no uniform feature of growth or reproductive cycle characterized the algae. In some, all generations were alike. In others, a series of asexual generations was punctuated occasionally by a sexual mode of reproduction. In still others, three or four different generations appeared, in the last of which the first form appeared again. Given von Sachs's definition of the alternation of generations as "the whole process of development which passes through . . . successive dissimilar generations, and finally returns again to the first form,"[33] all algae with dissimilar forms could be said to display it, whether in an ABAB, an ABBBA, or an ABCA life cycle. Not surprisingly, therefore, answering the question of which generations were homologous to each other remained difficult as far as the algae were concerned. Indeed," von Sachs concluded, "[it] is to a certain extent still impossible."[34]

With the publication of the fourth edition of his textbook in 1874, however, von Sachs redefined the criteria of alternating generations in such a way that he was able to claim that *all* sexual plants displayed them: "The development of all plants which possess sexual organs may be divided into two stages which correspond in all essential points to the two generations in the life history of a fern; . . . there is, therefore, in the whole vegetable kingdom, only one type of alternation of generations."[35] He came to this conclusion by redefining the term "spore" (Fig. 5.3). A spore, originally defined as any asexual reproductive cell, was now defined as a cell that arose from an asexual

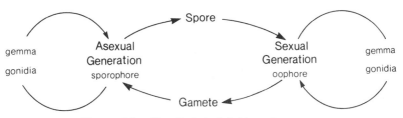

FIGURE 5.3. Von Sachs's definition of a spore

generation but germinated to produce a sexual generation. On the other hand, asexual cells that germinated to produce another asexual generation were to be termed "gemmae" or "gonidia," not "spores." "I begin," von Sachs wrote, "by designating as *spores* the reproductive cells which are produced in the sporangia of ferns and capsules of mosses,"[36] cells, in other words, that arise from an asexual generation and produce a sexual generation. Thus, he defined the two alternating generations as generations that produce sex cells and spores respectively: "Sexual cells and true spores indicate the turning points in the alternation of generations; they are not organs for direct reproduction, for each of them always produces something different from that form from which it immediately sprung."[37]

Von Sachs's conclusion that all sexual plants possess such a life cycle may have arisen from the continuing controversy over the nature of the *Oedogonium* life cycle. Since the 1850s it had been known that in *Oedogonium* (a filamentous green algae) and in other similar algae, the oospore (zygote) produced by fertilization did not germinate directly into a young asexual plant, but rather acted as a mother cell to swarming zoospores, which themselves formed new plants:

$$asexual\ filaments \rightarrow zoospores \rightarrow asexual\ filaments \rightarrow zoospores$$
$$\uparrow \qquad\qquad\qquad\qquad\qquad\qquad\qquad\qquad \downarrow$$
$$zoospores \leftarrow oospore \leftarrow gametes \leftarrow sexual\ filaments$$
$$(zygote)$$

In 1856 Nathaneal Pringsheim had suggested that such a life cycle was similar to that of mosses. The zygote of *Oedogonium,* he argued, was a "spore-building, fruiting cell"; it differed from the "fruit" of mosses only in that the moss zygotic cell produced tissue cells before producing spores, while the *Oedogonium* zygotic cell produced spores directly. In other words, the oospore, or zygote, of *Oedogonium* was equivalent to the spore-producing generation of mosses.[38]

This was how von Sachs now interpreted the life cycle of all algae: the zygote of a sexually reproducing alga was basically homologous to the entire asexual generation of the moss and fern. Thus, since sexual reproduction always produces a zygote, every sexual plant must display an alternation of generations, even if its asexual generation is limited to a one-celled zygote. If one were to imagine, von Sachs wrote, that "the act of fertilization did not result in the production of any vegetative structure or the second generation to be altogether suppressed, the fertilized oosphere would then itself become a spore."[39] This, in fact, was what occurred in *Oedogonium*. The zygote was equivalent to the entire asexual generation in the moss. But whereas in *Oedogonium* this zygotic cell produced spores directly, in mosses a stalk and capsule were formed before the spores were produced. Consequently, the whole of the plant kingdom could be unified by stressing the gradual increase in the importance of the asexual generation. Appearing first as a zygotic cell in the algae, the asexual generation gradually assumed a more dominant position in the plant kingdom.

Von Sachs's final position was therefore basically similar to Celakovsky's. Both realized that the theory of descent provided the key to the alternation-of-generations puzzle. The generations seen in the mosses and ferns represented the two generations that had been established during the early evolution of their algal ancestors, but along the long evolutionary road, the asexual generation had gradually assumed the dominant position. What seems strange is that these Central European botanists rarely asked why this progressive elaboration of one generation and the parallel atrophy of the other should take place. They deemed it sufficient merely to describe the sequence.

In 1874 Celakovsky produced a lengthy paper on the alternation of generations in plants in which he not only extended his phylogenetic perspective but also attempted to include in its domain the older morphological concept of plant, or shoot, metamorphosis. Hofmeister's alternation of generations, shoot metamorphosis, the sequence of sexual and asexual stages in the algae, and the more recent interpretation of the *Oedogonium*-like life cycles, Celakovsky argued, *all* were examples of alternating generations, since "only this is essential, that these generations differ in form and manner of propagation and that they alternate with each other."[40] He then subdivided the broad concept of alternation of generations into *Sprosswechsel,* or the alternation of shoots, and *Biontenswechsel,* or the alternation of "free-living beings."

Celakovsky included under *Sprosswechsel* not only shoot metamorphosis but also the production of leafy shoots from a thallus—in other words, the production of the leafy moss plant from the protonema. *Biontenswechsel* he further divided into two groups: the alternation seen in mosses and ferns, and the sexual-asexual generations seen in most algae. The former ("antithetical" or "contrasting") alternation of generations occurs, he wrote, when "a sexual ovum producer and an asexual spore-builder alternate with each other," whereas the sexual-asexual stages in algae represent a "homologous" alternation of generations.[41] The point of this distinction was to make explicit his belief that the antiphyte, or asexual generation, in the higher cryptogams was *not* homologous to the asexual generations of the algae. As Celakovsky argued, if one were to designate the zoospore-producing, asexual generations of algae as *A,* and the gamete-producing, sexual generation as *B,* then the two generations in mosses would have to be represented as *B* and *C, C* being a new "antithetical" generation which had little or no counterpart in the algae. "The greatest difference between the antithetic and the homologous alternation of generations," he wrote, "is that the asexual generation of mosses and vascular cryptogams is not identical to the asexual generation of thallophytes; that is to say their origin or phylogeny is essentially different."[42]

Arguing in a manner similar to Pringsheim and von Sachs, Celakovsky then pointed out that the rudiments of the new antithetical generation did make their appearance in some algae. In *Oedogonium,* for example, the zoospore-

and gamete-producing generations represented the homologous generations, *A* and *B*, while the oospore (zygote) directly produced a number of zoospore mother cells that represented the antithetical generation in a rudimentary state (generation *C*) (Fig. 5.4).

The same three generations, *A*, *B*, and *C*, were also present in mosses,

1. OEDOGONIUM

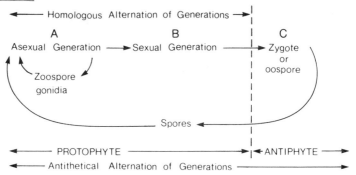

2. COLEOCHAETA

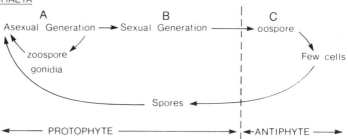

3. MOSS

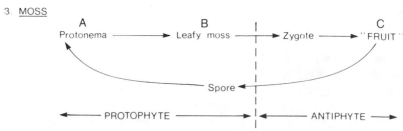

FIGURE 5.4. Homologous and antithetic alternation of generations according to Celakovsky

according to Celakovsky, but in this case *A* and *B* had been reduced to mere shoot generations representing the protonema and leafy moss plant respectively. "In the moss," Celakovsky wrote, "we can see that both vegetative *Bionten* have been reduced to mere shoot generations . . . for the sexless first generation of algae corresponds to the protonema of mosses, the sexual generation to the leafy moss plant, which together with the protonema still only make up a single *Biont.*"[43] Thus, generations *A* and *B* of algae and mosses represented the protophyte generation; generation *C,* the antiphyte generation.

Celakovsky then very elegantly incorporated his concept of *Sprosswechsel* into his definition of *Biontenswechsel.* As mentioned above, he denied that the alternation of generations *A* and *B* in the algae corresponded to Hofmeister's alternation of generations in the higher cryptogams and conifers. Rather, it corresponded to shoot metamorphosis. In his view both were asexual modes of propagation of a single *Biont,* and all such generations were "homologous": "Homologous alternation of generations (in algae) corresponds to the alternation of shoots in phanerogams. The former, being minor members of the protophyte generation, has the same significance as shoot alternation in the antiphyle generation."[44]

On the basis of this distinction between homologous and antithetical generations, Celakovsky presented a phylogenetic outline of the whole plant kingdom. The antithetical, "fruit" generation of mosses, he argued, was basically a new structure that was represented in some algae, such as *Oedogonium,* by a single zygotic cell. In this case "the *Anlage* of the fruit, which the egg cell is, transforms itself immediately into a spore."[45] With the elaboration of this antiphyte generation, the production of spores is delayed and the alternation found in mosses occurs.

No sooner had Celakovsky's ideas on the homologous and antithetical nature of the alternation of generations in mosses and ferns and the various stages in algal life histories been put forward, however, than a new set of discoveries was made which not only challenged these ideas but also seemed to suggest again the insignificance of sexual reproduction. In 1874, while working with Anton de Bary at the University of Strasbourg, William Farlow of Harvard University reported that young fern fronds could be produced by direct budding of the prothallus cells, and not necessarily following the act of fertilization. To Farlow, the antiphyte fern generation appeared to be "a direct continuation of the prothallus cells, and not a distinct organization temporarily attached to it."[46]

Two years later Pringsheim showed that if one cut the "fruit" of the mosses into pieces, it was possible to cultivate protonema threads from them without first producing spores. Thus, he concluded, both generations "possess an identical morphological character": "From their anatomical structure and their morphological character, therefore, the seta [the antiphyte generation] of the moss appears not to be a generation different in organization from the moss stem, but a less developed, leafless, sporangium-bearing axis morphologically equivalent to a leafy stem."[47]

As a result of his findings, Pringsheim developed a theory that was completely at odds with Celakovsky's concepts. The alternation of generations in mosses, he argued, was indeed homologous to the alternation of zoospore- and gamete-forming generations in the algae; both generations in the moss life-cycle were "relatively different, developed members of the same kind of organization, in which one bears the sporangia and the other the sex organs."[48] In other words, the alternation of generations in mosses and ferns was not marked by the appearance of a new "antithetical" generation but was strictly comparable to the so-called homologous alternation in the algae.

In this modern age of biochemistry, population genetics, and mathematical ecology, it needs to be emphasized that late-nineteenth-century botanists devoted much energy to deliberations like those just described. As pointless and confusing as their theories might appear to be today, to ignore these post-Hofmeister botanists in our modern histories would be to paint a false picture. They were basically concerned with uncovering ancestral family trees, and because the fossil record was so inadequate, they were forced to speculate about the past on the basis of what they could learn about the present: from the present they drew homologies, and in homologies they found ancestors. The question whether the zygotic cell of a filamentous green algae was or was not homologous to the asexual generation of mosses and ferns was thus fundamentally important; answering it would provide a clue to the origins of the whole plant kingdom. And that was the raison d'être of much post-Darwinian botany.

Since phylogenetic speculation was so widespread in nineteenth-century German, British, and North American biology, it is thus most curious that British botanists played no role in these post-Hofmeister, post-Darwin deliberations. British botanists during this period continued instead in the grand tradition of naming and classifying plants.

One reason for this lack of interest in the new botany lay in the long reign of the Linnaean system in English botany; Britain was the last nation to reject Linnaeus's system in favor of a natural system. Finally, in 1829, John Lindley published his *Synopsis of the British Flora,* the first application of a natural system to British plants. The Linnaean system, he wrote, "skims only the surface of things, and leaves the student in the fancied possession of a sort of information which is easy enough to obtain, but which is of little value when acquired."[49] Six years later he remarked: "Surely it cannot be denied that this school has acted as if the whole object of botany were naming and describing species, evidently mistaking the means for the end, and converting the study of the vegetable kingdom into a system of verbal trifling."[50]

Still, as influential as Lindley was in bringing about the collapse of the Linnaean system in England, he did not seek to change the fundamental goal of British botanists. Naming and classifying remained their primary tasks. The difference was that the new natural system adopted by the British botanists demanded a more profound knowledge of plant structure than the Linnaean system had. In addition, these botanists' rejection of the Linnaean system led them to deny the sexuality of the nonflowering plants and thus to remain indifferent to Hofmeister's claims.

In his numerous publications spanning the middle of the nineteenth century, John Lindley, professor of botany at University College, London, divided the plants into two groups: those with sex organs, spiral vessels, or cotyledons, and those without. Continuing to believe that sexual reproduction must be based on the presence of anther- and pistillike organs, he argued in 1847, for example, that although "it is true that such names as Antheridia and Pistillidia are met with in the writings of the cryptogamic botanists—these are theoretical expressions and unconnected with any proof of the parts to which they are applied performing the office of anthers and pistils."[51] Cryptogams, he believed, reproduced only by means of spores. Such views were, of course, commonly accepted among Continental botanists at that time. However, Lindley retained these views after the work of Hofmeister had been reported to the British by Arthur Henfrey, who occupied the chair of botany at King's College between 1854 and 1859. In 1853, in the third edition of his magnum opus, *The Vegetable Kingdom,* Lindley again denied the presence of sexual reproduction in mosses and ferns and thus remained indifferent to Hofmeister's scheme: "I confess that I am by no means satisfied with some of these opinions. The adoption of Steenstrup's theory of alternate generations seems to rise too much from *à priori* considerations, and the statements regarding the impregnating action of the spiral filaments in ferns appears to be wholly hypothetical."[52]

The Reverend Miles Berkeley, a Northamptonshire clergyman and graduate of Cambridge, is best known for his work on the Irish potato blight, but his writing during this period illustrates the difficulty the classical botanists had with Hofmeister's thesis. While agreeing that the prothallus of the fern was homologous to the moss plant, Berkeley denied any connection between these cryptogams and the phanerogams. "Although there is much resemblance between the formation of the archegonia and embryonic bodies in Lycopods, and that of the corpuscles and embryos in conifers," he wrote in his *Introduction to Cryptogamic Botany* (1857), "homology is out of the question, and we have, therefore, mere analogues to build upon, which, however curious, indicate no affinity."[53]

The triumph of the "new botany" in Germany and its virtual absence in England reflected, above all, the different status of science in the two countries. Science flourished in the newly revitalized German universities of the mid nineteenth century. From the beginning, these new universities rejected the utilitarian model of their English counterparts, and with it the attitude that universities existed to impart accepted knowledge to students and to examine them on it. On the contrary, at least in theory, German students were led toward independent thought and research. Offering only the Ph.D. degree, German universities became centers for research, not teaching. The new approach to botany was thus a necessary outcome of this change in emphasis. Plants were examined for their own sake and not as part of a medical curriculum; they were examined in the laboratory, not in the field or in herbaria; published studies emphasized their microscopic structure, their life histories, their development, and their physiology, not their names and classification. Botany became the discipline of professional botanists.

Lacking this new institutional base, British botanists continued in the old tradition. Indeed, if anything, the tradition was strengthened in the mid-nineteenth century by the rejuvenation of Kew Gardens, which had undergone a decline following the death of Joseph Banks in 1820. In 1840 the gardens were placed under government control and the directorship was given to William Hooker. The aim of the gardens was twofold: first, "gratifying the national love of gardening, and affording much popular information as to the appearance, names, uses, and native countries, etc., of an extensive series of useful and ornamental plants"; and second, "promoting the useful arts which depend on vegetable produce, supplying information to botanists, aiding their publications, and imparting knowledge of plants to travellers, merchants, and manufacturers, also . . . training plant collectors and gardeners for home, colonial and foreign service."[54] Under the direction of William Hooker (1841–65), his son Joseph Hooker (1865–85), and the latter's son-in-law W. T. Thiselton-Dyer (1885–1905), Kew Gardens continued to grow in prestige during the last half of the nineteenth century. As Colonial Secretary Joseph Chamberlain remarked in 1898, "I do not think it too much to say that at the present time there are several of our important colonies which owe whatever prosperity they possess to the assistance given by the authorities of Kew Gardens."[55] He may well have been thinking specifically of the growth of the tea industry in Sri Lanka and the Darjeeling Valley of India, the transporting of coffee plants from Arabia to the West Indies, and the introduction of the Brazilian rubber plant to Malaysia.

In many ways the state of English botany in the mid-nineteenth century resembled that of English physiology. Both disciplines remained indifferent to the new laboratory work being undertaken on the Continent; both had a descriptive anatomical and structural bias, much of it macroscopic; and both suffered as a result of their strong links with Britain's system of medical education. This education was centered increasingly at London's teaching hospitals and was narrowly utilitarian; it did not provide for such esoteric subjects as experimental physiology or the new botany. As Gerald Gieson remarked in his book on Michael Foster, the English physiologist, "Precisely because experimental science seemed so irrelevant to medical practice, it was scarcely represented at all in the curricula of the London hospital schools until very near the end of the century."[56] Similarly, experimental physiology and the new botany failed to find a home at Oxford or Cambridge, and the new universities in London, University College and King's College, concerned themselves only with the dissemination of existing knowledge; there was little room for laboratory research.

Botany's subservience to medicine at this time was stressed in William Thiselton-Dyer's historical article of 1925. It was restricted, he wrote, to teaching students the identification and properties of useful plants. "So limited, it was a compulsory part of medical training with its sequel *materia medica*. Huxley thought that this had become a mischievous encumbrance to study constantly more exacting."[57] In 1875 J. Hutton Balfour, professor of botany at Edinburgh University, wrote an article on botany for the ninth edition of the *Encyclopaedia Britannica*. Although obviously aware of the broad scope of botany, Balfour confined his essay almost exclusively to the structure and morphology of higher plants. "Balfour," Thiselton-Dyer complained, "simply tumbled into the Encyclopedia his class textbook for Edinburgh medical students," and so his article became "an illustrated enumeration of terms."[58] Plant development and physiology were dismissed in one and a half pages, while the structure and morphology of plants took up eighty pages! Not surprisingly, the alternation of generations was hardly mentioned.[59]

During the second half of the nineteenth century a group of young scientists took up the public championing of professional science, demanding both the inclusion of laboratory science in the education curriculum and government support for scientific research. In 1870, with the passage of the Education Bill and its demand for science teachers, Thomas Huxley organized a short summer course in laboratory biology at Kensington, "a course of instruction in Biology," he wrote, "which I am giving to Schoolmasters—with the view of converting them into scientific missionaries to convert the Christian Heathen of these islands to the true faith."[60] The course began in July 1871 with Michael Foster, E. Ray Lankester, and others acting as assistants. This course, in the words of Foster, "brought about a revolution in the teaching of biology in this country." Shortly afterward Foster and others introduced similar laboratory classes into the English universities. In 1873 Thiselton-Dyer became involved with Huxley's class and in 1875 he and Sidney Vines organized an eight-week summer course in laboratory botany. A year later Vines secured an appointment at Cambridge, where he began "a long-continued struggle for the advance of botanical teaching in the University." His struggle was exacerbated, however, by Charles Babington, who occupied the chair of botany at Cambridge and was uncompromisingly hostile to the new botany. "Now it is rare to find an undergraduate or B.A. who knows, or cares to know, one plant from another," Babington remarked in an address to the Ray Club in 1887. "I am one of those who consider this to be a sad state of affairs." It was said of Babington that he pitied the botanist who, never seeking living plants in their homes, armed with a microscope, ransacks their cell and fibre. A student of the first class in the Natural Science Tripos, observing a specimen of (what I shall call X) in his drawing room, on learning the name cried, "so that is really X? I know all about that; I guessed it would be set, and it was." Science which cannot see the wood for the trees, growing herb or animal for cell laid bare by scalpel, had for him no charm.[61]

Even today many biologists share Babington's qualms about the excesses of the new biology and botany; an effective compromise between the old and new still escapes us.

When Frederick Bower arrived at Cambridge in 1874, he found botany "moribund in the summer and actually dead during the winter," and Babington's lectures "wanting both in spirit and substance." The following year, however, he attended Foster's new class in laboratory biology and "learned what it meant to be taught science in a rational way." Thus Bower's early career in botany spanned those formative years when science slowly gained favor and status in England.[62]

Born in Ripon, Yorkshire, in 1855 and educated at Repton School, Bower was continually frustrated by the lack of any science in his education and by his awareness of the revolution that was taking place in Germany. What was a young man to do as "he felt the years of opportunity slipping from him," he asked, but go to Germany. Thus, in 1877, two years after von Sachs's textbook of botany appeared in English translation, Bower and Vines spent the summer in von Sachs's laboratory in Würzburg. There they were taught "the Hofmeisterian methods of the time." As Bower described it: "The short course of practical exercises through which Sachs put me in 1877 covered ground already traversed by Hofmeister, Pringsheim, and others. This was exactly what I required at the moment to give precision in preparation, observation, and record."

When he returned to Cambridge, Bower enrolled in Vines's practical class. "The excitement was intense," he wrote, "as we verified facts first revealed in German laboratories. We felt we were seeing things never observed before in Britain."[63] After graduating in 1879, Bower spent the following year in Strasbourg working under de Bary. There, with a host of other foreign students such as Vines and Farlow, he began his research career.

Returning to England in 1880, Bower worked in the Botany Department of University College, carried out research in the newly established Jodrell Laboratory at Kew Gardens, and taught botany at the South Kensington Normal School for science teachers. By this time Huxley's summer program had spilled over into the rest of the year and included more advanced classes. In 1884, for example, a thirty-day elementary botany course was offered in January and February, while from February to June a more advanced class was taught. Although the classes attracted fewer than a dozen students, Bower was pleased to note that botany was "for the first time recognized as an

FIGURE 5.5.   Frederick Orpen Bower. Reproduced from Karl Maegdefrau, *Geschichte der Botanik* (Stuttgart: Gustav Fischer Verlag, 1973), by permission of the publisher.

independent science at South Kensington."[64] Finally, in 1885, Bower was appointed to the chair of botany at the University of Glasgow, a position he occupied for forty years. It was a happy choice, for in Glasgow, Bower built up one of the finest botany departments in the English-speaking world. His students emigrated to take up university positions in Ireland, Australia, South Africa, the United States, and Canada, and Bower himself became one of the world's leading botanists.

Upon his arrival in Glasgow, Bower found that he was responsible for teaching a lecture course in botany to 185 medical students and a noncompulsory practical course to 42 of them.[65] Because of his strong leaning toward laboratory instruction, he felt some dissatisfaction with this program. Thus, in his address at the opening of the medical session in 1885, he warned that sciences taught only as part of a medical curriculum would be disadvantageous to the university, which was a seat of pure scientific learning. In addition, he argued that a laboratory-based botany course could be justified within a medical curriculum. "It is practical work joined with lectures, rather than the lectures alone, which give the science its true raison-d'être in the medical course."[66] Unfortunately, it was not until 1895 that Bower was able to offer a

FIGURE 5.6.   Anton de Bary's laboratory in Strasbourg the winter of 1879/80. Among those pictured are F. O. Bower (*far left*), A. de Bary (*seated, front row center*), and S. Vines (*center, back row*). Feproduced from Frederick Orpen Bower, *Sixty Years of Botany in Britain* (London and Basingstoke: Macmillan & Co., 1938), by permission of the publisher.

fully integrated lecture-laboratory class. The financial arrangements at Glasgow were mainly responsible for this delay. Before the Scottish University Act of 1889, a professor was forced to cover all course expenses with student fees, and this prohibited him from offering small, advanced laboratory classes, where expensive equipment was required. "The rich public," Bower wrote, "not the poor student, should finance what is so clearly for the public good." With the passage of the University Act, professors' salaries were fixed and the handling of all expenses was assigned to a central university body. Thus, equipment became the property of the university, not the professor, and money was made available for the appointment of junior staff and laboratory assistants. The significance of the 1889 act to the future of Scottish science becomes all too clear when one realizes that the University of Glasgow graduated only five honors students in science between 1836 and 1885.[67]

   In 1890 Bower published his monumental paper on the alternation of generations, and for the first time the alternation was explained in terms of environmental influences—in particular, the migration of plants from water to land.[68]

   The majority of English botanists had never been receptive to the German concept of a functionless morphology. Even their foremost morphologist, Richard Owen, argued that modifications of the ideal archetype were a result of adaptive requirements. In part, this English approach may have reflected the influence of the long line of field naturalists for which Britain is justly famous. As they assiduously made their collections from areas as different as Kent and Yorkshire, such naturalists could not help but be acutely conscious of the relationship between organisms and their environment. Their passion for collecting always involved more than mere naming and identification. When Alexander Irvine, for example, proposed that members of the London Botanical Society should produce a local flora, he requested that "the precise habitation of the species, the nature of the soil where it grows, [and] the attitude of its locality" be noted.[69] In part, too, this hostility to pure morphology reflected the influence of natural theology, in which the interrelationships between structure and function within an individual, as well as the intimate relationships between the individual and its environment, were used as evidence for design. Wherever we look in this world, Peter Roget wrote in his Bridgewater treatise of 1839, "we still meet with life in some new and unexpected form, yet ever adapted to the circumstances of its situation."[70]

Darwin himself was fully conscious that any theory that ascribed the origin of new species to natural causes must also explain the adaptiveness of these species. Any theory that claimed that species have descended from other species would be unsatisfactory, he wrote, "until it could be shown how the innumerable species inhabiting this world have been modified, so as to acquire that perfection of structure and coadaptation which most justly excites our admiration."[71] Likewise, Herbert Spencer argued that structure and function "are inseparable co-operators," and that "an account of organic evolution must be essentially an account of the interactions of structures and functions, as perpetually altered by changes of conditions."[72]

By the latter part of the nineteenth century, pure morphology was lessening its grip on German botanists also. Karl Goebel, for example, a pupil of Hofmeister's who, like Bower, had studied under de Bary at Strasbourg and von Sachs at Würzburg, became a leading critic of theories that separated physiology from morphology. As he wrote in his *Organographie der Pflanzen*, "All the phenomena of life have a definite relationship to environment, and therefore, the consideration of the configuration of the organs of plants is not merely a comparative historical criticism, but must take into account all the conditions of the environment which we find at the present day."[73] "Comparative historical criticism" was precisely the tool the German post-Darwinian botanists had used to approach the alternation-of-generations problem. It was enough to show how the life cycles of a wide variety of plants could be interpreted in terms of a universal alternation of generations. As Bower remarked, "There has grown up the idea that an alternation of generations is due to some quality inherent in many organisms . . . which leads them to pass through certain definite phases in the progress of their individual life.''[74] In other words, German botanists were usually content to describe the phenomenon rather than to explain it. Only Celakovsky had attempted such an explanation in his paper of 1874.

Celakovsky's explanation paid no heed to the environment, however. It was rather an amalgam of the rejuvenating effects of sexual reproduction, the concept of division of labor, and the biogenetic law. As Celakovsky expressed it, "The alternating of generations does not only allow separate species to renew themselves, but it is a memory of overall renewal in the plant kingdom in general." The first generation, Celakovsky claimed, becomes worn out, so it renews itself "while gradually engineering the antiphyte." Thus the protophyte generation atrophies while the antiphyte generation keeps on developing: "Even though the antiphyte has become the carrier of the development of the plant kingdom, the first generation has not *ipso facto* disappeared at once; according to the law of rejuvenation each form will return during its development back to the first generation."[75]

Bower's explanation, on the other hand, rested on the basic Darwinian assumption that the appearance and subsequent elaboration of the antiphyte generation must be based on some adaptive significance.

In 1887 Bower addressed the problem whether apogamy and apospory (the production of generations directly without gametes or spores) were "to be regarded as mere sports or as having a deep morphological meaning throwing light upon the origin of the two distinct generations." The work of Pringsheim and others, he noted, had shown that "no fixed and impenetrable barrier exists between the sporophore and the oophore; in fact, the formation of spores is not a *necessary* stage in the cycle of life." Bower regarded this as an example of "substitutionary growths," in that the normal alternation of generations may be extended by budding or cut short by apospory or apogamy (Fig. 5.7). Maintaining, then, that the sporophore generation was a new structure and not simply a homologue to the oophore, he concluded:

> In such a series of plants as *Oedogonium, Coleochaete, Riccia* and *Anthocercos,* there is ample indication of the formation of spores before the sporophore assumed its vegetative characters; in the later terms of the series that a differentiation of vegetative tissue of the sporophore from the true spore-forming tissue becomes apparent. Thus, on phylogenetic grounds, it appears improbable that apospory is a true reversion[;] . . . the phenomenon of apospory is a sport, not a reversion bearing pregnant interpretations with it.[76]

Having taken this stand in favor of an antithetical interpretation of the spore-forming generation (termed by Isaac Bayley Balfour the sporophyte generation, as distinct from the gametophyte generation), Bower then turned his considerable talents to an explanation of the phenomenon. From the very start it was clear that he believed it involved Darwinian adaptation; that the appearance and subsequent elaboration of the sporophyte generation indicated some form of environmental adaptation.

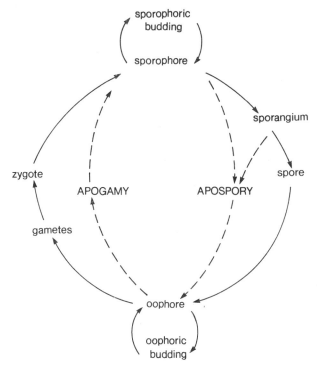

FIGURE 5.7. Apogamy and apospory according to Bower. Both are shown to be rare and insignificant events.

In 1887 Bower and the botanist Dukinfield Scott had visited W. C. Williamson in Manchester to receive instruction on coal-measure plants. In the notes made by Bower at this time one finds the first indication of what was to become his central thesis, that the sporophyte generation represents a terrestrial form:

> No exposed land before Devonian—∴ all veg. life (which must necessarily have existed to feed animals) must have been aquatic. Devonian gives us at once the three phyla of vascular cryptogams and the gymnosperms as well. Prob. evolution very rapid, subsequently to the [invention ?] of an alternation.[77]

Bower's papers of this period also contain a reprint of an 1889 paper by Walter Gardiner in which the latter stated not only that "all nature is in a perpetual state of desperate warfare—the care for self, the absolute disregard for others," but also that the production of swarm spores (zoospores) in *Oedogonium* was linked to unfavorable environmental conditions.[78]

By 1890 Bower had incorporated these ideas into his lectures to the students at Glasgow. From the lecture notes of William Lang, Bower's favorite student (who later became professor of botany at Manchester University), we learn that the lower plants were aquatic, that ferns were intermediate, that the phanerogams were mainly terrestrial forms, and that in these terrestrial forms the balance between the two generations altered. Bower also stressed that in the gymnosperms water was not required for fertilization, that instead the fertilizing element was transferred by a pollen tube, and thus, that fertilization had become truly terrestrial.[79]

In 1890 Bower published his famous explanation of the alternation of generations in plants. "I am convinced," he wrote, "that a merely formal comparison of different organisms, or of their successive stages one with another, will not suffice for the solution of the question as to the real nature of alternation."[80] The phenomenon can only be explained, he argued, by reference to the plant's external environment—in particular, the terrestrial environment. The prothallus of the fern, he noted, is delicate in structure and requires water for successful fertilization; thus, it is adapted to moist conditions. "It is, in fact, typically semi-aquatic in its nature, sharing its main characters with the Algae from which we have every reason to believe that the land-flora originated." The sporophyte, on the other hand, is robust, able to withstand drying, and requires dry weather to disperse its spores; it is adapted to terrestrial conditions. "Thus," he concluded, "the Fern is an organism with, so to speak, one foot in the water, the other on land" (the two feet being represented by the gametophyte and sporophyte respectively). The whole sequence of alternating generations—the process by which the sporophyte assumes greater and greater complexity and the gametophyte is reduced—"is to be correlated with a progression from the aquatic or semi-aquatic habit of the lower forms, to the very distinctly sub-aerial habit of the higher." Therefore, Bower claimed, the alternation of generations "is the result of adaptation of originally aquatic organisms to sub-aerial conditions of life."

The gametophyte, Bower further argued, is the older and preexistent generation, while the sporophyte is a later acquisition which, being terrestrial, must be absent or virtually absent from the algae. From this it follows "that the alternation must have been the result of an interpolation of a new development between successive gametophytes," or in other words the interpolation of a terrestrial phase between preexisting aquatic ones. The sporophyte, therefore, "cannot itself be a gametophyte which has undergone a change of form."[81] Given his environmental perspective, Bower argued that this antithetical alternation differed fundamentally from the sequence of sexual and asexual generations found in the algae. All such generations, he wrote, must "be considered as potential gametophytes, homologous one with another," and arise by the gametophytes' adapting to various environmental conditions.[82]

Four years later, in his presidential address to the Botanical Society of Edinburgh, Bower remarked that "a crisis has now arrived" in the explanation of alternation.[83] The distinction between the two generations had been shown by Eduard Strasburger to rest on minute nuclear differences—specifically, the different number of chromosomes in the two generations. This discovery was part of a series of revolutionary discoveries on cell structure and behavior that were made at the end of the nineteenth century, the period to which I now turn.

SIX

The Reemergence of Sex

During the last three decades of the nineteenth century, remarkable changes took place in the methods by which microscopists examined cellular material. New fixing and staining techniques made possible the detailed investigation of the cell nucleus, the importance of which was becoming increasingly obvious.[1] In 1885 the first edition of Arthur Lee's influential *Vade-Mecum* revealed that fixatives such as chromic, osmic, nitric, and picric acid and various carmine and hematoxylin stains were in general use.[2] Immersion objectives came into vogue, and at the end of the 1880s, cytologists began using the new Abbe-Schott apochromatic lenses, which produced images unclouded by either spherical or chromatic aberration. Together, these new techniques enabled microscopists to observe the intricate nature of nuclear division and, more to the point, the detailed cytological events surrounding fecundation.

With these discoveries, the definition and interpretation of sexual reproduction underwent a remarkable change. It became clear that sexual reproduction involved more than a simple stimulation of the egg by the entry of a fecundating principle contained within the sperm or pollen grain. Intricate *structural* components of the egg and sperm also were observed. Indeed, so intricate were these cytological events that not only did sexual reproduction come to be seen as a unique event, quite unlike asexual growth, but biologists began to define it in structural terms. An egg was not simply a sporelike cell, differing from the spore only in its mode of stimulation. In order to develop, an egg required the physical entry of a sperm nucleus and its chromosomes. Thus, by the early years of the twentieth century, sex had again become a central concern of biologists; no longer to be denied, it was recognized as more than a minor variation of asexual reproduction.

Such changes came about only after a period of intense controversy, however. It should not surprise us that a controversy developed, yet this aspect of scientific investigation needs to be stressed again. Scientists are brought up to believe that a sharp distinction exists between ''facts'' and their interpretation, and they are prone to ascribe changes in outlook solely to technological improvements. Such scientists would argue, for example, that once the new microscopes and chemicals came into use, cytologists would see exactly how the nucleus divided, how gametes and spores were really formed, and how chromosomes behaved during fecundation. Yet this is far from what actually occurred. Just as the invention of achromatic objectives in the 1820s generated rather than solved problems, so, too, in the last decades of the nineteenth century problems were generated and not immediately solved. Biologists in both periods may have shared a common technological tool, but it did not produce any agreement on what was seen. In both periods ''facts'' could not simply be observed; they had to be *interpreted,* and these interpretations reflected a host of differing assumptions on a wide range of biological issues.

In the 1870s it was still generally believed that fecundation involved the entry and dissolution of one or more sperm cells within an egg cell. In 1869, Eduard Strasburger had postulated that only one sperm cell of the fern was involved; that it passed down the neck of the archegonium, came into contact with the egg cell, penetrated it, and dissolved there.[3] Although it was generally assumed that the dissolution of sperm within the egg guaranteed that the offspring's characters were a blending of maternal and paternal characters, in the same way that white and black paints blend together to produce gray, it was more usual to stress the stimulating influence of this dissolution. In 1874, for example, the influential physiologist Wilhelm His reiterated the standard physicochemical argument that the sperm acted as a molecular stimulant for the egg to develop.[4]

The mode by which new cells arose was still a subject of controversy. Zoologists generally believed that new cells were formed by the division of preexisting cells, but botanists, faced with the complex cellular activities within the endosperm of the flowering plants, believed also in the occurrence of free cell-formation.[5] Moreover, the role played by the nucleus in these events was not clear. Some believed that a constriction and splitting of the nucleus into two halves preceded cell division, but others, such as Strasburger, reported that the old nucleus disintegrated prior to cell division and that the daughter nuclei appeared *de novo.*[6]

This uncertainty provided the background for the famous papers of Oscar Hertwig. Claiming in 1876 and 1877 that the ''cleavage nucleus [i.e., the zygote] arises from the conjugation of two different sexual nuclei, a female nucleus which is derived from the germinal vesicle and a male nucleus which is derived from the body of an entering spermatozoon,'' Hertwig posited a morphological theory of fecundation that challenged the generally held physicochemical interpretation.[7] Hertwig had been trained under Ernst Haeckel at Jena, and taught anatomy there until 1888. He had become interested in the process of fecundation in 1874, following the publication of Leopold

FIGURE 6.1.   Oscar Hertwig. Reproduced by permission of the Museum of Comparative Zoology, Harvard University.

Auerbach's findings. Auerbach had reported the *de novo* appearance of two "pronuclei" in the fertilized egg cell and their subsequent fusion to form the cleavage nucleus. "Suddenly the borderline between both nuclei disappears," he wrote, "and they become united to a single mass."[8] From his work on the eggs of the sea urchin, however, Hertwig reported that "at the period of maturation the germinal vesicle undergoes a retrogressive metamorphosis and is propelled to the yolk surface.... Its membrane is dissolved, [and] its contents decompose to be finally reabsorbed again by the yolk. The

Keimflecke, however, appears to remain unaltered and to continue in the yolk, where it becomes the permanent nucleus of the ripe ovum capable of fecundation.''[9] Hertwig also claimed that the other pronucleus represented the head, or nucleus, of the entering sperm.

At the same time, Hermann Fol, who, like Hertwig, had studied under Haeckel at Jena and was conscious of the important role Haeckel had bestowed on the nucleus, investigated fertilization in the starfish. He reported that the germinal vesicle underwent two rapid divisions and that only one nucleus, the female pronucleus, remained in the egg, the others being expelled. He also described the actual penetration of a *single* sperm into the egg, where, he explained, it fused with some egg protoplasm to form the male pronucleus. This pronucleus then traversed the egg to unite eventually with the female pronucleus.[10]

FIGURE 6.2.   Hermann Fol. Reproduced by permission of the Museum of Comparative Zoology, Harvard University.

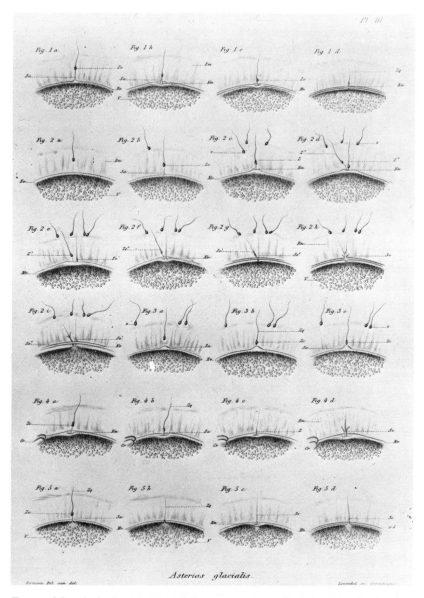

FIGURE 6.3.   A single sperm penetrating an egg. From Hermann Fol, ''Recherches sur la fécondation et le commencement de l'hénogénie chez divers animaux,'' *Mém. Soc. Phys. Hist. Nat. Genève* 26 (1879). Reproduced by permission of the Museum of Comparative Zoology, Harvard University.

The papers by Hertwig and Fol generated considerable criticism. Much of this stemmed from continuing opposition to the idea of nuclear continuity from cell generation to cell generation and also from hostility to the morphological interpretation of fecundation. Eduard Strasburger, like Hertwig and Fol a close associate of Haeckel's, denied nuclear continuity and argued in his text *Über Zellbildung und Zelltheilung* that "the contents of the pollen tube penetrate the egg in a dissolved form" and that although fecundation depends on the introduction of nuclear substance from the sperm, the substance enters as "physiological elements," not as morphological units of the sperm nucleus.[11]

A few years later, however, painstaking and exacting examination of the nature of cell division revealed the existence of formed nuclear elements, the chromatin, which, breaking into segments and splitting longitudinally, became distributed on the equatorial plate of the dividing cell. These pieces, or *Faden,* then migrated to the two poles of the cell, where they coalesced to form the nuclei of two daughter cells. Terming this process karyokinesis, or

FIGURE 6.4. Eduard Strasburger. Reproduced by permission of the Museum of Comparative Zoology, Harvard University.

"indirect nuclear division," Walther Flemming, professor of anatomy at the University of Kiel, argued that it was the only means by which nuclei divided in cell division. "Cell division," he remarked, "may be regarded as a process of asexual reproduction. In its course the chromatin separates from the achromatin, condenses into figures of a characteristic shape, [and] divides into two parts which provide the foundation of the daughter nuclei."[12] A year later Flemming agreed with Hertwig that fecundation involved the copulation of sperm and egg nuclei, but he argued that it was the chromatin threads of the male and female nuclei which united to form the cleavage nucleus.[13] Strasburger, meanwhile, changed his position on nuclear continuity. Writing in 1880 in the third edition of his *Über Zellbildung und Zelltheilung,* he noted: "In the plant kingdom the newly appearing cell nucleus can be traced back to an earlier existing nucleus. The new nuclei arise through the division of an older nucleus. The earlier concept in which the daughter nuclei were supposed to be newly formed owing to the dissolution of the mother nuclei, has proven to be false."[14]

At this time, however, although he accepted the continuity of nuclei, Strasburger denied the individuality and continuity of the chromatic threads (termed *chromosomes* by Waldeyer in 1888). Describing in detail the process of indirect nuclear division, he claimed in 1882 that the resting nucleus contained a *single,* long thread consisting of nucleomicrosomes embedded in a nucleohyaloplasm. By the fusion of adjacent microsomes, he argued, the thread became shorter and thicker, producing alternate thick and thin pieces of microsomes and hyaloplasm respectively. The single thread then broke within the thin areas, he reported, the number of breaks being more or less constant in any species. The microsomes then divided such that each chromatic element came to lie on the equatorial plate as a paired structure, which eventually pulled apart to form the nuclei of the two daughter cells.[15]

Strasburger's emphasis on the microsomal granules rather than on the chromosomes themselves sprang partly from his observation that cell division in *Hemerocallis fulva* deviated from the normal pattern: the cell sometimes divided into more than two daughter cells due to the retention on the equatorial plate of one chromatin element. This element, he noted, "forms a completely normal though very small cell nucleus," and thus "each segment of the nuclear thread shares the properties of the entire thread."[16] The role of indirect nuclear division, he therefore concluded, was to distribute *equal amounts* of the microsome material to each daughter nucleus:

> The significance of the complicated nuclear division may above all lie in the cell nucleus splitting into two completely equal halves. In the first segmentation of the nuclear thread the pieces are apparently of very unequal size and could become distributed often in a very unequal manner on both sides of the equatorial plate. Thus by splitting longitudinally and distributing the longitudinal halves to both daughter nuclei, the halves become true and equal.... *The*

simultaneous longitudinal division of the segments would be the surest means of distributing these substances equally to both daughter nuclei [italics added].[17]

This belief in the transient nature of the chromosomes and the primacy of their microsomal elements affected Strasburger's interpretation of fecundation profoundly. Arguing in 1877 that in plants the pollen tube introduced a "formless fecundating material," while in algae copulation consisted of the "fusion of similar parts of both swarmers with each other," he denied the unique role of the nucleus in the process of fecundation.[18] In 1882, however, although he still maintained that in algae "fecundation is a matter of the union of equivalent parts of both copulatory cells," he acknowledged that with gamete differentiation in higher plants, only the nuclear substance was involved.[19] He no longer argued that a formless fecundating material was liberated from the pollen tube. Instead, he spoke of small pieces of the nuclear thread, derived from the pollen nucleus, being passed through the walls of the pollen tube into the egg cell.

Two years later, however, Strasburger admitted not only that fecundation involved the fusion of complete egg and sperm nuclei but also that the nuclear segments remained preserved in the resting nucleus (although still joined end to end in a continuous thread). Upon fecundation, he argued, the nonnuclear material of the egg and sperm fuses, but the chromatin material lies in contact without actually fusing. Then the chromatin networks of the male and female nuclei divide into segments in the normal manner, distributing an equal amount of nuclear material to the two daughter nuclei so that each daughter nucleus will contain an equal number of paternal and maternal segments. Then and only then do the individual segments fuse (*verschmelzen*) end to end to form a single thread, "which is formed half from the nuclear segments of the father and half from the mother."[20]

By 1884, therefore, Strasburger had added his considerable prestige to those who believed that "the specific characteristics of the organism are based in the properties of the cell nucleus," and that sexual reproduction, through the agency of the nuclear segments, or "chromosomes," was the means by which the characteristics of both parents were passed to the offspring.

With the substitution of "nuclei" for "cells" and finally "chromosomes" for "nuclei," the definition of sexual reproduction obviously changed and the role of the male gamete was considerably enhanced. Apart from that, however, views on the significance of sexual reproduction were not altered by the discovery of its chromosomal base. Those who were concerned with problems of inheritance continued to view sexual reproduction as the means whereby male characters were incorporated into a female egg, and the offspring as a blending of both parental characters. Indeed, the fact that fecundation was seen to involve a complete fusion (*Verschmelzung*) of male and female cells, nuclei, and chromosomes lent a cytological basis to the blending of inherited characters, by which, in the words of Strasburger, "the permanence of the species as a whole" is maintained.[21] In addition, those who supported the physicochemical interpretation of fecundation could continue to do so, particularly after François Maupas showed that fatal aging occurs in cultures of ciliates unless conjugation occurs.[22] The view of fecundation as a process of rejuvenation held that the zygotic cell, once formed, continued to divide until its original energy was exhausted, at which time gametes were produced and a new burst of energy was generated by their subsequent fecundation. Thus, this view still had its adherents.

These interpretations of blending and rejuvenation were first fundamentally challenged by August Weismann, whose speculative writings of the 1880s generated a fierce controversy over the cytological events of gamete formation and subsequent fecundation, the biological implications of sexual reproduction, and the nature of the developmental processes. These differing reactions to Weismann's speculations led cytologists to see totally different events under the microscope; the "facts" did not speak for themselves.

THE PROBLEM OF MATURATION DIVISION

The commonly held view that germ cells arose in the same manner as tissue cells—in other words, by a typical, "indirect" nuclear division—began to be questioned in the 1880s. In 1884, for example, Strasburger stated that germ cells arose by a normal indirect division, but he also seemed to suggest that quite unique processes were involved. In the preparation of germ cells, he argued, it was necessary "that their idioplasm be reduced to half that mass which the embryonic nucleus possessed." But he also argued that it was necessary that the nature of the idioplasm or hereditary material change or be remodeled prior to fertilization, such remodeling perhaps being indicated by the "excretion" of polar bodies by the egg cell.[23]

It had long been known that close to the time of fertilization the egg cell buds off two minute bodies termed directive cells (*Richtungskörper*), or polar bodies, the significance of which had been a point of considerable controversy. One of the most popular explanations, and the one to which Strasburger took exception, was Edouard van Beneden's. At a time when the germ layers were being discussed widely with reference to the biogenetic law—specifically in terms of the embryonic ectoderm and endoderm of vertebrates constituting an embryonic recapitulation of the two-layered situation found in adult coelenterates (epidermis and gastrodermis)—van Beneden had discovered that the egg cells of coelenterates had an endodermal or gastrodermal origin, while the sperm cells were derived from the ectodermal or epidermal layer. "From the sexual point of view," he argued, "ectoderm and endoderm have an opposite significance," the ectoderm being the male germ layer, the endoderm the female.[24] Thus, since it always contains these two germ layers, the organism is hermaphroditic. In 1883 van Beneden had extended this hermaphroditic theory to include the individual cells; they, like the organism, also must be hermaphroditic. Thus, in order to produce a unisexual egg cell from a tissue cell, the male half of the nucleus must be removed as polar bodies, and to produce a male gamete the female half also must be removed.[25] As a result of these concepts, van Beneden concluded that just prior to fertilization, the germ nuclei were in fact only half nuclei, and believing them to be only half nuclei, he necessarily denied that fecundation involved a complete fusion of the two complete nuclear elements. Rather, he wrote, "Fecundation consists of a replacement, in the substitution of a half-nucleus furnished by the male and introduced by the sperm, for a half-nucleus eliminated by the egg in the form of polar bodies."[26] This view was generally disregarded. His conclusions that polar bodies represented an expelled male part of the nucleus, however, seemed to receive some approval. "The female sexual character of the egg," he wrote, "appears only after the expulsion of the polar globules; in fecundation the male elements of the egg are replaced by new elements supplied by the zoosperm."[27]

Van Beneden thus disagreed with Hertwig and Strasburger that fertilization involved the activities of complete egg and sperm nuclei. "One cannot give the name 'cell nucleus' to the clear spot nor the corpuscle contained in it," he wrote in 1876, "for the element looking like a nucleus which is formed near the surface of the yolk does not become the nucleus of the first cleavage sphere until after being united to another element having also the appearance of a cell nucleus. It is for this reason that I have called the peripheral bodies by the name 'pronucleus peripherique,' and the element which is formed in the center of the yolk by the name 'pronucleus central.' "[28]

Van Beneden's claim that the pronuclei were only half nuclei was further justified, according to him, by the significant discovery that each of the pronuclei in the nematode *Ascaris* contained only two "anses chromatiques." Further, he argued, upon conjugation the *anses* from each pronucleus *do not fuse* but line up on the equatorial plate as four *anses* which then divide, distributing four *anses* to each daughter nucleus.[29] Thus, van Beneden concluded that the embryonic cell must contain twice as many "anses chromatiques" as each pronucleus and that fecundation was essentially a mechanism of replacement or substitution. "Fecundation," he wrote, "appears to consist essentially in this reconstruction of the first embryonic cell, [a] cell revivified and provided with all necessary energy in order to transform into an individual like the parent."[30] A corollary to van Beneden's concept was the implication that since the gamete mother cell must divide in such a way as to produce a half nucleus and eliminate the other half as polar bodies, the mechanism of gamete formation must differ from the exact, quantitative, normal nuclear division.

Strasburger's early suggestions that the polar bodies were an excretion related to the remodeling of the idioplasm stemmed directly from the speculations of Wilhelm Roux. In 1883 Roux, expressing his interest in the mechanics of development, posed a fundamental question. If the characteristics of the organism were based on the properties of the cell nucleus, and if indirect nuclear division resulted in an exact duplication of nuclear material, how could a series of identical nuclei produced by the repeated and exact divisions of the embryonic nucleus induce those changes seen in ontological development? Roux concluded that although indirect nuclear division did ensure an equal distribution of the nuclear material in terms of *quantity,* nevertheless a qualitative change in the material took place during the nuclear divisions. Thus the nuclear chromatin was not a homogeneous mass of material but an aggregate of qualitatively different bodies that were distributed unequally during cell division. From this perspective Roux concluded that in order to re-form the original embryonic nucleus containing the full complement of these qualitatively different bodies, the nuclear material must be remodeled during the formation of the germ-cell nuclei.[31] Again, this suggested that the mechanism of germ-cell formation differed from the normal method of indirect nuclear division.

In 1887 Walther Flemming provided the first cytological evidence that the cell divisions involved in the production of sperm differed from the normal type of mitosis.[32] Spermatogenesis, he reported, involved two types of cell division. The first type differed little from normal mitosis, but the second, or "heterotypic mitosis," seemed to be unique in two ways. First, the chromosomes appeared as knots or rings. Second, and more significantly, the number of such ringed chromosomes was half the number that appeared in tissue cells—twelve rather than twenty-four in the salamander. However, for reasons that need not detain us here, Flemming saw nothing of very deep significance in these differences. We owe to August Weismann the statement that germ-cell production *must* be fundamentally different from tissue-cell production.

In a paper of 1885 Weismann had posed Wilhelm Roux's question and, like Roux, had concluded that the nuclear material of embryonic cells must change *qualitatively* during ontogeny, becoming less and less complex as development proceeds.[33] The fact that the longitudinal division of chromosomes during mitosis leads to daughter nuclei with an *equal quantity* of nuclear material does not prove, he argued, "that the quality of the parent nucleoplasms must always be equal in the daughter nuclei."[34] However, he went on to ask, if the cell nuclei change qualitatively during development, how can germ cells be produced again, for germ cells, unlike embryonic cells, contain within their nuclei "all the hereditary tendencies of the whole organism." Denying the possibility of any gradual or sudden retransformation of embryonic nucleoplasms into germinal nucleoplasm, he argued instead "that in each ontogeny, a part of the specific germplasm contained in the parent egg cell is not used up in the construction of the body of the offspring, but is reserved unchanged for the formation of the germ cells of the following generation."[35] However, he realized that the germ cells themselves were quite

FIGURE 6.5. August Weismann and the continuation of his germ plasm: his son, Julius. Photo, taken in Freiburg in 1909, reproduced by permission of the Museum of Comparative Zoology, Harvard University.

complex and must therefore contain in their nuclei a specific "histogenic nucleoplasm" to control their development. In order for embryonic development to begin, he speculated, it was necessary for this histogenic nucleoplasm to be removed, leaving the germ-cell nuclei with only pure germ plasm. This histogenic nucleoplasm, he concluded, was removed as polar bodies.[36]

Weismann's and van Beneden's notions on polar bodies had more in common than either man would admit. Both assumed that before fertilization could take place it was necessary to remove something physically from the germ nuclei, and that the maturation divisions leading to the production of germ nuclei were somehow related to this expulsion process. They differed, of course, in their views of the nature of this removed part, a difference that could be tested quite easily, both realized, by determining whether parthenogenetic eggs also removed polar bodies. From van Beneden's point of view, parthenogenetic eggs, since they were capable of development without fertilization, must be whole nuclei from which the male element had not been removed. In Weismann's view, however, it was still necessary for parthenogenetic eggs to have their own "histogenic nucleoplasm" and for this to be removed before development could proceed. Thus, it was something of a surprise to Weismann when he discovered later in 1885 that parthenogenetic eggs of *Daphnia* produced one polar body, not two. He concluded from this finding that the second polar body, expelled from normal eggs but not from parthenogenetic eggs, must have a different role than the first. The role of this second polar body provided the substance of Weismann's famous paper of 1887, "On the Number of Polar Bodies and Their Significance in Heredity," the reaction to which stirred the cytological controversy of the 1890s.[37]

Weismann believed that originally all organisms reproduced only asexually, in which case their germ plasms were completely homogeneous. When sexual reproduction took place for the first time, however, two different germ plasms necessarily came together (Fig. 6.6). In order to keep the quantity of total germ plasm constant in this first sexually produced generation, it was obviously necessary, he argued, that each individual germ plasm be halved *in quantity*. In each succeeding generation, therefore, to maintain a constant total quantity of germ plasm, the quantity of each individual or ancestral germ plasm had to be halved. In other words, in each subsequent sexual generation the total *number* of ancestral germ plasms, or *ids,* doubled while the *quantity* of each id was halved. Obviously, since each id was a very complex entity, there was a minimal size for each id such that the quantity of each could not be halved indefinitely. Since this stage was reached soon after sexual reproduction first appeared, the question remained how sexual reproduction could continue without a corresponding doubling of total germ plasm in each generation—the halving of individual ids being no longer possible. The answer was clear: there must be a mechanism which reduced *not* the quantity of each id but the *number* of ids.

Weismann visualized the ids as being arranged in rows along the nuclear

A) FIRST SEXUAL GENERATION

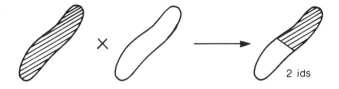

B) SECOND SEXUAL GENERATION

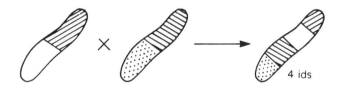

C) To produce eventually an idant (chromosome) with row
of ids (a to h).

FIGURE 6.6.   Reduction of the quantity of each id in the early generations of the first
sexually reproducing organism, according to Weismann. From Alice Baxter and John
Farley, "Mendel and Meiosis," *J. Hist. Biol.* 12 (1979). Reproduced by permission
of D. Reidel Publishing Co., Dordrecht, Holland.

threads. Clearly, then, in normal mitosis, where each thread, or chromosome,
splits longitudinally, the number of ids would be the same in both daughter
cells and equal to the number in the mother cell. Thus, in order for an id
reduction to take place during the production of germ cells, "there must be yet
another kind of karyokinesis, in which the [chromosomes] are not split longitudinally, but are separated without division into two groups, each of which
forms one of the two daughter nuclei."[38] Such a mechanism is shown in Fig.

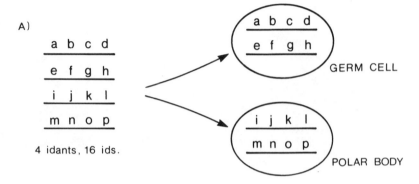

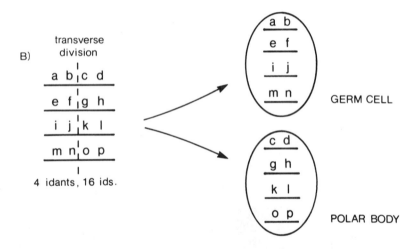

FIGURE 6.7. Reduction of the number of ids in modern sexually reproducing organisms (after Weismann). From Alice Baxter and John Farley, "Mendel and Meiosis," *J. Hist. Biol.* 12 (1979). Reproduced by permission of D. Reidel Publishing Co., Dordrecht, Holland.

6.7 (*A*). The four chromosomes do not divide longitudinally but are arranged singly on the equatorial plate of the dividing cell. Two chromosomes move to one pole to form the germ-cell nucleus; the other two are removed as the second polar body. Thus, Weismann concluded, "with the nucleus of the second polar body as many different kinds of idioplasm are removed from the egg as will be afterwards introduced by the sperm nucleus; thus the second division of the egg nucleus serves to keep constant the number of different kinds of idioplasm."[39] However, it was also possible for the number of ids to be reduced by a transverse division of each chromosome during maturation division (Fig. 6.7 [B]). In both cases the end result would be the same: the daughter nuclei would receive half the number of ancestral germ plasms, or ids, that were present in the mother cell. *Weismann termed this division, this reduction in the number of ids, a "reduction division."*

At the end of his 1887 paper Weismann made the very significant remark that since it was unlikely that the polar bodies would remove the same ids each time, those retained in the germ cells also would be different. In other words, the germ nuclei were probably different from each other. This conclusion was to become very important to Weismann in his later papers.

A few years later Weismann had to alter his theory slightly when Oscar Hertwig and Theodor Boveri (a professor at Würzburg and a former student of Richard Hertwig's) demonstrated that egg and polar-body formation was essentially the same as sperm formation. Boveri's interest in the polar bodies had been aroused by van Beneden's report of the reduced number of "anses chromatiques" in the germ cells. The express purpose of Boveri's first investigation was therefore to observe with his own eyes the process of fertilization and to verify van Beneden's description of the polar bodies.[40] Boveri did verify that the chromosome number was halved, but, disagreeing with van Beneden, he maintained that the division was a typical karyokinetic division and not simply an expulsion of the male element. In his study, Boveri also presented the first description of chromosome tetrads (*Vierergruppen*) and their behavior during reduction division. In *Ascaris megalocephala* var. *bivalens,* where the normal diploid number is four, Boveri observed two groups of chromosomes with four chromosomes in each group. The first division of the egg separated the tetrads into two dyads, one of which remained in the egg, the second of which entered the first polar body. The second division of the egg separated the two elements of each dyad, leaving a total of two chromosomes in the egg cell and two chromosomes in the second polar body. Meanwhile, the first polar body divided once. Thus the divisions produced a total of three polar bodies, each containing two chromosomes.

In 1890 Hertwig confirmed Boveri's findings. He also presented a complete description of sperm cell formation in which he showed conclusively that "the transformations of the nucleus which on one side take place in the transformations of sperm mother cells into sperm, and on the other in the production of polar bodies, show an extraordinary and very striking similarity."[41] Whereas in spermatogenesis one sperm mother cell divides into two spermatocytes, each of which quickly divides again to form four sperm, in oogenesis, the egg mother cell divides to form one oocyte and one polar body, and the oocyte quickly divides again to form one egg cell and the second polar body (Fig. 6.9). Thus, Hertwig concluded, the polar body does not represent a part of the germ plasm removed from the egg but has, in fact, "the morphological value of rudimentary egg cells."[42]

Thus Hertwig and Boveri confirmed that during the maturation process "there is a reduction by half of the originally existing number of chromosomes, and this numerical reduction is not therefore only a theoretical postulate but a fact."[43]

Two aspects of the Hertwig and Boveri papers need to be stressed at this stage. First, the reduction to which they referred was a reduction only in the number of chromosomes, not in the number of ids (i.e., chromosomal com-

FIGURE 6.8.   Theodor Boveri. Reproduced by permission of the Museum of Comparative Zoology, Harvard University.

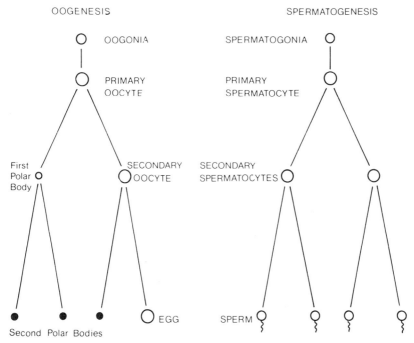

FIGURE 6.9.   Theodor Boveri's and Oscar Hertwig's interpretation of oogenesis and spermatogenesis. From Alice Baxter and John Farley, "Mendel and Meiosis," *J. Hist. Biol.* 12 (1979). Reproduced by permission of D. Reidel Publishing Co., Dordrecht, Holland.

ponents). Second, this numerical reduction took place during two sequential cell divisions and not by the bodily removal of part of the nuclear material in the form of polar bodies. Boveri termed this interpretation of polar bodies the "egg hypothesis," or "phylogenetic hypothesis," in that polar bodies are abortive eggs, which to him were nothing but a "phylogenetic reminiscence."[44] Also, contrary to Weismann's interpretation, Boveri claimed that both polar bodies "originate in the same way and have the same size and chemical composition"; since the first polar body actually divides once, "the four sperm arising from one mother cell correspond to the egg and the three polar bodies."[45]

Weismann viewed these papers by Hertwig and Boveri as a partial vindication of his own theoretical conclusions. Agreeing with them that the number of chromosomes and thus the mass of nuclear material was reduced by half, he also claimed that since the chromosomes were not alike "but are derived from the differing germplasms of various ancestors" (a claim with which, as we shall see, Hertwig will not agree), "it follows that a reduction of the ancestral germ-plasms is admitted."[46] On the other hand, Weismann was forced to admit that his previous interpretation of the polar bodies was in error; both were intimately related to the division process, but one of them was not a carrier of histogenic nucleoplasm. In addition, Weismann was led to question why there should be two sequential divisions when one could equally well fulfill the necessary goals of reduction. The answer came in the papers of Hertwig and Boveri: "because the number of rods is doubled before the process of reduction has begun."[47] As Weismann then argued, by having two divisions rather than one, "the highest possible number of combinations of germ plasms are offered for the operation of natural selection."[48] Thus the complex cytological events leading up to the formation of the germ cells were of fundamental importance to the interpretation of sexual reproduction. They represented, according to Weismann, "the attempt to bring about as ultimate a mixture as possible of the hereditary units [the ids] of both father and mother."[49] Therefore, sexual reproduction was the means of introducing variation into the population; it provided the raw material upon which natural selection acted. *Not only was sexual reproduction totally different from asexual reproduction, but it was absolutely essential to the whole evolutionary story.*

Weismann's interpretation of maturation division as a reduction in ids, or characters, formed the focal point of the controversy over germ-cell formation in the 1890s. The term "reduction division" was used in descriptions of spermatogenesis and oogenesis to refer to a qualitative reduction in characters. (A division that did not result in a reduced number of qualities was usually called an "equation division.") The controversy stemmed primarily from the question whether or not such a "reduction" took place. Many cytologists interpreted maturation division as a reduction in nuclear mass only, and not as a reduction in nuclear germ plasms. In addition—and this was to become crucial after the rediscovery of Mendel—they did not agree that each germ cell was uniquely characterized by different combinations of germ plasms. To Strasburger, Hertwig, Leon Guignard, and others, all germ cells of an organism were identical. Sexual reproduction was not a process leading to variations in offspring as Weismann maintained; rather, it was one which, by the fusion of parental nuclei, "tends to maintain the permanence of the species as a whole."

As a result of Weismann's papers, cytologists in the 1890s focused their attention on the problems of maturation divisions in eggs, sperm, and pollen grains. It should be stressed that the reduction-division controversy developed at the very time when the new apochromatic lenses were coming into use and at the time when a younger generation of embryologists was beginning to turn away from microscopic descriptions of embryonic stages to instead subject the embryos to experimental manipulation.

The basic position of those who opposed Weismann's hypothesis of reduction division was first clearly outlined in the 1890 papers of Hertwig and Boveri. The former worked on the two varieties of the nematode *Ascaris megalocephala;* the latter studied an echinoderm, a medusa, and five species of mollusk.

At that time it was known that in the normal mitotic division of *Ascaris megalocephala* var. *bivalens,* four chromatic threads appear. Each thread then divides longitudinally and four chromosomes are distributed to each daughter nucleus. During spermatogenesis, however, eight chromosomes eventually appear, not four. These are then distributed—two to each of four spermatids—by means of two nuclear divisions that follow immediately after each other without a resting stage.

The basic question, of course, was how these eight chromosomes were formed. Hertwig noted three possibilities (Fig. 6.10):

1. The chromatin network of the resting cell broke into eight segments instead of the usual four. Thus, maturation division occurred without any longitudinal division of chromosomes, and the eight chromosomes represented eight different individual threads.

2. The chromatin network of the resting cell broke into the usual four chromosomes, each of which then divided once longitudinally. Thus the eight chromosomes consisted of four identical pairs.

3. The chromatin network of the resting cell broke into only two elements, each of which then divided twice longitudinally. Thus the eight chromosomes consisted of two sets of four identical chromosomes.

On rather shaky grounds Hertwig opted for the third possibility. He argued that during the maturation process the chromatin network breaks into half as many chromatin threads as existed in a tissue cell and that *each* of these chromosome threads then divides *twice* longitudinally to form a "tetrad" of four identical chromosomes. Each member of a tetrad is then distributed to a gamete nucleus during two nuclear divisions, between which there is no resting stage. Clearly, as a result of this process, only a *mass* reduction of nuclear material and a *numerical* reduction in chromosomes occurs. There is no reduction in the number of *ids* nor is there any variability among the gamete cells.[50] Hertwig's position was backed by Boveri, who in *Zellenstudien III* reaffirmed his earlier belief that each tetrad was formed by the two longitudinal divisions of a single chromosome.

It should be noted at this point that both Hertwig and Boveri attributed the reduction of chromosome number in the gametes to the fact that only half as many chromosomal elements appeared at the initiation of the maturation divisions as appeared at the initiation of normal cell division. In *Ascaris,* for example, four chromosomes appeared in mitosis but, they assumed, only two appeared in gametogenesis. Each of these chromosomes then divided twice longitudinally to form two "tetrads," each containing four chromosomes. As Boveri wrote in 1890:

MITOTIC DIVISION

1   2   3

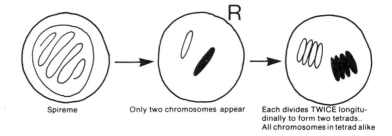

Spireme   Four chromosomes appear   Each chromosome divides once longitudinally to arrange themselves on equator

MATURATION DIVISION

HERTWIG'S INTERPRETATION

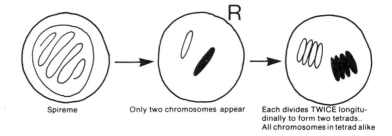

Spireme   Only two chromosomes appear   Each divides TWICE longitudinally to form two tetrads.. All chromosomes in tetrad alike

THE WEISMANN GROUP'S INTERPRETATION

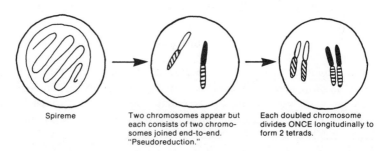

Spireme   Two chromosomes appear but each consists of two chromosomes joined end-to-end. "Pseudoreduction."   Each doubled chromosome divides ONCE longitudinally to form 2 tetrads.

4   5

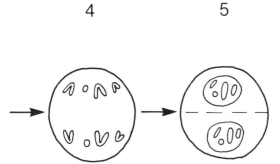

The chromosomes then move to each pole of the cell and the
cell divides. Each daughter cell then contains four chromosomes. Daughter cells identical to mother cell.

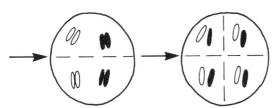

Two divisions without a resting stage. Each gamete receives
two chromosomes, half the somatic number. ALL CELLS
IDENTICAL.

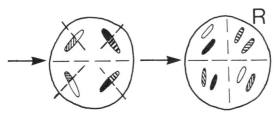

First division: Each tetrad moves apart longitudinally. Second
division: Transverse (shown by dotted lines). ALL CELLS
NOT IDENTICAL.

FIGURE 6.10.   The reduction division controversy. "R" denotes stage of reduction.

If we consider first of all the formation of the egg, then there is only one statement which can be put forth as certain and generally valid, that the reduction must take place, at the latest, in the primordial germ cell. Therefore, by the formation of the first maturation spindle the chromosomes already are present in the reduced number.[51]

Reduction (*R* in Fig. 6.10) took place before the cells began to divide.

It is clear that most of the more eminent cytologists of the 1890s supported the above position. Leon Guignard, professor of botany at the Paris School of Pharmacy, reported in 1891 that whereas the primordial pollen mother cells and epidermal cells of *Lilium* had twenty-four "batonnets chromatiques," only twelve segments, each consisting of two rows of chromatic granules, appeared in the definitive pollen mother cell. There is nothing to prove, he remarked, "that at the moment when the nucleus of the mother cell has taken birth, these 24 'batonnets chromatiques' are joined together in pairs, end to end, to give twelve." These twelve, longitudinally split chromosomes then moved apart at the first division to distribute twelve single chromosomes to each daughter cell. Each of these then split longitudinally again and twelve chromosomes passed to each of the four grains of pollen. Thus, as a result of two longitudinal splittings of each chromosome and two nuclear divisions, "the nucleus of each of the four grains of pollen is derived from a mother cell constituted of twelve segments." Apart from the fact that the cell began with twelve chromosomes rather than the typical twenty-four, Guignard could see nothing "étranger à la marche normale de la karyokinèse."[52] In 1898, after describing virtually the same process in *Naias major* (with the exception that both of the longitudinal chromosome splittings took place before nuclear division commenced), Guignard concluded:

> The second division simply has the object of distributing equally to the four sexual nuclei the chromosomes already formed during the first division. It reduces by half the *quantity* of nuclein which they receive, compared to the amount possessed by vegetative nuclei as a result of ordinary mitosis. But, no more than the first division, it does not reduce *qualitatively,* and *the four nuclei are equivalent with regard to hereditary properties* [italics added].[53]

In 1893 August Brauer, who like Boveri had studied under Richard Hertwig, repeated Oscar Hertwig's investigations on spermatogenesis in *Ascaris*. He observed that before nuclear division began, only half as many chromosomes appeared as appeared at the initiation of mitotic division, and that each chromosome then divided twice longitudinally. Thus, in his words, the only difference between normal mitosis and maturation divisions was that "one transverse division of the *Faden* does not occur, whereas one more division (nuclear) follows. The first causes the number of chromosomes to be reduced by half; the second division, together with the elimination of a resting stage between both divisions, produces a halving of the mass." Thus, he concluded, "a reduction division in the sense of Weismann does not occur."[54]

Similar conclusions were reached by Friedrich Meves, who was working on *Salamandra maculosa,* and by John Farmer, professor of botany at the Royal College of Science, who was working on the Hepaticae. Both recorded that maturation divisions of the nucleus were preceded by two longitudinal divisions of the chromosomes. Thus, Farmer concluded from his work on spore formation in the Hepaticae, "a study of these divisions affords not the slightest evidence in favour of any reduction division (in Weismann's sense) taking place. . . . The only reduction is a numerical one."[55]

Some cytologists interpreted maturation division in the terms demanded by Weismann's hypothesis, but as a group their prestige hardly matched that of their opponents. Indeed, three of them—Chiyomatsu Ishikawa, Otto vom Rath, and Valentin Haecker—were graduate students of Weismann's, and their goal, they stated, was to illustrate the validity of the Weismannian hypothesis rather than to test it.[56] The most impressive of those who supported Weismann's concept of reduction division was Johannes Rückert of Munich. In 1893 and 1894 he produced an objective summary of the whole problem together with factual evidence in support of Weismann's hypothesis.[57]

Admitting in his 1893 paper that previous work on *Ascaris* denied Weismann's law of reduction, Rückert pointed out that the work of Weismann's students on the maturation division of arthropods had provided factual proof of Weismann's theoretical concepts. Both schools of thought, he pointed out, generally agreed that prior to the first nuclear division the number of chromosomes seemed to be halved. They also agreed that each of these chromosomes eventually formed a tetrad, which was constituted of four parts. Then, pointing to agreement on a third issue, Rückert concluded: "Through two sequential mitotic divisions without a resting phase, the tetrad is divided in such a way that in each spermatid or ripe egg cell a single rod of each group arrives as a chromosome."[58] The questions at issue, of course, were the derivation of the reduced number of initial chromosomes and the derivation of the tetrad. To simplify matters, each question will be dealt with separately below.

Derivation of the Reduced Number of Initial Chromosomes

This question presented no problem to Guignard, Brauer, and Hertwig. To them, chromosomes were merely transitory bodies that dissolved into the nuclear sap at the end of each cell division; they simply assumed that prior to the onset of maturation divisions half the usual number of chromosomes segmented out.

However, to those who accepted that, in the words of Boveri, "the chromatic elements are independent individuals that retain this independence even in the resting condition,"[59] this decrease in chromosome number presented a difficult problem. Rather reluntantly, Boveri accepted the position that half the chromosomes degenerated sometime before maturation division commenced. In 1890 he observed, in addition to the tetrads, two small, darkly

stained bodies. "One could regard these two bodies," he wrote, "as degener-
ate chromosomes, and if this interpretation be correct, then the reduction
would be produced when half the chromosomes atrophied."[60] Strasburger, on
the other hand, argued that "the reduction of the number of chromosomes by
half is due to the fusion into one of two chromatic individuals."[61]

Although, as we shall see, this position seemed to be identical to the view
of Weismann's supporters, it was, in fact, fundamentally different. To Strasburger this union of two chromosomes into one was a complete fusion, or
Verschmelzung, such that in reality a completely new set of chromosomes
appeared. The Weismann group assumed that the initial reduction in number
was merely a "pseudoreduction," which appeared because two chromosomes
temporarily attached to each other end to end. Thus, each of these so-called
chromosomes was in fact a pair of dissimilar chromosomes. Weismann's
students believed that to ensure a combination of ids during the sexual process, it was necessary to retain the individuality of each chromosome. As
Haecker remarked of Strasburger's opinion, "I believe that by assuming such
a fusion the process of reduction is robbed of all theoretical significance, as
far as such significance bears upon the theory of heredity."[62]

Derivation of the Tetrad

Most of the more eminent cytologists of the day believed that each
chromosome divided twice longitudinally to form a tetrad with four identical
parts. "If it really exists," wrote Rückert, "then one must see the essential of
reduction to be the reduction of chromatin mass."

The Weismann group, however, believed that only one longitudinal division took place (Fig. 6.10). Each original chromosome, which they assumed
consisted of two chromosomes joined end to end, divided once longitudinally
to form the tetrad. Then, during the first nuclear division the two longitudinal
halves of each tetrad were distributed to each daughter nucleus and during the
second division the two chromosomes that had joined end to end separated by
means of a *transverse division.* "Such a division," Rückert remarked,
"would not only answer the requirements of all those who previously wished
to see explained only the empirically established numerical reduction of
chromosomes, but also would accomplish the reduction of ancestral plasms
postulated theoretically by Weismann."[63] The second nuclear division had the
effect of distributing half of the original and permanent chromosomes to one
daughter cell and half to the other, such that a reduction in the number of ids
took place. Moreover, unlike Hertwig and Strasburger, *Weismann's supporters denied that any fusion took place during fertilization.* To them the embryonic nucleus was not a newly generated entity but one which contained a
combination of permanent chromosomes, half from each parent.

An Overview of the Controversy

At this point it may be helpful to summarize the different positions taken in the reduction-division controversy. There were two points on which all the investigators agreed: (1) prior to the first nuclear division the number of chromosomes appeared to be halved such that the number of tetrads present was always half the normal somatic number; (2) through two sequential divisions, without a resting phase, the tetrad was divided such that each sperm or ripe egg cell received a single rod from each tetrad. There was no agreement, however, on the manner in which the reduced number of chromosomes initially appeared or on the manner in which the tetrad was formed and divided.

Derivation of the reduced number. Brauer and Hertwig, who believed the chromatin rods were temporary aggregates of granules, simply assumed that the spireme segmented into half the normal number of rods. Strasburger, on the other hand, believed that the reduction in number was caused by a process of fusion, or *Verschmelzung*. Boveri reluctantly set forth the hypothesis that half the chromosomes degenerated before the tetrads were formed.

Weismann's group assumed that the initial reduction was a "pseudoreduction" caused by the attachment of two individual, dissimilar chromosomes end to end. This interpretation differed from that of Hertwig and Brauer in that it stressed the *temporary* union of permanent individual chromosomes as opposed to the random grouping of chromatin granules into transitory rods. The Weismann group also differed from the majority by denying that any actual *Verschmelzung* took place during the process of fecundation.

Division of the tetrad. The Hertwig group, which included Strasburger and Boveri, believed that each chromatin rod divided twice longitudinally to form a tetrad with four identical parts. All the germ cells formed from the separation of such a tetrad would be identical since the reduction was only a reduction in mass.

The Weismann group believed that only one longitudinal division took place. Each rod, which was, in fact, two chromosomes joined end to end, divided longitudinally to form the tetrad. The tetrad was thus composed not of four identical parts but of two pairs of chromosomes. During the first nuclear division the two longitudinal halves were distributed to each daughter nucleus and during the second division the two chromosomes that had joined end to end were separated by a *transverse* division. The germ cells did not receive identical sets of chromosomes; half the permanent chromosomes went to one daughter cell and half to the other. Thus, a reduction in the number of ids occurred.

During the 1890s no special evidence was brought forward by either of the debating groups that would compel one to assent to its view. One could assume that the process of germ-cell formation varied from one species to the next, but this would be an exception to the general uniformity of cellular structure and processes throughout the plant and animal kingdoms. The different interpretations that were put forward reflected two facts: first, that even with improved techniques, it was difficult to evaluate what one saw through the microscope; second, that the investigators differed in their fundamental assumptions about the nature of the hereditary material and about the nature of development. In the 1880s and 1890s the old issue of preformation versus epigenesis reemerged with the questions, "Is embryonic development epigenesis or evolution? Is it the new formation of complexity, or is it the becoming visible of complexity previously invisible to us?"[64] The fact that the two groups also differed over this question illustrates again the major philosophical difference between them.

An epigenetic interpretation of the developmental process had become firmly established in the nineteenth century following the collapse of preexistence theories. As everyone realized, however, epigenesis was basically a statement of fact. It described what was seen to occur during ontogeny: a progressive and gradual appearance of parts. "There must be no mistake," wrote the Oxford zoologist Gilbert Bourne: "epigenesis is a fact, not a theory."[65] This was the basic problem, however. By the 1880s, mere formal descriptions of embryonic events no longer seemed adequate to many biologists; but causal explanations brought a return to preformationist-like doctrines. Thorough-going epigenesists simply had great difficulty explaining the causes of embryonic development. Such epigenesists, among them Hertwig and Strasburger, assumed that the embryonic nucleus consisted of basically homogeneous nuclear material produced each time afresh by the complete fusion of male and female germ-cell nuclei. According to Strasburger, Weismann's ids, visible "as discoid segments of the chromosomes," each contained all the hereditary characters—a view to which Strasburger had subscribed following his earlier work on *Hemerocallis*. Thus, he wrote, "the serially arranged ids in the chromosome are, in my opinion, repetitions of each other." The hereditary particles were not, as was later believed, bearers of discrete and different characters, and neither were they, as Weismann believed, bearers of different ancestral germ plasms. Instead, Strasburger argued, "the neoplasm of many and different ancestors enters into the formation of *each individual id*" (italics added), and their entry is brought about by "the fusion in pairs of the ids and therefore also of the chromosomes." "I do not," Strasburger concluded, "consider that these ancestral plasms exist isolated in the id; I regard them as completely fused into one."[66] "Neopreformationists" like Weismann and Roux, however, assumed that differentiation occurred because of the gradual appearance of a latent differentiation that was already present in the embryonic nucleus and in the germ-cell nuclei, which did not fuse. As we have seen, Weismann assumed that the

embryonic nuclei divided qualitatively such that the nucleus of the cells became progressively dissimilar. Cellular differentiation, then, was caused basically by this nuclear unlikeness. To Hertwig and Strasburger, however, the
embryonic nuclei always divided quantitatively to produce cells with identical
nuclei. In that case, differentiation could not be attributed to nuclear differentiation but came about through the interaction of cells with each other
and with their environment.

Both Hertwig (who was a Lamarckian) and Strasburger attacked the preformationist position from the same perspective. "Theoretical speculation,
which transcends the limits of experience," Strasburger argued, "must start
from definitely ascertained facts."

> Minute investigation of the longitudinal splitting of the chromosomes can but
> produce the impression of equal division; there is absolutely no foundation in
> fact for the assumption of unequal division. Hence, from the very beginning, I
> have taken the standpoint of epigenesis in forming my theoretical interpretation
> of the facts of development. The only conception of development that I am able
> to form is that it is a succession of stages, such that each stage determines the
> conditions for the succeeding stage and inevitably leads on to it.[67]

Hertwig, in his famous paper entitled "The Biological Problem of Today:
Preformation or Epigenesis?," vigorously attacked Weismann's position:

> To satisfy our craving for causality, biologists transform the visible complexity
> of the adult organism into a latent complexity of the germ, and try to express this
> by imaginary tokens. . . . Thus craftily, they prepare for our craving after cau
> sality a slumbrous pillow. . . . But their pillow of sleep is dangerous for biologi
> cal research; he who builds such castles in the air easily mistakes his imaginary
> bricks, invented to explain the complexity, for real stones. He entangles himself
> in the cobwebs of his own thoughts, which seem to him so logical, that finally
> he trusts the labour of his mind more than nature herself.[68]

By 1900 there was not enough evidence either to settle the preformation-
versus-epigenesis controversy or to resolve the question of chromosome individuality. Experiments on the former question yielded contradictory results for
different species, and the arguments for and against individuality were
based primarily on cytological observations—those of Rabl and Boveri, for
example. Such observations were no more convincing than the observations
of maturation division had been.

It is against this background that Mendel's work was rediscovered in 1900.
It is obvious that the favored interpretation of germ-cell formation—two longitudinal divisions—would not have led anyone to see the connection between
the chromosomes and Mendel's factors. If indeed chromosomes behaved in
the way the majority of cytologists described, and if Mendel's factors lay on
those chromosomes, then all gametes would be alike and there would be no
segregation of factors, no 3:1 or 9:3:3:1 ratios.

Nevertheless, Carl Correns, one of the codiscoverers of Mendel's work, did see the connection between Mendel's results and Weismann's interpretation of reduction division. To explain such results, Correns remarked,

> one must assume with Mendel, that after the union of sexual nuclei the *Anlage* for the "recessive" character, in our case for the green, becomes suppressed by the other "dominant" character, the gold. The embryo becomes all gold. The *Anlage* however remains; it is only "latent," and before the definitive formation of the sexual nuclei there always occurs a separation of both *Anlage* in such a manner that *half* of the sexual nuclei contain the *Anlage* for the recessive character, for green, and *half* the *Anlage* for the dominant, for gold. The separation takes place at the earliest in the formation of sperm and pollen *Anlage*. The 1:1 ratio very much supports nuclear division taking place by Weismann's reduction division.[69]

It should be emphasized again, however, that Weismann's hypothesis was little supported in 1900 and that those who did support it had worked almost exclusively on insects. Mendel's botanical colleagues were unanimously opposed to any form of reduction division. "There is no reduction division in the plant kingdom nor anywhere else," wrote Strasburger in 1894, and this was a view to which botanists subscribed even after 1900.[70]

By the turn of the century, therefore, sexual reproduction had become an issue of major importance to biologists. Although its definition had changed and now emphasized the behavior of male and female chromosomes, just how these chromosomes arose and how they acted during fecundation remained points of controversy. As a result, the significance of the sexual act remained obscure. Cytologists had moved away from ascribing the phenomenon to an act of sperm stimulation and toward a concern with heredity. Still, it was not clear whether sexual reproduction involved simply the conservative act of blending male and female characters, thereby conserving the characters of the species, or whether it was an act by which variation was introduced into the population so as to ensure continuous change or evolution. Sex had reemerged, but what it involved and what it signified remained unclear.

SEVEN

Sex, the *Sine Qua Non* of Evolution

Although the work of late-nineteenth-century cytologists revealed many of the morphological characteristics of sexual reproduction and showed that it could no longer be dismissed as a minor variation of the normal asexual processes, the details of the process and its biological significance remained obscure, the focus of controversy. By 1905, however, the speculation had ended. By then cytologists were interpreting sexual reproduction as a process without which evolution could not have occurred, a process which introduced variations into a population. Variations arose, they argued, because the gametes produced during maturation division were different from each other and because their chromosomes did not fuse during fecundation—just as Weismann had claimed.

This rapid end to the reduction-division controversy can be traced to two major turn-of-the-century developments. First, the work of Hugo de Vries and the rediscovery of Mendel's work provided cytologists with a theoretical base from which to reinterpret their findings. Mendel's theories were not accepted because they could be fitted into an already established cytological picture; on the contrary, the cytological picture was quickly realigned to fit the new theoretical framework. The absence of such a framework had perpetuated the cytological controversy of the 1890s; the presence of a unified framework solved the problem in the first decade of the twentieth century. Second, the anti-Weismann position was fatally undermined when many of its advocates gave up cytology or changed their opinions.

The acceptance of both the new theory and the new cytological facts that sprang from it was in turn a result of a profound change that began to take place in some branches of the biological sciences at the end of the nineteenth century. Such changes impinged directly on cytologists and embryologists. First in Germany, and later in the United States, a totally new outlook on embryology arose to replace the old phylogenetic emphasis. Mere description, it was argued, should be replaced by the study of "developmental mechanics," or *Entwicklungsmechanik*. Such a program formed the raison d'être of a new journal, *Archiv für Entwicklungsmechanik der Organismen,* the first volume of which appeared in 1894 under the editorship of Wilhelm Roux. "Developmental mechanics," Roux wrote, "is the doctrine of causes of organic forms, . . . [and] the causal method of investigation is experiment."[1] In the final analysis, he argued, the processes of development would have to be reduced to physical and chemical phenomena, but initially such an approach would not be possible. The importance of descriptive and comparative embryology is denied, he remarked, by authors "who see the *present* task of developmental mechanics in the immediate reduction of organic formative processes to purely inorganic, physico-chemical components." If, however, "we limit ourselves to that which is *possible* at present, we can regard *this* task only as a final goal, which for the present, and even for some time to come, we shall approach in a *direct* path only at a relatively slow pace."[2] Initially, therefore, the developmental mechanists undertook a series of experimental manipulations of developing embryos.

The results of these early experiments were, however, as indecisive in resolving the Weismann controversy as cytology had been. Roux's discovery in 1888 that one blastomere of a two-cell-stage frog embryo (in which the other blastomere had been killed) formed only half a larva seemed to support his and Weismann's theories and the cytological work of Rückert and others, but a discovery by Hans Driesch a few years later appeared to support Hertwig's, Strasburger's, and Boveri's findings. Driesch found that the isolated blastomere of some animals actually formed complete larvae. These seemingly profound differences between epigenetic and preformationist interpretations of development were quickly resolved, however, by the American cytologist and experimentalist Edmund Wilson. In turn, Wilson's experimental findings influenced the cytological descriptions of maturation division and fecundation. Thus the answers seemed to come not from cytology alone but from a subtle combination of descriptive and experimental work.

In 1891 Edmund Wilson was appointed to the chair of zoology at Columbia University, but before taking up this appointment he spent a year in Europe, working part of the time with Theodor Boveri in Munich and part of the time at the famous zoological station in Naples. In Naples he was exposed to the new experimental embryology and to the controversy over embryonic development, and thus he concentrated on this area in his own work during the 1890s.

In 1893 Wilson rejected the Roux-Weismann hypothesis on the basis of his work on *Amphioxus*. He found that the isolated blastomeres from the two-cell-stage embryo formed complete larvae. He also discovered, however, that the capacity of a blastomere to form a complete larva diminished with age; a blastomere from a two-celled embryo formed a larva, but one from an eight-celled embryo formed only a blastula. Thus, he concluded, "the ontogeny assumes more and more the character of a mosaic work as it goes forward. In the earlier stages the morphological value of a cell may be determined by its location. In later stages this is less strictly true and in the end the cell may become more or less completely independent of its location, its substance having become finally and permanently changed."[3] As such change occurs, of course, the isolated blastomeres become less and less capable of forming a complete larva, he argued. Given the fact that "the mosaic character of the ontogeny emerges from the indifferent condition of the early stages at different periods in different animals, and in many cases appears more or less distinctly from the beginning," Wilson felt that the conflict between the experiments of Roux and Driesch had been reconciled. He therefore concluded that "the phenomena of regeneration are not incompatible with a modified form of the mosaic theory, in which the hypothesis of qualitative division is repudiated."[4]

Wilson received additional support for his views in 1896 when his student Henry Crampton, who was working on the snail *Ilyanassa obsoleta,* showed for the first time that in some animals an *isolated* blastomere is capable of only partial development. In this mollusk the isolated blastomeres never formed a complete embryo but executed "a typical partial development, in Roux's sense of the term."[5] Wilson, who added an appendix to this paper, realized that organisms could be arranged in a series with regard to the time at which mosaic development began, the gastropods being at one extreme and *Amphioxus* at the other. Although the gastropods seemed to support Roux's mosaic theory, and *Amphioxus* backed Hertwig's epigenetic theory, in reality, Wilson argued, both extremes could be explained from one viewpoint. "It is certain," he commented, "that development must be fundamentally of the same nature throughout the series, and the differences between the various forms must be of a secondary moment."[6] Such degrees of partial development, Wilson concluded, were due to varying conditions of the *egg cytoplasm,* not the egg nucleus.

In Naples, during the summer of 1902, Wilson, assisted by Walter Sutton, began to test this cytoplasmic hypothesis in his work on isolated blastomeres and *fertilized egg fragments* of the nemertean worm *Cerebratulus.* He reported in December of that year that although the isolated blastomeres were capable of only partial development, the pieces of egg fragments formed complete embryos. Thus, he concluded, although cleavage and differentiation were caused by cytoplasmic localization, such localization was not complete until *after* the first cleavage of the embryonic cell. Before that time the egg was equipotential.

> The topographical arrangement of the egg materials is a leading factor in the determination of the form of cleavage, but . . . this arrangement is itself not a primary feature of the egg organization but a secondary result attained by a progressive (i.e. epigenetic) process.[7]

But what, asked Wilson, controls this cytoplasmic epigenesis? Not surprisingly, he concluded that "the primary determining cause of development lies in the nucleus, which operates by setting up a continuous series of specific metabolic changes in the cytoplasm."[8]

In the second edition of the *The Cell in Development and Inheritance,* which was published just before the rediscovery of Mendel's work, Wilson also noted: "In the later stages differentiation may occur in the nuclear as well as in the cytoplasmic substance."[9] If the nuclear material itself underwent differentiation during ontogeny, however, it was necessary to explain how this took place. Once again Wilson's only recourse lay in assuming some sort of nuclear preformation as the ultimate cause of embryonic differentiation, but it was not until his address to the New York Academy of Sciences in December 1904 that he made this clear, pointing out at the same time the importance of Mendel's work to his own studies of embryological development. In order for the nucleus to be the ultimate cause of differentiation, he argued, the nuclear material, the idioplasm, must be "built into a complex fabric having a definite architecture" (a view that was incompatible with Hertwig's claim of a complex, but homogeneous, idioplasm). Mendel's work, Wilson continued, was explicable only if one assumed that characters are born on discrete bodies that can be mixed and "shuffled like a pack of cards." "One of the most significant and remarkable discoveries of modern biology," he added, "is the fact that such entities exist"—namely, the chromosomes. Wilson then concluded his paper with the following important statement:

> The germ consists of two elements, one of which undergoes a development that is essentially epigenetic while the other represents an original controlling and determining element. The first is represented by the protoplasm of the egg. The second is the nucleus, which must apparently be conceived as a kind of microcosm or original preformation, consisting of elements which correspond, each for each, to particular parts or characters of the future organism.

Thus, he stated, "if we admit such a distribution of characters among the chromosomes . . . to just this extent have we admitted the principle of preformation as applied to the nuclear substance or idioplasm."[10]

However, Wilson had previously drawn attention to the parallel between chromosomes and Mendel's factors in a short paper written just four days before his *Cerebratulus* paper appeared. The Weismann interpretation of the maturation processes, he pointed out there, "gives a physical basis for the association of dominant and recessive characters in the cross-bred, and their subsequent isolation in separate germ cells, exactly as the Mendelian principle requires."[11] This paper was Wilson's response to a report by his student Walter Sutton, who in his cytological investigations of the chromosomes of *Brachystola magna* had realized that they behaved in a way that was exactly equivalent to Mendel's factors.[12] In turn, Sutton's work was an attempt to illustrate, by cytological means, Theodor Boveri's experimental findings and

Thomas Montgomery's more theoretical pronouncements. Montgomery had proposed that a pairing of male and female chromosomes occurred during maturation division (which explains the reduced number of tetrad groups in reduction division).

Thomas Montgomery, at the University of Pennsylvania, had argued in 1898 and 1899 that maturation involved only a reduction in the total mass of the chromatin. Like Hertwig, he rejected the concept of chromosome individuality and believed that the reduced number of tetrads was caused by fusion of the hereditary particles.[13] By 1901, however, he had completely revised his description of maturation division. In that year he published an extensive article detailing his observations of many different species and presenting his new interpretation of maturation division. He explained that his earlier interpretation had been the result of faulty preparations in which the material had stained too deeply to show a longitudinal cleavage. However, it is clear that his new interpretation was due, to an even greater extent, to his recent conversion to the theory of chromosome individuality.

According to Montgomery's 1901 article, the process of maturation division was as follows: the spermatocytes entered a growth phase, which was followed by the first maturation division. During this growth phase a reduction in chromosome number occurred due to a temporary union of the chromosomes into pairs (a pseudoreduction). The two members of the pair were connected end to end, thereby giving rise to U or V shapes. Montgomery also postulated that each "arm" of the U or V (actually separate chromosomes) split longitudinally. The first maturation division was transverse and separated the two different chromosomes; the second division was equational and separated the two halves of each chromosome.[14]

In the course of a year, Montgomery had completely converted to the theory of chromosome individuality. This change was due in part to his studies of the sex chromosome (which he called the chromatin nucleolus, not being aware of its function in sex determination). Since the chromatin nucleolus was visible throughout the cell cycle, it was easier to accept its individuality. The other reasons he gave for accepting the persistence of chromosomes as individual structures also were based on morphological observations, including the fact that the same number of chromosomes always emerged from a resting nucleus as entered into it.

Montgomery's main contribution in his 1901 paper, however, was not the idea of pairing or his support of chromosome individuality. Rather, it was his suggestion that each pair of chromosomes that joined to cause a pseudoreduction was composed of one chromosome of paternal origin and one of maternal origin. Montgomery seems to have arrived at this vitally important conclusion after following two separate lines of reasoning, one based on observations, the other on theoretical considerations. Because the former were a rather weak basis on which to make such a sweeping generalization,[15] it is probable that his theoretical views played a more important role. Montgomery noted that in the germ cells of the Metazoa there were regular cycles of generations. In each cycle, one found a stage of conjugation of maternal and paternal cells—that is, fertilization. This stage, he said, constituted a process of rejuvenation. Such a view was, of course, commonly held, but Montgomery took the idea one step further by arguing that the synapsis stage was a delayed part of the conjugation of the germ cells and served to rejuvenate the chromosomes. "The reason for the final union of these chromosomes is obvious," he wrote.

> It is evidently to produce a rejuvenation of the chromosomes. From this standpoint the conjugation of the chromosomes in the synapsis stage may be considered the final step in the process of conjugation of the germ cells. It is a process that effects the rejuvenation of the chromosomes; such rejuvenation could not be produced unless chromosomes of different parentage joined together, and there would be no apparent reason for chromosomes of like parentage to unite. At the same time the so-called "reduction in the number" of the chromosomes is effected, but this is probably not primal, but rather a necessary result of the conjugation of the chromosomes.[16]

Although elegant, Montgomery's argument had little to support it. If one rejected his idea of chromosome rejuvenation, little remained. As the debate over maturation division clearly indicates, theories based solely on observations of chromosomes were weak and open to dispute. What was needed was more information based on experiments.

This information finally came in 1902, when Theodor Boveri published the results of a brilliant series of experiments on multipolar mitosis. By this time it had been shown that the cells of one species were characterized by a particular *number* of chromosomes. Boveri demonstrated that a specific *combination* of chromosomes was essential for normal development to occur. According to Weismann's theory (noted earlier), any *one* chromosome (idant) was sufficient to ensure normal development; how they combined was immaterial.

By 1902 it had also been established (through studies of natural and artificial parthenogenesis and the development of nonnucleated egg fragments) that one set of chromosomes, either maternal or paternal, was sufficient to ensure normal development. Given this information, some investigators might have proposed that all the chromosomes of a nucleus were essentially the same. Boveri, however, was not so inclined. The explanation struck him as inefficient; why would Nature provide four or more identical chromosomes to do a job that a single chromosome could fulfill. Boveri believed that the only reasonable explanation for the multiplicity of chromosomes was a multiplicity of functions. "It seems to me probable," he wrote, "that the multiplicity of chromosomes are only of significance in their individual differences."[17]

The work of Hans Driesch on sea urchin eggs provided the impetus for Boveri's experiments. Driesch had bred a large number of doubly fertilized eggs (eggs containing two sperm), all of which developed abnormally. Boveri remarked: "Under the assumption that all the chromosomes were of equal value, I could think of no grounds for this development."[18] Using his own supply of doubly fertilized eggs, Boveri observed that four spindle poles were formed. They contained three sets of chromosomes—eighteen from each nucleus, or a total of fifty-four. Since each of these fifty-four chromosomes divided longitudinally, and four centers, not two, competed for each pair, Boveri realized that the chromosomes would be distributed randomly among the four poles and that a large variety of combinations was possible at each one. Thus, each of the four cells produced by the cleavage of a dispermic egg contained the same protoplasm but different chromosomal arrays. If the pathological development of dispermic eggs depended on a disturbance in the protoplasm, one would expect all four blastomeres to exhibit the same type of pathology. If, on the other hand, the pathological development was brought about by the abnormal distribution of the chromosomes, all four blastomeres should develop differently. Boveri's results supported the latter theory. When the four blastomeres produced by a dispermic egg were separated, each developed a different type of pathology.[19]

However, a second abnormal condition found in nature lent even stronger support to Boveri's theory. Under certain conditions, a doubly fertilized egg developed only three poles, not four. Thus, when the fifty-four pairs of chromosomes arrived at the equator, they were randomly arranged at three poles instead of four. Boveri calculated the probabilities for different chromosome distributions at each pole as well as the percentage of blastomeres that would receive the normal complement of chromosomes in both the four-poled and three-poled eggs. According to his results, no pole in the four-poled egg should receive a normal complement, and only 11 percent of the three-poled blastomeres should do so. Boveri's experimental results were strikingly close to theoretical prediction. From 1,170 tetrasters, he obtained only 10 normal larvae, although none was completely normal; of 695 triasters, 58 (or 8 percent) produced normal larvae.[20] Since in parthenogenesis normal larvae developed with only half the usual number of chromosomes, Boveri ruled out the possibility that the number of chromosomes alone affected the development of the abnormal larvae. He concluded that a particular combination of individual chromosomes was the key. "Thus all that remains is that not a definite number but a definite combination of chromosomes is necessary for normal development, and this means nothing other than that the *individual chromosomes must possess different qualities.*"[21] This physiological individuality precluded the dissolution and intermixing of the chromosomes as separate granules between divisions. Boveri therefore had to forsake his earlier interpretations of maturation division.

Proof that the chromosomes were qualitatively different enabled Boveri to offer a more convincing explanation of the events of reduction division than Montgomery had. As noted in the last chapter, Boveri originally believed that half the chromosomes atrophied prior to maturation division and that a tetrad represented one chromosome divided twice longitudinally. His own dissatisfaction with this explanation, however, is reflected in the fact that he did not pursue the idea and readily changed his view. In 1904 he cited the work of Rückert and especially that of Korschelt as important keys to his conversion. Korschelt's pictures of *Ophryotrocha*, whose somatic cells contain four chromosomes, clearly showed the process of pairing, because pairing took place after the formation of the spindle. Boveri's own work on multipolar mitoses convinced him that Montgomery's hypothesis of maternal-paternal pair formation during synapsis was correct.

> We now reach a new and very important turn in the question of reduction. It relates to the proof of the different values of the chromosomes of one and the same nucleus. If all the chromosomes of one nucleus, both those of paternal and of maternal origin, had essentially the same value, the reduction could be effected by any arbitrary conjugation of pairs. The various combinations which the germ cells would thus contain would cause nothing else but individual variation. If, however, the chromosomes of a nucleus played different roles in the life of the cell and thus in the entire organism, so that only a definite combination enabled normal functioning, then the chromosome pairing (conjugation), which is a preparation for reduction, must possess a very definite regularity.[22]

Furthermore, to Boveri, the nature of this regularity was obvious: studies of parthenogenesis had shown that the egg and sperm nuclei were equivalent and that each contained all the hereditary qualities. If one labeled the chromosomes of the egg nucleus *a, b, c, d,* then one must also label the chromosomes of the sperm nucleus *a, b, c,* and *d.* If the chromosomes paired in an arbitrary manner, such as *a:b, a:c, b:c, d:d,* then the germ cells would contain arbitrary combinations of chromosomes depending only on chance. One would lack *a,* another would lack *c.* The germ cells would thus be defective with regard to certain qualities, as is evident from the results of multipolar mitoses. It is immediately obvious, Boveri wrote, that only one mode of pairing is possible. *Homologous chromosomes—one maternal in orgin, the other paternal—must pair with each other.*[23]

Both Montgomery and Boveri reached their conclusions about maturation division without relying on the newly rediscovered Mendelian laws. (Montgomery, in fact, denied that the chromosomes were the bearers of the Mendelian traits.) Rather, their fresh interpretations of the process of maturation division served to connect the chromosomes and Mendel's factors. The connection was made very quickly, however.

Boveri, for example, appended a brief footnote to his paper on multipolar mitoses as follows: "It is clear that therefore the simple considerations which Weismann developed for the reduction division also require at least considerable modification since random distribution of the chromosomes into two groups should in general be equally harmful as a multipolar mitosis. These and related problems, as well as the relevance of this to the results of botanists in studies of hybrids and their descendants, will be discussed separately." In his vague reference to "botanists," it is certain that Boveri had Mendel in mind.[24]

Boveri's paper on multipolar mitoses in turn prompted Walter Sutton to address the question whether the different chromosomes could be identified by their morphological features. Sutton, who spent the summer of 1902 in Naples working with Wilson and Boveri, chose to study the chromosomes of the grasshopper *Brachystola magna* since they were large and easy to identify. He claimed that the chromosomes could be classified by pairs into eleven sizes and that these size relations were maintained in all the cells of the species for at least eight generations. Each pair consisted of one maternal and one paternal chromosome, which, he said, agreeing with Montgomery, paired during the synapsis stage of maturation division. Furthermore, this pairing suggested to him an explanation of Mendel's laws: "I may finally call attention to the probability that the association of paternal and maternal chromosomes in pairs and their subsequent separation during the reducing division as indicated above may constitute the physical basis of the Mendelian laws of heredity."[25]

One might wonder at Sutton's ability to distinguish chromosomes of eleven different sizes since, as previously indicated, it was extremely difficult to interpret cytological preparations in those days. It is clear, however, that Sutton was looking for such differences, influenced as he was by Boveri. It should also be pointed out that Sutton's explanation was based on the study of insect cells—a group in which Weismannian reduction division had long been suspected to occur.

In 1903 Sutton published his paper "The Chromosomes in Heredity." Here he linked the chromosomes to the Mendelian factors by pointing out the equivalence of the maternal and paternal sets of chromosomes, the pairing of maternal and paternal homologues during maturation division, and the individuality of chromosomes. He also noted that the mechanism of reduction division generated a large number of chromosome combinations: a germ cell with four chromosomes could be constituted in sixteen different ways, theoretically, and upon fertilization, 256 offspring combinations could arise. In other words, as Weismann had speculated twelve years earlier, sexual reproduction seemed to be a means by which variations were introduced into the population, both at the time of fertilization and during gamete formation.[26]

In 1904 Boveri demonstrated how the chromosomes could account for Mendel's laws. He asked the reader to imagine a trait with the dominant allele *D* and recessive allele *R*, like those Mendel had described in his theory. *D* and *R* would be localized on a pair of homologues. "Through reduction in oo- and spermatogenesis, these two homologues, bound together at the reduction spindle, would be separated into different egg and sperm cells. Exactly half the sperm cells would contain *D*, the other half *R*, and likewise for the egg cells." The random union of these germ cells would produce Mendel's ratio of 1:2:1.[27]

RESOLUTION OF THE REDUCTION-DIVISION CONTROVERSY

Once the correspondence between the chromosomes and Mendel's factors had been pointed out, biologists had a new framework within which to interpret their cytological preparations. There was some resistance to the chromosome theory, but it is surprising that the resistance was not stronger, considering the nature and the duration of the controversy the cytologists had engaged in over reduction division. All the evidence supporting the link between chromosomes and Mendel's factors came from animal studies, most of which were restricted to insects. Mendel's data, on the other hand, came from plant studies, and according to Strasburger and other botanists, there was as late as 1904 an obvious discrepancy between the behavior of plant chromosomes during maturation division as described by plant cytologists and the behavior that Mendel attributed to his "factors."

Strasburger and his students continued to describe the occurrence of two longitudinal divisions in the maturation process and thus denied that the tetrad was formed by one longitudinal division of a pair of homologous chromosomes. Although in 1897[28] Strasburger reported a transverse division during maturation of the pollen cells of the lily, and thus agreed with Rückert and Haecker that a Weismannian reduction division took place in the plant kingdom, he hastily retracted this view in a paper of the same year: "We return again to the concept we had temporarily abandoned ["we" referring to his American student David Mottier and himself], that the existence of reduction divisions cannot be supported by the presently known facts in the plant kingdom."[29] It was a view to which he subscribed also in 1900.[30] In 1903 and again in 1904, Mottier, who by then had moved to Indiana University, repeated his claim that no Weismannian reduction took place and that there was no basis for considering chromosomes to be qualitatively different. The purpose of fecundation, he believed, lay in "the blending of two lines of descent and possibly the restoration of the power of growth and cell-division."[31]

Given, then, the obvious discrepancy between the behavior of Mendel's factors and the behavior of plant chromosomes as it was generally understood, it is puzzling that those who opposed the chromosome theory never made the devastating and legitimate claim that Mendel's factors could not occur on the chromosomes: the behavior of these chromosomes was not such as would lead to an independent assortment of factors. The fact that opponents of the chromosome theory did not make this claim reflects a curious development among plant cytologists *after* the rediscovery of Mendel and publication of the work of Hugo de Vries. Between 1900 and 1905, opposition to the Weismannian interpretation of reduction division collapsed completely.

In the first place the anti-Weismann position was badly undermined when many of its advocates gave up cytology. Leon Guignard, for example, after discovering double fertilization in angiosperms at the end of the nineteenth century, turned his attention to the study of plant secretions. Similarly, Oscar

Hertwig played no part in the chromosome debate of the early twentieth century, although he did continue to discuss problems of fertilization.

More significant was the rediscovery of Mendel, which led to a fresh burst of activity in the field of maturation division. A new series of papers appeared, most of which argued in favor of the Weismann interpretation. In 1904 R. Gregory, part of the Bateson group at Cambridge, reported reduction division in the ferns, and a year later reduction division in *Ascaris* was reported for the first time.[32] Among these papers were some by Farmer, Strasburger, and, as we have already seen, Boveri. Between 1903 and 1904 all three changed their minds completely. According to Haecker, the reversal in the thinking of Strasburger and Boveri was chiefly responsible for the triumph of Weismann's interpretation.[33]

In 1903 John Farmer reported that as a result of recent investigations it was necessary to "considerably modify the conception of the process as already set forth by ourselves" and to conclude that a qualitative reduction division took place.[34] A year later Strasburger changed his opinion for the last time. He reported that prior to maturation division in *Galtonia candicans,* the twelve chromosomes joined in pairs to form six and that subsequently these paired chromosomes split apart by a transverse division. "As a result," he concluded, "the much disputed question whether the maturation division is a reduction division would be decided in favor of Weismann." Similarly, he reported "an undoubted reduction division" in both *Lilium* and *Tradescantia.*[35]

Strasburger's change of heart was clearly influenced by the writings of Hugo de Vries. In *Mutationstheorie,* de Vries claimed that new species arise through the appearance of "progressive mutations," or new pangenes, while varieties arise through a change in the position of already existing pangenes. Such a change, he argued, comes about as a result of sexual reproduction, which "can unite characters by exchange of elements in every possible kind of combination."[36] Elaborating on this claim in his paper "Befruchtung und Bastardierung," which was read at the 1903 meeting of the Dutch Society of Science, he stressed what he called the "principle of duality": that every being is a double being, inheriting one part of its nature from its mother and the other half from its father. He noted, however, that though the bearers of these parental characters are intimately connected, "they are not, by any means, fused into a new indivisible entity. They form twins, but remain separate for life."[37] Thus, he continued, when two sex nuclei unite at fertilization, very little happens. "A penetration or fusion of their substance does not take place. They remain separate in spite of the union."[38] On this basis de Vries argued that the numerical reduction of chromosomes "means nothing but the separation of two nuclei which had so far worked together for a period. It is like the parting of two parents who have walked along together for a while."[39] Before this separation, de Vries explained, the parental chromosomes lay either side by side or end to end, a difference which explained much of the previous confusion over transverse division of the chromosomes. However, de Vries stressed, the separation of chromosomes was not a simple matter; it involved the creation of a lasting "reciprocal influence."

> Shortly before their separation, their leave taking, they are still the same as before. But now they exchange their individual units, and thus cause the creation of those countless combinations of characters, of which nature is in need in order to make species as plastic as possible, and to empower them to adapt themselves in the highest degree to their ever changing environment. . . . This increase of variability and of the power of individual adaptation is the essential purpose of sexual reproduction.[40]

De Vries' publications profoundly influenced Strasburger's reinterpretation of fertilization. As Strasburger wrote in 1908, they furnished "the basis for our conceptions of fertilization and heredity."[41] They provided for Strasburger and his school a new explanation for what they saw under the microscope. The linking and separation of the paired chromosomes now appeared to be more than the temporary "pseudoreduction" described by their opponents. Of course, a complete *Verschmelzung* did not occur, as Strasburger had stated previously, but there was an exchange of units such that, as Strasburger always maintained, the postmaturation chromosomes were in a sense new formations. Strasburger, too, had long stressed that the male and female chromosomes remained separated, fusing not at fecundation but during the initial stages of maturation division, and thereby producing a cell with half the normal number of chromosomes. Furthermore, the work of de Vries (and Mendel too) emphasized discrete hereditary units, or pangenes, rather than the chromosomes per se, and thus substantiated Strasburger's claim that chromosome number was only secondarily significant in the reproductive process.

Thus, after the work of de Vries and Strasburger's final acceptance of Weismann's view of reduction division, it was necessary only to deny that a complete fusion of chromosomes took place at the initiation of the maturation divisions. Strasburger, however, never accepted the theory that the chromosomes paired only in a temporary end-to-end "pseudoreduction." Rather, he believed that the chromatin particles (termed "gamosomes") "unite [*vereinigen* not *verschmelzen*] to form a single zygosome, out of which two new chromosomes arise anew." Moreover, putting aside his earlier belief that the male and female chromosomes retained their complete individuality both before and after maturation, and that the postmaturation chromosomes represented *totally* new structures, he now stated: "These chromosomes do not contain exclusively gamosomes springing only from the father or mother, but rather a partial exchange [*teilweise Answechselung*] takes place."[42]

By emphasizing an exchange rather than a fusion of chromatic material, Strasburger fundamentally changed his position on the significance of sexual reproduction. Instead of a *Verschmelzung* ("fusion") and a device to ensure the stability of the species, or as he expressed it in a paper of 1901, "the equalization of species characters,''[43] he now spoke of an *Answechselung* and a device to ensure massive variation and advantages in the struggle for life. Such variations arose not only because of the assortment of chromosomes present but also because of "the exchange of homologous hereditary units [that take place] between the pairing chromosomes.''[44]

Charles Allen of the University of Wisconsin, another of Strasburger's American students, was still unconvinced, however. "It seems very difficult,'' he wrote in 1905, "to bring them [Strasburger's 1904 findings, etc.] into harmony with the best known facts as to the heterotypic figures in the lily.'' He still reported seeing two longitudinal divisions and the *fusion* of chromosomes into pairs. "All appearances indicate,'' Allen insisted, "that the object of the whole process is the fusion of the chromosomes.''[45] He admitted, however, that this intimate contact of chromosomes need not necessarily be a true fusion from which only one type of germ cell developed; "there may be more or less mixture of the microplasms as a result of the contact of the two threads, but no actual fusion.'' Such a mingling, he wrote, "would naturally lead to a purely Mendelian distribution of parental characters in the pure-bred offspring of the next generation.'' Allen quite correctly pointed out that the Mendelian ratios occurred only in "isolated qualities of plants or animals whose other qualities either have not been shown to follow the law or have been shown to deviate from it in a marked degree.'' The hereditary units may be distributed in accordance with Mendel's laws, he wrote, but this distribution is negated by the cytological facts in the lily and by the other cases in which a double longitudinal splitting occurs.'' "The questions whether there ever occurs a complete fusion, or whether in any case no fusion at all occurs,'' he concluded, "must be left for future investigation.''[46]

By 1905, however, Allen's doubts about the process of maturation division were not shared by the vast majority of cytologists. In that year Farmer and Moore produced their famous summary paper in which the term "maiosis" (Gk *meiōsis* = "reduction") was introduced for the first time. The importance of Mendel's work to the cytologists' change of heart was made most explicit. Pointing out that the nature of maturation division as understood by the botanists "is equivalent to a denial of the permanence of the chromosomes from one generation to another,'' they went on to say:

> Investigations on the behaviour of hybrids militate strongly against the assumption that during fertilization any real fusion of the parental substances responsible for the expression of particular features occurs. . . . How could one account for the segregation of ancestral characters in interbreeding hybrids, if the individuality of the original chromosome become really obliterated during each generation.[47]

It is not hard to understand why the cytologists changed their interpretations of maturation division so easily. Their differing cytological interpretations rested on events that took place during the very early stages of cell division. It was, and still is, very hard to observe these events through a light microscope and be at all sure of interpreting them correctly. As late as 1961 M. M. Rhoades noted correctly that "cytological observations on the double-ness or singleness of leptotene chromosomes might be questioned because of the subjective nature of interpretations of fine cytological detail."[48] It was virtually impossible to decide whether a double chromosome thread was two different threads lying side by side or the product of a longitudinal division of one thread. It was also impossible to determine by cytological means whether the coalescence of two threads into one was a real fusion, a mixture of chromosome particles, or merely a temporary union. These facets of matura-tion division were, of course, the very points at issue in the controversy.

Another complicating factor in the debate was the universal assumption up to 1900 that the chromosome pairs would be connected end to end. In that year, however, Hans von Winiwarter finally suggested that chromosomes might lie side by side.[49] In 1925 E. B. Wilson pointed out that some of the confusion over the nature of reduction division was due to this widespread assumption about the significance of longitudinal and transverse clefts. "From the start," he wrote,

> investigation of the problem has been confused by the emphasis that Weismann laid upon the supposed theoretic significance of a transverse division of the chromosomes, thus implying that a longitudinal division is *ipso facto* an equation-division. Some of the leading early investigators, such as O. Hertwig, Boveri, and Brauer, found both maturation divisions to be longitudinal, and were thus led to deny the occurrence of a reduction-division. The fallacy of this was demonstrated as it gradually became clear, especially through the work of Winiwarter and his successors, that what seems to be a longitudinal split may in reality be the separation of two chromosomes that have been associated side by side; and conversely, that a division that seems to be transverse may be only the final separation of such a pair.[50]

Many cytologists admitted these difficulties. William Cannon, for example, remarked that the microscope could never resolve whether Hertwig or Weis-mann was correct; "the optical effect in the dividing sex nucleus would be the same in either case."[51] In 1904 Strasburger admitted that the facts were difficult to ascertain and that views on maturation division were determined by prevailing theoretical precepts.[52]

The Weismannian interpretation of sexual reproduction became established very rapidly. In 1917, for example, Bower wrote to Macmillan Press suggest-ing that he and two of his colleagues publish a book on sex and heredity, a book that would be "written in such a way that any boy or girl of 18 could read it without a blush."[53] He stressed that the word "venereal" would be avoided at all costs (a reference to the fact that Bower had been lecturing on sex to teachers and social workers in Glasgow at the request of the Glasgow Committee for Combating Venereal Disease). The publishers need not have worried, however; the book dealt only with theoretical principles and life cycles. After publication, Isabel, Marchioness of Ailsa, wrote from her home in Culzean Castle that she had been "thrilled with the sense of romance and beauty" of the text and expressed her belief that the facts presented—many of them botanical—"must prove the surest means to dispel the darkness and ignorance which prevail."[54] Bower's lectures and book gained "rapt attention" and were a "huge success." "Sexual propagation," Bower wrote, "is not a mere matter of increase in number, but one which touches the very springs of evolution and progress in living beings."[55] Although the parallelism and convergence observed in ferns had convinced Bower that the inheritance of acquired characters did occur, in his book he nevertheless accepted the reality of an immortal and unspecialized germ plasm.

Weismann himself, aware that pangenes and Mendelian factors were not identical to his ids and ancestral germ plasms, remained somewhat indifferent to the rediscovery of Mendel. "We must postpone the working.of this new material into our theory until a much wider basis of facts has been supplied by the botanists," he remarked in 1904. By 1913, however, he looked upon the Mendelian theory as proof of the existence of his "determinants" and as a result completely changed his definitions. He termed the totality of chromosomes in a germ cell (the collected determinants for the whole individual) *Vollide*, and each individual chromosome (which contained qualitatively different determinant groups) *Teilide*, or "partial ids." The chromosomes were thus no longer a collection of ancestral germ plasms or idants; they were ids made up of the determinants.[56]

Thus the stage was set for the birth of classical Mendelian genetics and the overriding importance it gave to the process of sexual reproduction. Twentieth-century biologists came to view sexual reproduction as perhaps the single most significant event in the biological world. Not only was it the vehicle by which maternal and paternal traits were passed to offspring, but it introduced into a population the variations without which evolution could not occur.

But what about the other functions of sexual reproduction? Did it serve only to generate new genetic combinations, or were the Victorians correct after all? Did not fertilization also serve to rejuvenate the life of the ovum, without which it would die? Indeed, did not fertilization also rejuvenate the individual? Charles Brown-Séquard, the famous French physiologist who had succeeded Claude Bernard to the chair of experimental medicine at the Sorbonne, clearly thought that it did. In 1889 he injected himself with saline extracts of dog testes, believing that since a loss of semen leads to spermatorrhea, an intake of it could have nothing but beneficial results.[57] He was, he pointed out, seventy-two years old and he felt it; he had much less energy than he had had in earlier times, he needed much rest, and he was usually completely exhausted by early evening. But after the injections, he proclaimed, "everything has changed; I have regained at least all the energy I had

FIGURE 7.1.   Charles Brown-Séquard. Reproduced by permission of the Museum of Comparative Zoology, Harvard University.

years ago.'' He was no longer tired, he reported; he did not sit down all day, his assistants were astonished, and he could even work after dinner. ''I can without difficulty and even without thinking about it, run up and down stairs.'' His feces were longer and more solid than before, and, most amazing of all, he reported a 25 percent increase in the distance he was able to shoot his jet of urine!

On a less esoteric level, Maupas had shown in 1888 that without conjugation a ciliate protozoan "deteriorates, uses itself up, simply by the prolonged exercise of its functions."[58] Colonies of such ciliates eventually do conjugate, however, and in so doing rejuvenate themselves to begin a new cycle of asexual fissions. If this was true for ciliates, Maupas argued, why would it not be true for the higher organisms? In them, Harvard-based embryologist Charles Minot speculated, "the union of two sexual unicellular individuals, of different origin, form[s] an asexual cell [the zygote], which then goes on dividing asexually for many generations until the original energy is exhausted."[59] All metazoans, Minot argued, pass through a cycle of youth, old age, and death, but "before senescence conquers, the sexual products are thrown off and effect the process of *rejuvenation*."[60]

At the turn of the century a group of American biologists once again subjected these old ideas to a series of experimental tests. In 1902 Gary Calkins of Columbia University, a protozoologist of some repute, kept individuals of *Paramecium* (a ciliate) alive for more than 500 generations in the absence of conjugation. Nevertheless, he reported, these clones became sluggish every three months and abnormalities appeared in them. Although such colonies sometimes recovered after conjugation, he noted that a more dramatic recovery occurred if he changed their environment, either by agitating them mechanically, changing their nutritive medium from hay to beef extract, or by suddenly increasing the temperature of the medium. The fact that periods of low vitality could be overcome by such artificial means suggested to Calkins that Weismann's theory of infusorial immortality might be basically valid. Later the same year he reported that vitality could also be restored by adding traces of potassium, sodium, or magnesium salts. "The organism," he wrote, "may be compared with a storage battery which gradually runs down as its charge is used. Like the battery, the infusorian cell is capable of being recharged and of inaugurating a new cycle of generations."[61] In nature, Calkins noted, such recharging seems to occur naturally; the environment is constantly changing, and every rainfall brings in traces of mineral salts. In a Woods Hole lecture during the summer of 1902, Calkins reported that one clone of *Paramecium* had been kept alive for 665 generations simply by altering the ciliates' environment at regular intervals. Drawing a parallel with Jacques Loeb's work on artificial parthenogenesis (see chap. 9), he concluded: "We are justified in regarding the artificial rejuvenation of *Paramoecium* as a phenomenon of the same order as the chemical fertilization of the egg."[62]

Two years later, however, Calkins changed his mind. Three separate series of *Paramecium* clones had died out after 379, 570, and 742 generations respectively, despite the addition of beef, pancreas, brain, mutton, lecithin, pineapple and apple extracts, mineral salts, galvanic stimuli, and even nitroglycerine to their culture media. Despite these environmental changes, however, the race died from exhaustion! Under the experimental conditions used,

Calkins concluded, "continued life was impossible, and if this conclusion be granted to obtain in nature, then we must agree with Maupas and others that the indefinite continuance of life without conjugation is improbable."[63]

Improbable perhaps, but not impossible. Calkin's student Lorande Woodruff took up the cudgels. By constantly changing the growth medium, he succeeded in cloning 490 generations by 1908, 1,230 generations by 1909, and 2,012 generations by 1911. Not surprisingly, Woodruff concluded that under suitable conditions, cultures of *Paramecium* were immortal, and that conjugation was not essential.[64]

In 1908, however, Minot published his classic treatise, *The Problem of Age, Growth, and Death*. In it he reported that immediately after fertilization the rabbit embryo's metabolic rate began to increase rapidly, and that this increase was manifested in many ways, including at least a 1000 percent daily increment in body weight. Very rapidly, however, this daily weight increase dropped—to only about 6 percent twenty-three days after birth, and to less than 1 percent after two and a half months. Finally, gametes were produced; these cells, which lacked any growth capacity, were in the last stages of senescence. Basically, Minot postulated, senescence resulted from an increase in the protoplasmic content of body cells and their differentiation, and could be counterbalanced only by a fresh start every generation, a start for which fertilization was responsible:

> The ovum is a cell derived from the parent body, fertilized by the male element, and presenting the old state to us, the state in which there is an excessive amount of protoplasm in proportion to the nucleus. In order to get anything that is young, a process of rejuvenation is necessary, and that rejuvenation is the first thing to be done in development. The nuclei multiply; they multiply at the expense of the protoplasm ... Then begins the other change; the protoplasm slowly proceeds to grow, and as it grows, differentiation follows, and so the cycle is complete.[65]

"The process of the segmentation of the ovum is that [which] we must call rejuvenation," Minot concluded, and is initiated by the action of a sperm cell on an egg cell. This theory, which like so many valued scientific theories combined raw data with a certain beauty, seemed at odds with Woodruff's teeming clone of *Paramecia*, which by 1912 was in its twenty-five hundredth generation of immortality.

Was this yet another stalemate between a beautiful theory and ugly facts? Not so, Herbert Jennings reported. "Age and death, though not inherent in life itself, are inherent in the differentiation which makes life worth living."[66] Whereas it had been shown that fertilization had two functions—to rejuvenate the egg and to produce variation—conjugation had only one. However, the function of conjugation was not to rejuvenate, as everyone seemed to have believed before, Jennings noted; it was to produce variation. Jennings had studied under E. L. Mark at Harvard, and after receiving his doctorate in 1896, had spent a winter with Max Verworn in Jena studying the behavior of protozoans. In 1906 he moved to Johns Hopkins University, where he began his studies on the genetics of protozoa.

After an exhaustive series of experiments, Jennings concluded that the average rate of reproduction decreased after conjugation and that mortality was higher among the protozoa that had conjugated than among those that had not. The most striking effect of conjugation, he noted, was "the great *increase in variability.*" Variability, however, was no longer expressed in vague, subjective terms. The era of quantitative and statistical biology had arrived; variability was measured as the coefficient of variation. Measuring the rate of fission, for example, Jennings presented the data seen in Table 1 (I have extracted some of his figures from a larger table).[67]

Because of the increased variability, Jennings explained, many more lines of conjugants died out than survived. Conjugation, therefore, was not a process of rejuvenation; it was "an ordeal." Its benefits exactly mirrored those of normal sexual reproduction: "Conjugation does not rejuvenate in any simple, direct way. What it does, is to produce variation, to produce a great number of different combinations, having different properties. Some of these are more vigorous, others less vigorous. The latter die, the former survive."[68]

The real biological significance of sexual reproduction and conjugation lay in the production of variation (Fig. 7.2). Sexual reproduction was not, as previously assumed, simply a method of procreation. Indeed, as Maynard Smith wrote in 1971, "at the cellular level sex is the opposite of reproduction; in reproduction one cell divides into two, whereas it is the essence of the sexual process that two cells should fuse to form one."[69]

Although this disassociation of sex from procreation seems somewhat academic in organisms that reproduce only by sexual means, in plants and many lower animals sexual and asexual reproduction exist side by side, playing, presumably, very different roles. In a very different context this disassociation of sex from procreation came to have enormous social repercussions; it provided the biological backcloth for the "sexual revolution" of the twentieth century.

TABLE 1

Experiment	Time	Conjugants		Nonconjugants	
		N	Coefficient	N	Coefficient
1	2 weeks	56	53.103 ± 4.232	59	12.975 ± 0.819
1	4 weeks	42	42.870 ± 3.689	59	27.743 ± 1.850
3	4 days	36	29.369 ± 2.528	16	12.756 ± 1.546

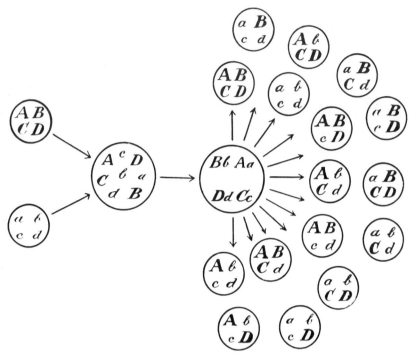

FIGURE 7.2. The essence of sexual reproduction. Gametes with factors *ABCD* and *abcd* conjugate to form the zygote *AaBbCcDd*. These factors can then recombine during maturation division to form sixteen possible recombinations. Reproduced from Thomas Hunt Morgan, *Heredity and Sex* (New York: Columbia University Press, 1913), by permission of the publisher.

EIGHT

Cytologists Have All the Answers

By the early years of the twentieth century, cytologists would have been justified in believing that they, above all others, had at last solved the major puzzles of sexual reproduction. They had shown that fertilization involves the entry into the egg of discrete morphological entities, the chromosomes, and in so doing had redefined the nature of sexual reproduction, clearly distinguishing it from asexual reproduction. It was true, of course, that their new interpretations had been made possible only with the help of experimental work. Nevertheless, the cytologists could legitimately claim credit for providing the first precise account of fecundation. Descriptions of distinct, visible chromosomes had replaced the earlier emphasis on nondescript matter osmosing into the egg. Sperm were no longer viewed as peculiar cells that dissolved within the egg, but were defined simply as vehicles of chromosome transfer.

Cytologists had provided answers to other related problems as well. They had redefined the nature of the gametophyte and sporophyte generations in plants in terms of the number of chromosomes they possessed. As Eduard Strasburger noted in 1907, "All individuals that have the same chromosome number belong to the identical generation, no matter what their form might be."[1] These same cytologists had also provided answers to three long-standing mysteries: Why is an unfertilized egg usually incapable of development? How does a sperm initiate development? and How is sex determined? Cytologists, not physiologists, seemed to have provided answers to all these basic questions.

CYTOLOGICAL ALTERNATION OF GENERATIONS

That chromosomes are essential to an understanding of the alternation of generations in plants was first suggested in 1893 by Charles Overton, a native of England and a distant relative of Charles Darwin's who was then *Dozent* in biology at the University of Zurich. Overton not only made the obvious remark that "the essential difference between vegetative and reproductive cells is indicated by the smaller number of chromosomes in the nuclei of the latter," but he went on to suggest "that the reduced number of chromosomes in the nucleus is a feature which is peculiar, not to the reproductive cells, but to the whole sexual generation (gametophyte)." He reported that the sporophyte cells of *Ceratozamia* (a gymnosperm) contained sixteen chromosomes, while the gametophyte cells carried only eight. "It will be a matter of great morphological as well as physiological interest," he concluded, "to establish beyond the possibility of doubt that the alternation of generations . . . is dependent on a change in the configuration of the idioplasm; a change, the outward and visible sign of which is the difference in the number of the nuclear chromosomes in the two generations."[2]

At that time little data existed on chromosome *numbers*. Counting chromosomes presumably required that the counter accept that they were permanent structures. As we have seen, most botanists were concerned only with chromatic granules and believed that chromosomes were temporary structures set up to ensure equal distribution of these granules during cell division. Even when chromosomes were counted, however, the results did not seem to support the idea of constancy within each generation. In 1888, for example, when Strasburger concluded for the first time that "the segments remaining in the daughter nuclei are the same that existed in the mother nucleus,"[3] and thus that "the separate segments of the *Anlage* retain their morphological individuality permanently,"[4] he nevertheless reported that the chromosome number was not the same in all tissue cells of the lily. It seemed to be constant in all pollen mother cells, however, and from this he concluded that there is "a tendency to reduce the number of segments in generative cells."[5] Even though Strasburger now accepted the permanence of chromosomes during vegetative cell divisions, he thought that in order to produce pollen it was still necessary for a true fusion to take place; through a complete fusion (*Verschmelzung*) of chromosomes a reduced number of entirely new chromosomes arose each time. As a result, Strasburger concluded, "to this number as such I should like to attribute only a subordinate significance, which perhaps may be related to the mechanics of the division process."[6] The constant and reduced number of chromosomes in generative cells, he noted, "is in fact the simplest way by which an equality of mass of nuclear threads is obtained to be united in fecundation."[7]

In 1891 Leon Guignard noted that whereas the pollen mother cells of the lily contained twenty-four chromosomes, the pollen cells only twelve, the cells of the embryo sac carried nuclei with either twelve, sixteen, twenty, or twenty-four chromosomes. "The chromatic segments do not conserve their autonomy," he naturally concluded.[8] A year later Strasburger changed his mind again and agreed with Guignard: "The morphological identity of the chromosomes in the resting cell nucleus is not maintained."[9] In 1894, however, Strasburger announced that in ferns, mosses, and perhaps also in some algae and fungi, "the nuclei of the sexual generation contain only half as many nuclei as do those of the asexual generation."[10] This constancy in number, he now maintained, "suggests . . . that though the chromosomes may lose their morphological individuality in the resting nucleus, they do not lose their physiological individuality. . . . In one word, it must be assumed that the individuality of the chromosomes persists in the resting nucleus, and determines the breaking up of the nuclear filament into the corresponding number of chromosomes in the succeeding prophase." He continued to believe, however, as he had earlier, that "the reduction of the number of chromosomes by half, at the initiation of the sexual generation, is due to the fusion into one of two chromosomatic individuals."[11]

The discovery that the reduced number of chromosomes was characteristic of the complete gametophyte generation rather than simply of the generative cells led Strasburger to conclude that "reduction is not to be regarded as a preparation for the sexual act"; rather, "it really marks the beginning of the new generation which comes into existence with the primitive number of chromosomes."[12] On the surface, such an assumption seems peculiar, but one should remember that in vascular plants the process of chromosome reduction is separated by a whole generation from the act of fecundation. Reduction division produces spores; the entire gametophyte generation produced from these spores contains the reduced number of chromosomes. Gametes are produced by the normal mitotic process, and only with the fusion of these gametes is the double number of chromosomes produced again so that all cells of the sporophyte generation contain this double number. What, then, was the cause of this reduction if, as Strasburger maintained, it was not related to the sexual act? Strasburger argued that since the gradual development of the asexual generation could be traced step by step from its beginning in *Oedogonium* and *Coleochaete,* through mosses and ferns, and on to the higher plants, then "the morphological cause of the reduction in the number of the chromosomes and of their equality in number in the sexual cells is . . . phylogenetic. I look upon these facts as indicating a return to the original generation from which offspring were developed having a double number of chromosomes."[13] "In one word," he concluded, "it is the repetition of phylogeny in ontogeny . . . the return of the most highly organized plants, at the close of their life-cycle, to the unicellular condition."[14]

The biogenetic law, that most gripping and romantic of concepts in the whole of biology, and long-favored research topic among zoologists, had at last found a new home in the plant kingdom.[15] The law exercised a profound influence on generations of zoologists in Britain, Germany, and North America, most of whom considered the goal of their discipline to be to describe the embryonic stages of animal development and thus to evaluate the animals' place in the phylogenetic tree of life. In 1890, in his presidential

address to the British Association for the Advancement of Science, Arthur
Marshall remarked that the biogenetic law "forms the basis of the science of
embryology and . . . alone justifies the extraordinary attention this science has
received," and in 1896, botanist Dukinfield Scott urged others in the field to
consider their subject from "this point of view."[16]

Strasburger was one of the biogenetic law's most fervent champions. In
1874, when Haeckel announced his famous gastraea theory in the pages of the
Jenaische Zeitschift für Medicin und Naturwissenschaft—the theory that the
gastrula stage in ontogeny represents a recapitulation of a two-layered gastraea organism, the ancestor of all Metazoa—Strasburger, who was then
professor of botany at Jena, made the coincidental announcement (in a paper
immediately following Haeckel's) that "only through phylogenetic treatment
is biological knowledge to be enhanced." Since we cannot trace the true
phylogenetic development of the Metazoa, Strasburger went on to say, indirect methods must be used. The key was to be found by examining ontogenies, which "we have reason to interpret as a repetition of phylogenetic
development."[17] Thirty-four years later, when Strasburger received a Gold
Medal from the Linnaean Society, the occasion being the fiftieth anniversary
of the Darwin-Wallace papers, he stressed again his support for this
phylogenetic approach. "The seed I received at Jena," he said, "sprang up
early. I took the path of phylogenetic speculation and have pursued it faith-
fully."[18]

The same phylogenetic reasoning lay behind Strasburger's claim in 1894
that the gradual reduction in the gametophyte stage from mosses to gymnosperms could be explained in terms of "the general law which determines the
phylogenetic disappearance of organs which have become useless." However, whereas Bower explained this uselessness in terms of the gametophyte's
nonadaptability to the terrestrial environment, Strasburger explained it
phylogenetically. Initially, organisms were asexual, he argued, and then sexual reproduction appeared, "at a certain stage of phylogenetic evolution in
virtue of certain properties possessed by organized matter as such." As the
sexual replaced the asexual, however, it became necessary for "the numerical
conditions of multiplications" to be maintained by other means. Among these
means was that adopted by the mosses, ferns, and phanerogams. "There
sprang up an altogether new generation from the product of the sexual act,"
Strasburger wrote, "the function of which [was] to reproduce asexually a large
number of individuals."[19] It is no wonder, then, that Bower had complained
in 1890 that such descriptions "will not suffice for the solution of the question
as to the real nature of alternation."[20] The English Darwinists were seeking
explanations in terms of adaptability, not in terms of phylogenetic causes.

Nevertheless, Bower was influenced by Overton's and Strasburger's work
on chromosomes. He agreed with Strasburger that the fusion of ids that took
place in the spore mother cell, producing thereby a reduction in chromosome
number, was the last step in the sexual act, and that in plants the first step (the

fusion of gamete cells) and the last step were widely separated. The sporophyte, he wrote, "would throughout its vast body, developed through long years, retain the individuality of the ids of the two parents; the final fusion only taking place in the spore-mother cells." As a result, he realized, migration to land could explain the elaboration of the sporophyte stage but not its origin. That, he wrote, "is to be sought in connection with the steps of sexual coalescence and subsequent reduction of the doubled number of chromosomes."[21]

With so many botanists committed to the view that the object of their studies was to discover the ancestor of vascular plants, and thus ultimately to construct a genealogical tree of the vegetable kingdom, there was, at the end of the nineteenth century, considerable argument over the relative merits of the antithetic versus the homologous theory of alternating generations. Dukinfield Scott, for example, a paleobotanist and student of von Sachs's, praised Bower for his theory of "amphibious alternation" and agreed with him that this explained the development of the sporophyte stage, but he did not share Bower's views on the alternation's origin. "All Darwinians owe him a debt of gratitude," Scott wrote, clearly referring to Bower's claim that all Darwinists must "see that all the characters which the morphologist has to compare are, or have been, adaptive."[22] Morphologists could not ignore function, Scott argued. "Morphological characters pure and simple" do not exist.[23] However, Scott did not share Bower's interpretation of the origin of the terrestrially adapted sporophyte stage. "Nature is conservative," he wrote, "and when a new organ is to be formed it is, as everyone knows, almost always fashioned out of some preexisting organ. Hence, I feel a certain difficulty in accepting the doctrine of the appearance of an intercalated sporophyte by a kind of special creation."[24] The new sporophyte arose, Scott argued, from the already existing asexual stages of the thallophytes. As Bower correctly noted two years later, however, the antithetic theory merely stated that the sporophyte arose from a zygotic cell, not from "nothing." The real question was: "Has the neutral generation or sporophyte been the result of change of any other part of the sexual generation than the zygote itself? If so, the alternation is of homologous generations; if not, then the alternation is what is styled antithetic."[25]

The answer to Bower's question rested not only on the interpretation of the significance of apospory and apogamy and of the historical relationships between thallophytes, vascular cryptogams, and phanerogams, but now also on chromosome studies, particularly those pertaining to algal life cycles. Scott argued that apospory and apogamy were vitally important phenomena and "that there is no such hard and fast distinction between the generations as the antithetic theory would appear to demand."[26] Bower, on the other hand, continued to argue that their rareness precluded any such significance. Scott, the specialist in fossil plants, took exception also to the progressive feature of the antithetic theory, that having arisen in the algae, the sporophyte slowly developed through the bryophytes to the ferns and beyond. Fossil evidence clearly mitigated against any such progression. "There is no reason to believe," he stressed, "that the Bryophyta, as we know them, were the precursors of the vascular cryptogams at all,"[27] to say nothing, one might add, of the origin of the angiosperms—which even today is shrouded in mystery. Instead, Scott pointed out, both ferns and bryophytes seemed to have originated independently from some algalike precursor, a fact which is much easier to explain from the viewpoint of the homologous theory. Clearly, all the traditional arguments had not served to alleviate the mystery, but the new chromosome discoveries seemed to offer a more hopeful line of attack. Such histological differences between the two generations, Scott admitted, "would appear to militate against their homology."

On the other hand, Lloyd Williams had by then described the *Fucus* life cycle as cytologically identical to that of animals; the plant had the double number of chromosomes, and meiosis took place during gametogenesis. While Scott admitted that *Fucus* was far removed from any direct line of descent, he nevertheless noted: "It is a striking fact that the only direct line of evidence we have goes dead against the idea that the sexual generation necessarily has the reduced number of chromosomes."[28] Bower naturally pointed out that *Fucus* was "peculiar," and he stressed, wisely, that one could expect to find great variation in algal life cycles before "settling down" to the fixed and constant position that appeared in terrestrial plants. However, he did argue that should a reduction division be found to occur in the production of zoospores from the zygote of *Oedogonium,* "we should be right in recognizing in these small cell bodies [the zygotic cells] the rudimentary correlative of a sporophyte—the sort of beginning from which a neutral generation may have sprung in land-living plants."[29] Bower therefore appeared to suggest that the debate over antithetic versus homologous alternation could be resolved with this one piece of cytological information. Scientific papers at this time called for investigations into the life histories of algae and their chromosomes, and once again cytologists appeared to be the likely solvers of the mystery.

Partly because of these "requests," and partly as a result of the fresh outburst of cytological studies that followed the rediscovery of Mendel, the information was not long in coming. In the first place, Farmer's, Moore's, and Digby's work on the cytology of apogamy and apospory suggested that indeed these were abnormal cytological events. In the former, where a gametophyte produces a sporophyte without the intervening production of gametes, it was found that the cells producing the sporophyte directly carried two nuclei that ultimately fused to form a double chromosome generation, just as is the pattern in normal postsexual stages. Apospory, on the other hand, arose because the plant contained double the necessary number of chromosomes throughout, in the absence of reduction and fertilization.

In 1900 David Mottier, professor of botany at Indiana University, reported a chromosome reduction in the tetraspores of the brown alga *Dictyota,* and four years later Lloyd Williams announced that this plant clearly manifested an alternation of generations—the plant producing the tetraspores carried thirty-two chromosomes, while the sexual generation produced from the tetraspores carried only sixteen chromosomes. Similarly, J. Wolfe at Harvard described the cytological changes in the life cycle of the red alga, *Nemalion*, in which "the cystocarp (= zygote) of the red algae is held to be the homologue of the sporophyte in higher plants."[30] Then, in 1905, Charles Allen, like Mottier a student of Strasburger's, reported that the zygote nucleus of *Coleochaete* divided *immediately* by means of a reduction divison *before* it actually produced the spores (see Fig. 5.4). Thus, he concluded, "the assumption of an homology between the sporophyte of archegoniates and the spore-mass originating from the repeated division of the *Coleochaete* zygote is untenable."[31] According to Bower's thesis, the zygotic cell would divide initially by normal mitotic division to produce the rudiments of the sporophyte generation, and only after that would reduction division occur to produce the zoospores.

By this time, Strasburger had reinterpreted his views of maturation division in light of de Vries' work. He defined the visible bodies on the chromatic threads as "pangenosomes," describing them as complexes of numerous invisible pangenes—the ultimate bearers of hereditary properties. It was his belief that during maturation division a complex group of these pangenosomes, or ids as he termed them, fused together, and that "during their fusion a possible exchange of pangenes takes place." Because the gametophyte and sporophyte generations differed (the latter carried each id, or pangenosome, doubled), Strasburger introduced into the literature the terms *haplo-id* and *diplo-id*: "It would perhaps be desirable if the terms 'gametophyte' and 'sporophyte,' which can be applied only to plants with single and double chromosome numbers, should be set aside in favor of terms that apply to the animal kingdom. To this purpose, I permit myself to introduce the words 'haploid' and 'diploid.'"[32]

Then, in 1908, Bower finally published his magnificent overview of the whole subject. Entitled *The Origin of a Land Flora,* it reiterated Bower's basic theme: aquatic algae reproduce sexually. On land, however, they could have reproduced in this way only if water were present or if a nonaquatic means of fertilization were invented. Thus, less dependence would necessarily have been placed on sexual reproduction and an alternative method of multiplication would have been substituted. The sporophyte arose from the zygote, which initially divided immediately into zoospores. In proportion as the plants spread to drier climes, however, more and more emphasis was placed on these spores, so that more nutrition was demanded by the sporophyte. As a result, the spore-forming and nutritional aspects of sporophyte life were enhanced. "This [development] for the first time stamped the sporophyte with a character of independence and permanence, while the number of spores produced might now be practically unlimited." At the same time, heterospory arose, and the dependence on water for fertilization (and thus on free-living gametophytes and motile sperm) was replaced by the formation of seeds and pollen tubes. As Bower described it, "We see the sexual process accommodated to that sub-aerial life which has led to the dominant position of the sporophyte."[33] Finally, the climax was reached in angiosperms, "where the gametophyte is found to have dwindled away to an exiguous residium of a few ill-defined cells, with virtually no vegetable characters at all."[34]

By this time, however, Bower had come to accept that the fundamental criterion by which these alternating generations were to be recognized was cytological. No longer defining the generations in terms of external form, internal anatomy, or the possession or absence of sex organs—all terms by which earlier botanists had attempted to deal with the phenomenon—Bower wrote:

> The gametophyte, or haploid phase, will then be recognized as extending from the spore to the zygote in each cycle, and shows "n" chromosomes normally in all its nuclear divisions: the sporophyte, or diploid phase, is recognized as extending from the zygote to the spore, and it shows "2n" chromosomes in all its normal nuclear divisions. . . . They provide a structural basis for the distinction of the two generations more exact than any other, a distinction which runs parallel with those less accurate criteria on which the recognition of the generations was first founded.[35]

Indeed, this definition seemed to provide a clear distinction between the Hofmeisterian alternation of generations, shoot metamorphosis, and Steenstrup's alternation of generations in animals. It also enabled Bower to distinguish cytologically between the antithetic and the homologous alternation of generations in the algae. The former, occurring in *Dictyota* and some brown algae, was, according to Bower, obligatory, indifferent to fluctuating environmental changes, and typified by a $2n$ to n cycle. Homologous alternation, on the other hand, occurring in desmids, the Conjugatae, *Coleochaeta,* and similar algae, was dependent on external conditions, was not obligatory in its succession, and above all stood "on a basis of cytological unity." *Both generations of the homologous type were either diploid or haploid;* no cytological alternation of generations occurred. In both homologous and antithetical types, it should be pointed out, the two generations could also be morphologically identical.

Bower stressed again that the *origin* of the antithetic alternation of generations must be sought "in those post-sexual complications which are so frequently the consequence of nuclear fusion."

> The initial factor appears to have been "sterilization", that is, the delay of reduction by the conversion of cells which are potentially, and were ancestrally, sporogenous, into cells which serve no longer a propagative but a vegetative function. . . . On a biological theory, the nutrition of the increasing number [of

spores] was secured by the conversion of some of the potential germs to form a vegetative system, which should provide for nutrition and protection. . . . Accordingly, the time of spore formation was deferred, and a vegetative system, ultimately of great extent, was intercalated.[36]

Although the origin of these generations depended on "sterilization," their subsequent elaboration required migration to land. Where antithetic generations occurred in the algae, he pointed out, the form of the two generations appeared to be the same since both lived in relatively the same environment. In other words, Bower seemed to be suggesting that the migration to land did explain the origin of any *structural* differences between the sporophyte and the gametophyte.

THE CYTOLOGICAL TRIGGER

By 1900, cytologists had also provided an answer to the perplexing question, Why is an egg usually incapable of development until it is fertilized by a sperm? Prior to that time, Bischoff and other physiologists had assumed that the sperm or the spermatic fluid exerted some physicochemical effect upon the egg. Cytologists, however, had assumed that an unfertilized egg lacked some structure or structures that could only be provided by the sperm at fertilization. Since the formation of the achromatic spindle and associated structures seemed to initiate division of the zygotic cell, cytologists naturally looked to these structures for an answer to the problem. An answer came in 1887. Theodor Boveri discovered that a sperm carried not only its nucleus, but also a single centromere, into the egg. This centromere then divided to form the two active centers of cell division from which the asters and spindles arose. As a result of Boveri's discovery, a completely morphological picture of fertilization emerged:

The ripe egg possesses all of the organs and qualities necessary for division except the centrosome, by which division is initiated. The spermatozoon, on the other hand, is provided with a centrosome, but lacks the substance in which this organ of division may exert its activity. Through the union of the two cells in fertilization, all of the essential organs necessary for division are brought together; the egg now contains a centrosome which by its own division leads the way in the embryonic development.[37]

Thus, as Edmund Wilson pointed out in 1896, although the end result of fertilization was the union of two germ nuclei, "the active *agent* in this process is the centrosome."[38] Once again the discipline of cytology had provided the answers. Perhaps the most dramatic example of the cytologists' ability to provide answers, however, was their work on the question of sex determination.

THE CYTOLOGY OF SEX DETERMINATION

What factors cause an embryo to develop into a male or a female individual? By the 1890s the issue was thought by many to have been resolved. "The determination of sex," Edmund Wilson concluded in his half-page discussion of sex determination in the first edition of *The Cell in Development and Inheritance,* "is not by inheritance but by the combined effect of external conditions." The issue hardly warranted a more lengthy discussion, at least not by one with Wilson's interests. Past work had, according to Wilson, rendered "it certain that sex as such is not inherited."[39]

Had Wilson chosen to discuss the issue at length, however, his readers would have concluded that the solution to the problem was not as clear-cut as Wilson suggested. As François Maupas had remarked only a few years previously, "Knowledge of the conditions and causes which determine the production of the sexes constitutes one of the most difficult and obscure questions in biology."[40] Nevertheless, Wilson was justified in noting that most theories of sex determination began with the premise that the fertilized egg was sexually undetermined, sex being subsequently acquired by the influence of environmental factors on the developing embryo. Moreover, by the end of the century, this influence had been traced to an extraordinary number of factors.

One popular idea, which had been quashed by Hertwig's and Fol's discoveries, was that sex was determined by the number of sperm that penetrated the egg: the higher the number, the greater the likelihood of a male offspring. Other workers had suggested that sex was related to the time of fertilization: the earlier it occurred, the greater the likelihood the offspring would be male. Rather questionable statistical studies had suggested to others that sex was related to the age or vigor of the parents: the sex of the offspring was identical to that of the oldest or more vigorous parent. Patrick Geddes and J. A. Thomson, whose publications on sex were widely read in the late nineteenth century, suggested that a male offspring resulted when catabolic or disruptive changes occurred in the embryo's cytoplasm.[41] By far the most popular explanations, however, were those supported by experimental evidence. Many such studies had shown that a higher proportion of males resulted if the embryos received insufficient nourishment, while well-fed embryos tended to become females. Maupas, in a widely quoted series of experiments, subjected rotifer eggs to differing temperature regimes: 95 percent of the eggs that were exposed to low temperatures became females, while 97 percent of those that were subjected to high temperatures became males.[42] Thus, S. Watasé concluded in 1892, "the differentiation of sex in the developing embryo is purely an ontogenetic phenomenon . . . in the hands of external forces."[43] As another author pointed out in 1873, however, the fact that a higher proportion of males resulted when embryos were malnourished was equally "attributable to the fact that the females, being largest and requiring more nourishment, succumb most readily under such treatment."[44]

The issue of sex determination was again thrown into considerable confusion in 1899 when Lucien Cuénot's lengthy treatise on the subject appeared. Cuénot, who was later to become one of the few French biologists to accept the validity of Mendelianism, discussed sex determination in a wide variety of parthenogenetic and normal forms and concluded that "the sex of an animal is in all cases determined in the egg or, at the latest, at the moment the egg is fertilized."[45] In some animals, he argued, sex is determined in the egg, in which case the sperm has no influence; in others, sex is determined at fertilization by the reciprocal activity of egg and sperm. Thus, he wrote, "it is necessary to abandon the generally admitted opinion that represents the fecundated egg or even the embryo for some period as indifferent, that is to say capable of developing in any direction whatever."[46]

Cuénot's paper was a portent of events to come: a year later Mendel's work was rediscovered. William Bateson saw immediately that Mendel had described a mechanism for maintaining discontinuous variations in nature, and in his first report to the Evolution Committee of the Royal Society, Bateson suggested that sex was a prime example of such a variation.[47] Just as pea seeds could be either round or wrinkled, so could individuals be either male or female. And just as pea shape was determined by a pair of single alleles, so too was sex. What could be simpler or more obvious? In 1903 William Castle presented the first paper in support of this totally Mendelian interpretation of sex differentiation.

Castle was a member of the first generation of American-trained geneticists. After graduating from Harvard in 1893, he took a position as an assistant to C. B. Davenport and two years later gained his doctorate under E. L. Mark. Following Davenport's lead, Castle became an early convert to Mendelianism. In 1903 he published four papers on the subject, including one entitled "The Heredity of Sex."

"Certain phenomena of sex," Castle argued, are "found to have their almost perfect parallels in recognized Mendelian phenomena"—namely, the principle of dominance and the principle of segregation.[48] There were, in other words, two alleles for sex: one for maleness and the other for femaleness. Since every individual received one allele from the mother and one from the father, it was necessary to assume that all individuals were heterozygous for sex, the character for maleness being dominant in the male and recessive in the female, and vice versa for femaleness. Thus, Castle argued, "all animals and plants are *potential* hermaphrodites, for they *contain the characters of both sexes,* but ordinarily the characters of one sex only are developed, those of the other sex being latent or else imperfectly developed."[49] The presence of the other sex in at least a latent condition was proven, according to Castle, by the occurrence of rudimentary organs peculiar to the other sex, by the ability of some parthenogenetic females to produce offspring of both sexes, and by the occurrence of true hermaphrodites—"sex mosaics," as Castle termed them.

Castle then came to the crux of his argument. If true Mendelian segregation took place during the formation of gametes, then "all possible combinations of gametes are formed in fertilization, and in the frequencies demanded by the law of chance." Half the offspring would therefore be homozygous for sex:

$$\text{Male} \times \text{Female}$$

Gametes　　($\male$) ($\female$)　($\male$) ($\female$)

Offspring　1 $\male$ $\male$: 2 $\male$ $\female$: 1$\female$ $\female$

However, since individuals homozygous for sex do not exist, "it is highly probable that an egg bearing the character of one sex can unite in fertilization only with a spermatozoon bearing the character of the opposite sex."[50] The necessity of "selective fertilization" thereafter became a central issue in the debate over sex determination.

Thomas Hunt Morgan, for one, was highly critical of Castle's theory. Since a 1:1 ratio of the sexes occurs, not the 1:2:1 ratio predicted on the basis of Mendel's principles, Morgan argued, it was "absurd to attempt to apply Mendel's law to the problem of sex." He further noted that there was no evidence to support the idea of selective fertilization.[51] Moreover, since Castle assumed that every individual was a "sex hybrid," he had not answered the question, What actually determines the sex of an individual? "Castle's theory appears needlessly complex," Morgan argued,

> and the whole attempt to apply the Mendelian principle to the question of sex does not appear to have been very successful. . . . It fails to account for the very problem that a theory of sex should explain, namely, the problem of what it is that determines whether an egg that contains both potentialities becomes a male or a female.[52]

This, of course, was the crux of the argument against a Mendelian interpretation of sex determination. The problem of sex determination clearly lay within the confines of embryology, not genetics. The possession of sex alleles, as postulated by Castle, did not explain how a zygotic cell developed into a male or a female. Mendelian interpretations simply did not provide the answer demanded by embryologists like Morgan.

Although he was prepared to admit that the sex of an embryo was determined no later than the time of fertilization, Morgan was unwilling to concede the necessity of separate gametes bearing either maleness or femaleness. For example, he was not willing to accept the findings of C. E. McClung, who in 1902 had reported that half the sperm in a species of orthoptera carried an additional "accessory chromosome," which, McClung argued, "is the bearer of those qualities which pertain to the male organism."[53] It was impossible to substantiate this conclusion, Morgan argued, "as long as we know nothing at all in respect to the conditions in the egg." McClung himself, it should be noted, was very circumspect in his concluding remarks. "It must accordingly be granted," he stated, "that there is no hard and fast rule about the determination of sex, but that specific conditions have to be taken into account in each case."[54] Morgan clearly supported this statement; sex might well be determined early, but to conclude from that premise that male and female primordia were separated in the germ cells was invalid. Both sexes are "potentially present in the nucleus," he wrote. Moreover, the egg "appears to be in a sort of balanced state, and the conditions to which it is exposed . . . may determine which sex it will produce."[55]

Arguing in support of the Mendelian interpretation of sex determination, Bateson in England and Castle in the United States attempted to convince their peers that "acceptance of Mendel's principles of heredity as correct must lead one to regard discontinuous (or sport) variation as of the highest importance in bringing about polymorphism of species and ultimately of the formation of new species."[56] Their interest in sex determination sprang from their assumption that sex was a classic example of such polymorphism. They were not concerned, however, with associating these Mendelian characters with any material object within the cell. At Columbia University, on the other hand, a rival school of cytologists was working under the leadership of Edmund Wilson. By this time convinced that the behavior of Mendel's factors paralleled that of chromosomes, Wilson and his students were examining further connections between Mendelianism and cytology. William Sutton's famous paper had appeared in 1902, and that same year McClung, another of Wilson's students, reported the existence of accessory chromosomes.

Thus, Wilson began his investigations into the possible existence of sex chromosomes. Work over the past summer, he reported in 1905, showed "constant and characteristic differences in the chromosome groups, which are of such nature as to leave no doubt that a definite connection of some kind between the chromosomes and the determination of sex exists in these animals."[57] Examining species of hemiptera, he found that in some the female tissue cells carried one more chromosome than existed in the male cells (Fig. 8.1, Type I); in others the cells of both sexes carried the same number of chromosomes, but one chromosome was much smaller in the male than in the female (Fig. 8.1, Type II). Wilson also reported that in the case of the former species half the sperm carried an additional "accessory," or "heterotropic," chromosome, while in the latter species half the sperm carried a small "idiochromosome." In all cases, on the other hand, the eggs carried the same number of chromosomes. The different crosses were as follows:

Type I. *Protenor* sp.
Eggs (7 chromosomes)× Sperm (6 chromosomes + 1 heterotropic
 chromosome) → Female (14 chromosomes)
Eggs (7 chromosomes) × Sperm (6 chromosomes) → Male (13 chromosomes)

Type II. *Euschistus* sp.

Eggs (7 chromosomes) × Sperm (7 chromosomes) → Female
(14 chromosomes)

Eggs (7 chromosomes) × Sperm (6 chromosomes + 1 small
idiochromosome) → Male (13 chromosomes + 1 idiochromosome)

"It is evident," Wilson remarked in his major review paper of 1906, "that a substantial basis now exists for the Mendelian interpretation of sex-production worked out by Castle."[58] According to this interpretation there

Type I.

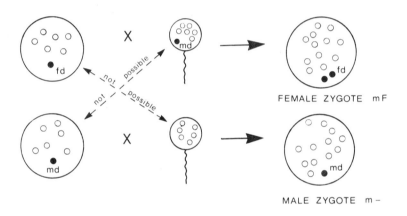

Type II.

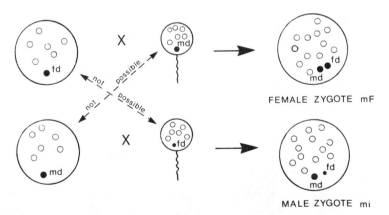

FIGURE 8.1.   Edmund Wilson's early interpretation of sex determination based on selective fertilization: *fd*, female determinant; *md*, male determinant.

were male- and female-producing eggs and sperm, and fertilization was selective.

Such conclusions were necessary, Wilson argued, because male individuals in the Type I species carried one unpaired heterotropic chromosome. This chromosome, he theorized, "[must] be the male-determinant which exerts its effect uninfluenced by association with a female-deterimant" (Fig. 8.1, *md*). But since any sperm that carries this heterotropic chromosome must produce a female, Wilson concluded that "the maternal mate or fellow, with which it becomes associated on entering the egg, is the dominant female-determinant" (Fig. 8.1, *fd*). On the other hand, he reported, the sperm that lacked this heterotropic chromosome produced males, which showed that the male determinant was derived from the egg nucleus. There must therefore be two types of egg, Wilson argued—one with the male determinant and one with the female determinant—as well as two types of sperm. We must assume, Wilson wrote, that

> there are two kinds of eggs that contain respectively the male- and the female-determinant, and that the former are fertilized only by spermatozoa that lack the heterotropic chromosome (i.e., the male determinant) and *vice versa,* giving the combinations mF (female) and m-(male). *Such a selective fertilization is therefore a* sine qua non *of the assumption that the heterotropic chromosome is a specific sex-determinant* [italics added].[59]

Basically the same situation obtained in the Type II species. There, male individuals carried not only the male-determinant but also a minute idio-chromosome, which, Wilson argued, "[could be] regarded as a disappearing, or even vestigial, female-determinant that is recessive to its larger fellow." As in Type I, the eggs were of two types, containing either a male determinant or a female determinant. Thus again, selective fertilization was essential; a sperm carrying the vestigial chromosome could combine only with an egg bearing the male determinant.

At this time Nettie Stevens, a research assistant at Bryn Mawr and a former graduate student of Thomas Hunt Morgan's, also began to investigate the accessory chromosomes. In May 1905 she submitted a monograph on spermatogenesis in the mealworm to the Carnegie Institution, which was funding her work. In her report she noted that the cells of the female mealworm contained twenty chromosomes, while those of the male contained nineteen large ones and one small one. "The spermatozoa which contain the small chromosome," she wrote, "[determine] the male sex, while those that contain 10 chromosomes of equal size determine the female sex."[60]

Wilson, unlike Stevens, was not convinced that a strictly Mendelian interpretation of sex determination was as yet justified. In 1905, for example, while he agreed that the facts led "irresistibly" to the conclusion that the accessory and idiochromosomes were sex determinants, he nevertheless added the rider that sex determination was more probably due to "differences of degree or intensity, rather than kind, in the activity of the chromosome groups in the two sexes."[61] Wilson became much more explicit on this issue in his major paper of the following year. There he pointed out that sex *determination* needed to be distinguished from sex *inheritance*. Sex chromosomes, he argued, might well be the vehicles of Mendelian sex inheritance, but sex determination might be due to factors other than the chromosomes. Moreover, Wilson felt so uneasy about the assumption that fertilization was selective that he concluded his paper with a non-Mendelian chromosomal explanation. Selective fertilization would not be deemed necessary, he wrote, if the idea that the sex chromosomes were specifically male- and female-determining were abandoned. His findings could be explained equally well, he pointed out, by assuming that the important factor was quantitative differences in the chromatin: whenever two sex chromosomes occur in a zygotic cell, a female results; if only one, or one plus the tiny, vestigial idiochromosome, appears, then a male develops. This interpretation differed from the Mendelian theory in that it attributed to chromosomes a difference in the degree of their activity:

> Under this assumption the facts might receive a general formulation in the statement that the association of two more active chromosomes of this class produces a female, while the association of a more active and a less active one (or the absence of the latter, as in the case of the heterotropic chromosome) produces a male. . . . If it could be adopted the necessity of selective fertilization would be avoided, for the observed results would follow from the fertilization of any egg by any spermatozoön.[62]

Despite these remarks, however, Wilson clearly favored the Mendelian interpretation. Having been in the forefront of research that linked specific Mendelian characters with individual chromosomes, it would now make little sense for him to argue that sex was somehow controlled by mere quantitative differences in the chromatic material. Thus he concluded his 1906 paper with the following words: "These relations unquestionably afford a concrete basis for an interpretation of sex-production that assumes a Mendelian segregation and transmission of the sex-characters and to this extent, they accord with the general assumption of Castle."[63] Indeed, that same year Wilson took issue with Richard Hertwig's claim that the nuclear-to-cytoplasmic ratio determined the sex of an organism, and that this ratio could be altered by environmental conditions. The basis of sex determination, Wilson once again reiterated, was nuclear constitution, not mass.[64]

Wilson's doubts were not shared by Nettie Stevens. Although once a student, and now a colleague, of Thomas Hunt Morgan's, she did not share Morgan's belief that sex determination was an embryological problem. She attacked Wilson's suggestions that the sex chromosomes differed only in activity, not in kind, and that they merely transmitted rather than determined sex characters. "On the whole," she concluded, "[the theory] which brings the sex determination question under Mendel's Law in a modified form, seems most in accordance with the facts."[65]

In December 1906 the American Society of Naturalists met to discuss "The Biological Significance and Control of Sex." Wilson, one of the participants, again argued that as a working hypothesis, sex characters must be treated as Mendelian alternates, but again he expressed doubts. Only in insects, he pointed out, had it been shown that male- and female-producing sperm exist, and only in some rotifers were male- and female-producing eggs known to occur. "Here," as he put it, "our actual knowledge ends so far as fertilization is concerned. . . . Until we can be sure on this point, it is almost idle to speculate on the subject; for if such a double predestination exists there must obviously be a selective fertilization, such that each form of egg is fertilized by the appropriate form of spermatozoon; and if this be so, sex is not determined by fertilization, but fertilization by sex."[66]

Most speakers at the conference did not support the Mendelian interpretation, which may account for the extreme caution of Wilson's remarks. Robert Harper of the University of Wisconsin, for example, concluded his paper on sex determination in plants by stressing that "no simple differential distribution of chromosomes on Mendelian principles at the period of chromosome reduction could in any way account for the conditions presented by the higher plants."[67] The occurrence of the alternation of generations in plants rendered the issue extremely complex, for not only were the factors of sex differentiation to be faced, but so were those that determined whether sexual or asexual reproduction should take place. In addition, Harper noted, the time at which sexual differentiation seemed to take place varied greatly. In the heterosporous ferns and fern allies and in hermaphroditic seed plants, it was feasible to believe that sex determination was associated with reduction division and spore formation, but in homosporous ferns and dioecious seed plants, sex differentiation presumably occurred in the gametophyte stage—or in what was left of it.

Thomas Hunt Morgan, who at that time was a member of Wilson's Zoology Department at Columbia, also was critical of Wilson's views. Taking much the same approach as he had in his earlier paper on the subject, Morgan argued that sex might well be determined by internal factors, but that did not mean that the sex characters were "preformed" in primordia lying in different germ cells. Every egg and sperm carried the potentialities of both sexes, he believed. "The sex of the embryo is determined by an internal condition that is present in the egg or sperm, which leads to the domination of one of the two possible alternatives. This is modern epigenesis as I understand it." Morgan objected to any interpretation that treated alternative characters like sex "as entities in the germ cells that may be shuffled but seldom get mixed. With each new deal, the characters are separated, one germ cell getting one character, and another the contrasted character."[68] Like many embryologists, Morgan realized that reducing a character such as sex to a particulate, segregated unit explained nothing. To Morgan, sex—indeed every character—represented the outcome of a long developmental process. Wilson had made the same point earlier when he remarked that sex determination should be distinguished from sex inheritance. The former, Wilson had argued, may well result from factors that are entirely different from those concerned with the latter. Nevertheless, this distinction was immediately clouded over by his statement that "for the purpose of analysis it will, however, be convenient to speak of the idiochromosomes or their homologues as "sex determinants," this term being understood to mean that these chromosomes are the bearers of the male and female qualities, respectively."[69]

A great deal of work on sex determination in plants and animals took place during these years, as a result of which both Wilson and Morgan changed their opinions—curiously in opposite directions. Wilson moved away from the Mendelian interpretation of sex while Morgan moved toward it.

In 1907, having discovered that there were male- and female-producing pollen grains, Carl Correns suggested that only the male was heterozygous for sex. Since the females were assumed to be homozygous, the required 1:1 sex ratio would arise without selective fertilization.[70] In a critique of Corren's paper, Wilson noted that, cytologically speaking, the male insects Correns had studied were heterozygous, since they carried the unpaired accessory chromosome or the unequally paired idiochromosomes. Wilson noted a fatal flaw, however, in the assumption that chromosomes actually carry characters. Alone in the male, the accessory chromosome carried the male characters, yet according to Corren's hypothesis, the accessory chromosome was derived from a homozygous mother and presumably there carried the female characters! This was a *reductio ad absurdum,* he noted. The conclusion seemed obvious: "Either the females of these insects must be physiologically heterozygotes (as I assumed), or the so-called 'sex chromosomes' do not bear the sexual tendencies."[71]

Meanwhile in England, Bateson and Reginald Punnett realized that Doncaster's earlier work on varieties of *Abraxis* (a lepidopteran), which Doncaster had attempted with difficulty to fit into the Wilson-Castle framework, could be explained more simply by assuming that the male was homozygous for sex and the female heterozygous.[72] Their explanation was simpler, of course, because it did not call for selective fertilization. Faced with two explanations that were at odds with his own, Wilson realized that some experimental proof of selective fertilization was needed. Such proof never materialized, however. Examining the entry of sperm into the egg, Morgan and his students saw no evidence of selective activity. Instead, they reported, the first sperm to enter was the one that stood with its longitudinal axis vertical to the egg's surface.[73]

On the other hand, Wilson's thesis derived considerable indirect support from Elie and Emil Marchal's work on *Bryum,* a species of moss that forms separate male and female moss plants (gametophytes). The existence of separate gametophytes suggested that the sporophyte generation was a sexual heterozygote which during reduction division produced spores carrying either the male or the female character. "Such is indeed the fact," Wilson stated, arguing that the Marchals had shown that gametophytes produced artificially by outgrowths of the moss sporophyte without spore formation were often hermaphrodites, which produced both archegonia and antheridia. "Translating this into cytological terms," Wilson reported, "cells that contain only a single or haploid series of chromosomes bear but one tendency, male or female; while those that contain the double or diploid series bear both tendencies."[74]

Faced with these conflicting data, Nettie Stevens retreated a little from her earlier position. Having shown by 1908 that species of coleoptera, orthoptera, and diptera had two types of sperm, which differed in their chromosome structure, she now admitted that the material "does not throw any further light on the question whether the dimorphic spermatozoa are themselves in some way instrumental in determining sex in these insects; or whether sex is a character borne by the heterochromosomes and segregated in the maturation of the germ cells of each sex."[75]

Perhaps the most significant development, however, in terms of its effect on Wilson's and Morgan's views, was the reinterpretation of "Mendelianism" that surfaced at this time. Before the work of William Johannsen, professor of plant physiology at the University of Copenhagen, Mendelians had assumed that *characters* were inherited from parents and that they were transmitted by means of Mendelian factors, which, cytologists argued, were actually physical bodies located on the chromosomes. Thus, they had theorized that the chromosomes carried the characters of "maleness" and "femaleness." As Johannsen argued, however, personal qualities or characters were *not* inherited and transmitted, and "genes," a word he invented, were not character-carrying particles.[76] Thus, as Wilson argued in 1912, the characters, or phenotype, were not borne by factors but required the cooperation of many factors. "The entire germinal complex," he wrote, "is directly or indirectly involved in the production of every character . . . [and these characters] appear as responses of the germinal organization operating as a unit-system."[77]

In an address to the American Association for the Advancement of Science in 1908, in which he first introduced the terms "X-chromosome" for the heterotropic, or large, idiochromosome, and "Y-chromosome" for the small idiochromosome, Wilson posed these fundamental questions: Can sex be treated as a form of Mendelian heredity, in which the gametes bear male and female tendencies or factors that correspond to those which represent the dominant and recessive members of a pair of allelomorphs? Should we think of maleness and femaleness as due to the presence in the egg of specific male and female determinants that disjoin in maturation and combine in fertilization?"[78] A strong case could be made for this interpretation, he argued, although three different possibilities had been suggested: (1) both sexes were heterozygous, (2) only the male was heterozygous, or (3) only the female was. There were, however, two principal difficulties with the first possibility, he admitted. One was the necessity of assuming selective fertilization, "which, though possible, seems *a priori* improbable." The other difficulty was the pattern of fertilization in the bee: unfertilized eggs produced males and fertilized eggs produced females. "Under this hypothesis, the female tendency must be derived from the spermatozoon. But this is a *reductio ad absurdum*," Wilson wrote, "for the male is derived from an unfertilized egg which has by this hypothesis eliminated the female tendency." In other words, how could this male then pass a female tendency to the egg? "This difficulty," he concluded, "seems to me to constitute a formidable obstacle not only to Castle's hypotheses, but to the whole Mendelian interpretation."[79]

As Wilson also pointed out, however, although the other two interpretations rendered selective fertilization unnecessary, they faced equally difficult problems. Thus, he concluded, "I think it must be admitted that until these and various other specific difficulties have been satisfactorily met, the Mendelian interpretation will fall short of giving an intelligible or adequate explanation."[80] Many of the difficulties of the Mendelian interpretation would disappear, Wilson pointed out, if it were naïvely assumed that a single X-chromosome determines the male tendency and two X-chromosomes create the female tendency. Most of the data could then be explained as follows:

Unfertilized egg (X)	—fertilized by Y- or O-bearing sperm → XY or XO male
	—fertilized by X-bearing sperm → XX female
	—parthenogenesis in haploid state → X male
	—parthenogenesis in diploid state → XX female

Thus an egg would produce a female not because of the introduction of a dominant Mendelian female character but because of the introduction of a second X-chromosome. Although Wilson regarded this "naïve formulation" as provisional, "a possible guide to inquiry," it is clear that it gained support as the character interpretation of Mendelianism was replaced by the genotype interpretation and as the question of sex inheritance gained preponderance over the question of sex determination.

Morgan, meanwhile, continued to oppose both the Mendelian and the chromosomal interpretation of sex determination. As Garland Allen has pointed out in his biography of Morgan, "From the point of view of an embryologist, the most important question [to him] was not so much how hereditary information was *transmitted* from one generation to the next, but rather how that information was *translated* into adult characters."[81] Nevertheless, around 1910 Morgan accepted the chromosome theory of sex determination.

By that time, Morgan had completed the investigations into the problem of sex determination in aphids and phylloxerans which he and Nettie Stevens had begun in 1903. This work clearly convinced Morgan that sex was indeed "connected with" or "produced by" chromosomal differences in the egg and sperm.[82] Phylloxerans and aphids, whose fertilized eggs produce both parthenogenic females and sexual males and females, had long presented a stumbling block to the Mendelian interpretation of sex determination, requiring the formulation of complex speculative systems to "save the phenomenon." By following the chromosomes through a complete life cycle, Morgan found that the fertilized winter egg produced only females because only the sperm with six chromosomes (in *Phylloxera fallax*) fertilized the egg (also with six chromosomes), whereas the sperm with only four chromosomes degenerated. It seemed clear that the four-chromosome sperm would in fact have produced a male because winged females, produced parthenogenetically from the original "stem mother," produced two kinds of eggs. Those with a full complement of 12 chromosomes produced parthenogenetically small sexual females and those with only 10 chromosomes produced small sexual males (see Fig. 8.2). Clearly, he concluded, both eggs and sperm contain factors that determine sex, and such factors are linked to the presence or absence of two accessory chromosomes.

In an exhaustive review of the whole question, Morgan pointed out that two contrasting theories existed, one of which was qualitative, the other quantitative. "By a qualitative interpretation," he explained, "I understand

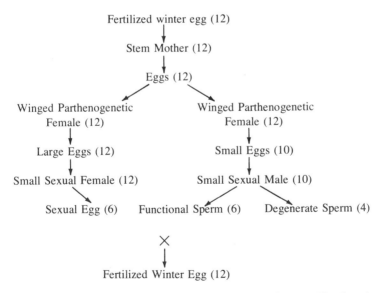

FIGURE 8.2. The life cycle of *Phylloxera fallax* (after Morgan). Numbers in parentheses indicate the number of chromosomes at each stage.

that there may exist in the gametes certain bodies or substances containing the materials either for a male or for a female.'' The quantitative interpretation, on the other hand, assumes ''that male and female are two alternative possibilities of the living material, which possibility is realized depending on quantitative factors.'' In the latter case, therefore, the gametes are not male and female but ''contain certain factors which when combined, give rise, in an epigenetic fashion, to one or the other alternative.''[83] Morgan naturally favored the quantitative interpretation and criticized, for example, Castle's attempt to reconcile the two views.

Castle realized by this time that male and female could no longer be interpreted as antagonistic members of a Mendelian pair. Instead, he argued, ''female is the male condition plus something else.'' This something else— the ''differential factor,'' as Castle termed it—''is inherited as a Mendelian character dominant over its absence.'' He further argued that Wilson's ''X-factor'' might well be this differential factor. There were difficulties inherent in this explanation, of course, as Morgan pointed out, and it required that one make numerous hypothetical assumptions.[84]

Nevertheless, two years later Morgan reported that ''sex is now treated by the same methods that are used for Mendelian characters in general,'' and to make the point doubly clear he claimed that ''the heredity of sex can be best understood when one sex is regarded as the pure line, or homozygous, and the other sex is treated as heterozygous.''[85] This rather dramatic change in Morgan's opinions resulted in great part from some of his early work on the fruit fly *Drosophila,* which he had begun in his Columbia laboratory around 1908 or 1909. In 1910 he discovered male flies that had white eyes rather than the normal red eyes. Upon crossing these males with red-eyed females, he found that the F_1 generation was totally red-eyed, while the F_2 generation consisted of the typical 3:1 ratio of red- to white-eyed offspring. To his surprise, however, all the white-eyed flies were male, and the red-eyed flies occurred in a ratio of two females to one male.

Morgan saw immediately that these and other crosses involving red- and white-eyed flies could be explained only by assuming that color was a typical Mendelian trait and that indeed the sperm did carry sex factors, as Wilson and Castle had claimed.

> Assume all of the spermatozoa of the white-eyed male carry the single ''factor'' for white eyes ''W''; that half of the spermatozoa carry a sex-factor ''X'' and the other half lack it, i.e., the male is heterozygous for sex. Thus, the symbol for the male is ''WWX'' and for his two kinds of spermatozoa WX-W. Assume that all the eggs of the red-eyed female carry the red-eyed ''factor'' R; and that all of the eggs (after reduction) carry one X, each, the symbol for the red-eyed female will be therefore RRXX and that for her eggs will be RX-RX.

An F_2 cross between a red-eyed female (RWXX) and a red-eyed male (RWX-) produced 25 percent red-eyed females (RRXX), 25 percent red-eyed

F₁ Female (RWXX)

		RX	WX
F₁ Male (RWX–)	RX	RRXX	RWXX
	[R–]	[RRX–]	[RWX–]
	[WX]	[RWXX]	[WWXX]
	W–	RWX–	WWX–

FIGURE 8.3. Sex-limited inheritance in *Drosophila* according to Morgan. The gametes and individuals that did not arise are noted in brackets.

females (RWXX), 25 percent red-eyed males (RWX–), and 25 percent white-eyed males (WWX–). But why were there no homozygous red-eyed males (RRX–) or homozygous white females (WWXX), which would normally occur were this a typical dihybrid Mendelian cross? Morgan realized that the answer lay in the absence of R– and WX male gametes: in other words, R and X must be transmitted together, thereby producing only two classes of sperm, RX and W–. "The fact is," he wrote, "this R and X are combined, and never have existed apart" (Fig. 8.3).[86]

Morgan's switch to a Mendelian interpretation of sex determination became very apparent in a paper he published in 1911. Pointing out that in *Drosophila* and man the male is heterozygous (♀ ♂, or XO) and the female is homozygous (♀ ♀, or XX), and that in other organisms the reverse is the case, he argued that in both cases "the chief drawback to these formulae, is, in my opinion, the absence of any character to stand for maleness." Maleness, according to these formulae, Morgan argued, rested on the absence of femaleness, and that "does not appeal to me as a sufficient explanation of the development of a male; for the male is certainly not a female minus the female characters." "It seems to me," he concluded, "that if we are to succeed in bringing sex into line with Mendelian methods we must be prepared to grant that there are representative genes for the male condition and others for the female; and we must so shape our formulae that the female carries the genes for the male and the male carries those for the female." This scheme differed from earlier ones in that two sets of genes were assumed to be involved and carried by all gametes:

			Adult organism	Gametes
Female	♀ ♀	XX	Fm Fm	(Fm)
Male	♀ ♂	XO	Fm fm	(Fm) and (fm)

Sexual dimorphism arose, Morgan claimed, "either because one female gene has become stronger than the others, or because one has become weaker."[87] He claimed that this new interpretation was little more than a development of his earlier views "in a form [that is] more in accord with the Mendelian treatment of the problem." He avoided any reference to dominance and recessiveness, substituting instead vague references to larger and smaller, stronger and weaker, genes. He referred to the F or f gene as a "quantitative factor [in which] the mass relation between these female genes and those of the male [somehow] turned the scale" toward or away from femaleness. In addition, Morgan now assumed that in the case of *Drosophila,* "the weakened female gene is contained in the so-called Y-chromosome."[88]

Given this extraordinary confusion and the rampant spread of speculative systems, it is not surprising to see a fairly widespread rejection of the Mendel-like explanations and a return to the pre-1900 explanations. What had occurred, of course, particularly in the case of Morgan, was not so much a rejection of previous explanations as a total change in the questions asked. Previously, Morgan's training had led him to see the question of sex determination in embryological terms: What factors caused a zygote cell to differentiate into either a male or a female? Now, however, heavily involved as he was in studies on *Drosophila* breeding, Morgan saw the question in genetic terms: How was sex inherited? He no longer posed the question, How do these factors, once inherited, cause the zygote to differentiate? It is not surprising, therefore, that those who had not made the switch from embryology to genetics could not accept the new explanations.

None other than Thomas Montgomery argued forcibly in 1910 that "there is no valid reason to interpret sex as an immutable unit character resident in or presided over by particular chromosomes, and sorted out and distributed by Mendelian segregation with all the complex mechanisms of dominance and determiners." Sex, Montgomery argued, was "the result of a labile process which may be changed by a variety of influences."[89] The Mendelians had changed the question. No longer able to see any solution to the problem of how sex developed, they limited themselves to the easier problem of how sex was inherited. Montgomery and others who were still interested in the developmental question, but who were not able to distinguish the developmental question from the inheritance question, remained skeptical.

Montgomery also pointed to the confusing data that seemed to deny the validity of a simple link between sex chromosomes and sex. His main thrust, however, was directed against the belief that chromosomes can be considered independent units. They are, he protested, "but parts of a larger whole," and thus "[it] is erroneous to speak of the chromosomes as automatic units, for they are but parts of the cell or cell complex."

> The organism acts as a whole, not simply as the sum of many parts; it is the interrelation of the activities of the many parts, added to these, that constitutes the behaviour of this major unit. Now to assume that particular chromosomes

alone are sex-determinants is to disregard this complex inter-activity. At the most, we are justified in concluding only that the chromosomes have a share in the establishment of sex. . . . The hypothesis is too naive, it assumes too great simplicity of the cell, it tastes too strong of rigid predetermination.[90]

Other critics went even further. Wilson's revised thesis, in fact, was one of many that interpreted sex determination in terms of a quantitative relation between chromatin and cytoplasm. Achille Russo, in a lengthy attack on the Mendelian system, argued that the character of hybrids, including sex, "depend in greatest part upon a special metabolism or chemical condition residing in the egg cytoplasm" and that "by artificial modification of the chemical state" the sex of an individual could be changed. After reporting that eggs of lecithin-fed rabbits gave rise almost exclusively to female offspring, Russo restated the old maxim that nutrition was one of the most important factors in determining sex. Thus, he concluded, sex and other characters "depend upon a suitable modification of the germ-cell, related in some way to the age, the physiologic state and other conditions of the particular individual."[91] According to Harvey Jordan, who cited the evidence presented by Russo and others, it "appears unequivocal that external conditions can determine the kind of sex and entirely vitiate the Mendelian scheme of ordinary crosses." Perhaps the most influential critic was Richard Goldschmidt, whose work would span the 1920s. Beginning his investigations of sex determination at this time, he, too, found the Morgan-Wilson scheme simplistic.[92]

Likewise, Neo-Lamarckians could not accept the Mendelian, or simple chromosomal, theory of sex determination. The leaders of the American School, Edward Cope, Alpheus Hyatt, and Alpheus Packard, died only a few years after the rediscovery of Mendel, so it was left to the French Neo-Lamarckians to articulate the Lamarckians' hostility to the new genetics and to theories of particulate inheritance.

Maurice Caullery, a student of the most preeminent French Neo-Lamarckian, Alfred Giard, agreed that sex often seemed to be determined at the moment of fecundation, but he could not accept the simplistic assumption that chromosomes carry sexuality. "Sexuality," he pointed out, "is a property of the total organism: one can no more imagine a chromosome representing sexuality than it does excretion or locomotion."[93]

Despite these problems and criticisms, the chromosomal theory of sex determination rapidly gained favor. In his book, whose title, *Heredity and Sex,* clearly indicated the new direction he had taken, Morgan provided perhaps the classic account of the situation which has since found its way into all first-year university texts on biology. The solution to the problem of sex determination, he wrote, "is very simple." In *Protenor,* for example, "any egg that is fertilized by a sperm carrying six chromosomes produces an individual with thirteen chromosomes. This individual is a male. Any egg that is fertilized by a sperm carrying seven chromosomes produces an individual with fourteen chromosomes. This individual is a female."[94] Other insects, he noted, had two sex chromosomes, not one: one large, X; the other, Y. Thus, "any egg fertilized by a sperm carrying an X chromosome produces a female that contains two X's or XX. Any egg fertilized by a sperm containing a Y chromosome produces a male that contains one X and one Y, or XY."[95] Similarly, according to Morgan, the inheritance of red or white eye color, which he had previously described as sex-limited, since white eyes were limited to the male sex, could now be explained by "a very simple hypothesis." "Any character," he pointed out, "that is dependent on the sex chromosomes for its realization will show sex-linked inheritance." The emphasis was now on the chromosomes; the idea that sex was determined by a single pair of alleles was no longer supported by either Wilson or Morgan.

> What, then, have the chromosomes to do with sex? The answer is that sex, like any other character, is due to some factor or determiner contained in these chromosomes. It is a differential factor of such a kind that when present in duplex, as when both sex chromosomes are present, it turns the scale so that a female is produced—when present in simplex, the result is to produce a male. In other words, it is not the sex chromosomes as a whole that determine sex, but only a part of these chromosomes.[96]

Neither, however, was sex determined by one or more pairs of Mendelian alleles carrying either maleness or femaleness. And neither did Morgan pose the question, How is sex determined in an embryological sense? For as Wilson so rightly noted in 1911, "as to the operation of the quantitative factor represented by the X-chromosomes, nothing is really known." One could conclude only that development is "turned" in either the male or the female direction "by some specific but unknown action induced by the presence in the zygote of more or less of the specific X-material."[97] And so the puzzle remained.

Whatever the shortcomings of these new answers, however, the cytologists had obviously found in the chromosome a structure of extraordinary significance. The behavior of these structures was the key to the sexual act; their quality determined an individual's sex; their quantity determined whether a plant belonged to the sporophyte or gametophyte stage. In addition, another structural component of the cell, the centromere, seemed to be responsible for initiating egg development. Thus, despite the rising prestige of the experimentalists—in particular, the new breed of physiologists who called themselves "biochemists"—the descriptive biologists, with their sophisticated new microscopes, microtomes, fixatives, and stains, could still claim to be among the elite of the biological profession.

NINE

Against Cytology

Experimentation and statistical research may well be indispensable, Edmund Wilson noted in 1909, "but it is well to remember that the sex problem was first attacked by such methods, and that they long gave inconclusive or wholly misleading results." On the contrary, he reminded his readers, "the most fruitful suggestions for its solution were first given by morphological studies . . . [and] the newer experimental work is bringing complete demonstration to these suggestions."[1] Wilson was referring, of course, to the experimental work of Boveri, Morgan, and himself; he was not alluding to Jacques Loeb's experiments, which called into question the whole morphological approach to the problems of sexual reproduction.

Those who looked for physical and chemical explanations of sexual reproduction did not accept the answers provided by the cytologists. Somewhat eclipsed by the cytological discoveries of the late nineteenth century, they reemerged at the turn of the century to seek once more a true physicochemical explanation of the events surrounding fertilization. Such an outlook, for example, permeated a series of articles entitled "Biological Problems of Today," which appeared in the 1898 volume of *Science*. Among the contributors to this series were Jacques Loeb and Thomas Hunt Morgan.

Jacques Loeb was motivated by the desire to account for all living phenomena in physicochemical terms. Physical chemistry—in particular, stereochemistry, van't Hoff's theory of osmotic pressure, and the theory of electrolytic dissociation—he argued in his 1898 paper, all provide tools with which to effect an understanding of the constituents of living matter.[2] Born in 1859, Loeb had studied at the Universities of Berlin and Strasbourg before moving to the University of Würzburg in 1886 to become Adolph Fick's assistant. Fick had been a student of Carl Ludwig, who in 1847 had joined forces with Hermann von Helmholtz, Emil du Bois-Reymond, and Ernst Brücke to form the Physical Society. These four had "imagined that we

235

should constitute physiology on a chemico-physical foundation, and give it
equal rank with Physics.'' They took as their premise von Helmholtz's statement that ''physiologists must expect to meet with an unconditional conformity to the law of the forces of nature in their inquiries respecting the vital
processes; they will have to apply themselves to the investigation of the
physical and chemical processes going on within the organism.''[3]

In the late 1880s Loeb joined the ranks of the developmental mechanists
and made the acquaintance of the American pioneers in the field. Feeling that
his career would be seriously jeopardized in Germany because he was a Jew
and a socialist, he emigrated to the United States and took up a position at
Bryn Mawr in 1891. Thomas Hunt Morgan was appointed to Bryn Mawr that
same year, bringing with him a classical training in descriptive embryology
under William Brooks at Johns Hopkins University. In 1894 Morgan spent ten
months at the Naples Station, met Hans Driesch, and returned to Bryn Mawr
committed to the new experimental, analytical embryology.

During the 1890s these embryologists discovered that when placed in
various salt solutions, unfertilized sea urchin eggs actually began to develop.
Thomas Hunt Morgan described such events in 1899, but he refrained from
taking a totally mechanistic view of the process. To reveal that physicochemical stimuli trigger an egg to divide, he argued, ''is not in itself a mechanical
explanation of the principal changes that take place.''

> To speak of a mechanical explanation of development, when we mean only a
> study of the responses to stimuli from without, gives an exaggerated and errone-
> ous idea of the entire problem. It is the vital structure of the egg on which the
> result largely depends, for if one kind of stimulus is as capable as another of
> starting the development of the egg, then we have accomplished very little in
> way of explanation if we have only determined what these stimuli may be.[4]

Morgan was conscious that a complete explanation involved both an understanding of fertilization per se and an analysis of development and inheritance.

No such qualms were felt by Loeb, however. Totally committed to the
belief that physical chemistry provided the answer to all biological riddles, he
announced in 1899 that ''former researchers [have] led me to suspect that
changes in the state of matter (liquefactions and solidifications) might play an
important role in the mechanics of life phenomena.'' Arguing that some
constituents of normal sea water may prevent parthenogenesis, he set out to
show that ''by making two changes in the constitution of sea water the eggs of
the sea urchin might be able to produce perfect embryos without being
fertilized.'' He reported that unfertilized eggs left in a mixture of magnesium
chloride and sea water for two hours before being returned to normal sea water
formed normal larvae. This mixture, he concluded dramatically, ''was able to
bring about the same effect as the entrance of a spermatozoon.''[5]

Loeb's conclusions were sweeping and totally countered Boveri's morphological theory. The unfertilized egg lacked nothing; all the "essential elements" for complete development were present in it. Parthenogenesis was prevented under normal circumstances only by the unfavorable constituents of sea water (or blood in the case of mammals). The sperm, whose role in fertilization had only recently been resurrected after two centuries of neglect, was once again relegated to a minor function. "All the spermatozoon *needs* to carry into the egg for the process of fertilization are ions to supplement the lack of one or counteract the effects of the other class of ions in the sea water or both." As Loeb triumphantly exclaimed, ions, not nuclei, were essential to fertilization, "which may interest those who believe with me that physiologists ought to pay a little more attention to inorganic chemistry."[6] "I consider the chief value of the experiments on artificial parthenogenesis," he concluded in a lecture given at Woods Hole in 1899, "to be the fact that they transfer the problem of fertilization from the realm of morphology into the realm of physical chemistry."[7]

The following year, having realized that the effect of the magnesium chloride could equally well be attributed to the higher osmotic pressure of the medium rather than to the individual ions per se, Loeb found that similar osmotic concentrations of sodium and potassium chloride and even urea produced the same effect on the eggs as did magnesium chloride. Substituting a theory of "osmotic parthenogenesis" for his earlier ionic theory, Loeb wrote: "The development of the unfertilized egg is produced through an increase in the concentration of the surrounding solution," and "there can be no more doubt that the essential feature in this increase in the osmotic pressure of the surrounding solution is a loss of water on the part of the egg."[8] Then, moving quickly from artificial fertilization to normal fertilization, Loeb claimed that an entering sperm was hypertonic to the egg (contained a higher concentration of ions than the egg), and thus water osmosed from the egg into the sperm, thereby stimulating the egg to develop. Once again he argued: "There is certainly no reason left for defining the process of fertilization as a morphological process."[9] Expanding his work shortly thereafter to include work on annelid eggs, Loeb found that these eggs, too, could be induced to develop, both by "osmotic fertilization"—increasing the osmotic concentration of the surrounding fluids—and by "chemical fertilization"—changing the concentration of individual ions without changing the total ionic concentration.[10]

Loeb's findings were greeted with rapturous praise in the popular press. The December 13, 1902, issue of *Harper's Weekly* carried a front-page picture of Loeb with the caption, "Americans of Tomorrow: Jacques Loeb," and described him as "one of the three or four greatest biologists living." Articles in *Cosmopolitan* and the *Fortnightly Review* even linked his experiments to the creation of life in the laboratory! "It was near to a realization of the dreams of Berthelot and Claude Bernard, aye, of every chemist who ever bordered the mysteries of life, the manufacture of life in the laboratory. In some ways, it was the most vital discovery in the history of physiology."[11]

By this time, however, Loeb was theorizing that the role of sperm in fertilization was more fundamental than merely acting as carriers of ions and hypertonic solutions. In his annelid paper in 1901 he suggested that sperm carry into the egg a "catalytic substance" which accelerates a process that would begin of its own accord, "but much more slowly."[12] This interpretation was far more "biological" than any offered previously; it implied that sperm manufactured and carried specific enzymes into the egg. Similar hypotheses had been proposed earlier in France. J.-B. Piéri had argued that a liquid extract from echinoderm sperm contained a soluble ferment, ovulase, which upon diffusion into the egg "has the property of bringing about segmentation of ovules." In 1900 Raphael Dubois argued that fermentation involved the reciprocal activity of "spermase" from the sperm and "ovulose" from the egg, but that because the former was incapable of direct diffusion into the egg, it needed the mechanical activity of the sperm in order to gain entrance. That, Dubois concluded, was the "raison d'être" of the spermatozoa.[13] A year later, however, William Gies, a student of Loeb's at Woods Hole, tested these enzyme theories and found them wanting. Studying liquid extracts from the sperm of *Arbacia* in which care had been taken to minimize osmotic differences, he found that they did not cause the eggs to divide.[14] Despite his attacks on morphological theories, Loeb now admitted that the sperm were needed to transmit hereditary qualities. "We must in future consider," he wrote, "the possible or probable separation of the fertilizing qualities of the spermatozoa from the transmission of hereditary qualities through the same."[15] This, of course, was a restatement of Newport's and Meissner's earlier views. It was also the opinion of Victor Hensen, who in 1881 had argued that fertilization involves both a physicochemical impetus and a morphological "moment."[16] In his 1902 review of the whole process of fecundation, Yves Delage took the same approach.

Born in 1854, Delage was by this time professor of zoology at the Sorbonne. He, like Loeb, had investigated artificial parthenogenesis and merogony (the artificial development of nonnucleated egg fragments containing sperm nuclei). Fecundation, he argued, had a double purpose: to put a ripe egg into a state capable of developing (embryogenesis), and to introduce male hereditary material into the egg (amphimixis). Studies on fecundation and artificial parthenogenesis, he noted, had shown that "the morphological phenomena of fecundation—in particular nuclear copulation—are essentially related to amphimixis, and that embryogenesis depends on concomitant physicochemical phenomena."[17] Such phenomena, he pointed out, "include the removal of cytoplasmic water by the male pronucleus, which absorbs it, dehydrates the cytoplasm, and thus communicates to the egg the capacity to divide."[18]

Thus, the explanation of fertilization proposed by the physiologists was at odds with that offered by the cytologists. Both groups agreed that the curious mobile spermatozoon, discovered over two centuries before, had two basic functions: to transmit through its nucleus and chromosomes the inherited male qualities and to initiate egg development. However, whereas the cytologists believed that egg development was initiated by the introduction of a missing centromere, the physiologists believed that the initiation of development involved some simple chemical reaction—a reaction so simple, in fact, that changing the osmotic pressure or ionic concentration could initiate egg development in the absence of spermatozoa.

Oscar Hertwig completely misunderstood these explanations of the role of sperm. Physicochemical accounts of fertilization, he protested in a passionate rebuke of Loeb's work, do not increase our biological understanding. "Fertilization is an amphimixis," he exclaimed, "a mixing or fusing of the properties of the two parental generators. How can the properties of the father be transmitted to the egg through osmosis or through ions or through catalytic substances? In [the] face of this fact, how can one speak of osmotic or chemical fertilization? Fertilization is a biological process which, at this time, one cannot expect to be resolved into a chemical-physical process by the concepts and experimental methods of chemists and physicists."[19]

It quickly became apparent, however, that Loeb's simple explanations were inadequate. Eggs induced to develop by these means did so abnormally and, unlike eggs fertilized by sperm, failed to produce a fertilization membrane. (In normal fertilization the outer membranes of the egg become separated from the protoplasmic contents by a clear space.) In addresses to the International Zoological Congress and the International Medical Congress in 1907 and 1909, respectively, Loeb reported that a fertilization membrane could be induced by treating sea urchin eggs with a monobasic fatty acid, and that normal development would occur if these eggs were subsequently placed in hypertonic sea water. "We see," he reported, "that the formative stimulus in the artificial activation of the egg of the sea urchin consists of two phases, . . . first the artificial causation of the membrane formation and, second, the subsequent short treatment of the egg with a hypertonic solution."[20] Membrane formation, he continued, could be caused by any cytolytic agent, and "from this we draw the inference that the membrane formation depends upon the cytolysis of the surface layer of the egg," which, being dissolved, gives the appearance of a membrane becoming separated from the protoplasmic layers beneath.[21]

According to Loeb, the spermatozoa contained two essential chemical components that together activated the egg to develop normally. One, a cytolytic chemical, dissolved away the superficial layers of egg protoplasm, while the other acted like hypertonic sea water. Further experiments convinced Loeb that the spermatozoon "causes the development of the egg through two agencies; one of these agencies is a cytolytic substance, a so-called lysin,'' which is situated at the surface of the sperm; the other, situated in the interior of the sperm, acts in a corrective manner to prevent general lysis.[22] Thus, sperm of the same species as the egg will cause the egg to develop normally because both substances are introduced into the egg. Sperm of a foreign species, however, will cause the egg to disintegrate because only the surface lysin is introduced. This hypothesis enabled Loeb to account for egg-sperm specificity, a problem that his earlier "osmotic" and "chemical" theories had failed to explain. What is more, Loeb's hypothesis suggested that the egg-sperm interaction might be related to the wider phenomenon of immunity.

In the early years of the twentieth century, scientists believed that blood contained lysins that destroyed the blood cells of foreign species but not those of the same species. Living cells, including egg cells, were thus thought to be immune to the lysins produced by cells of the same species. In fact, Frank Lillie later realized that fertilization was an example of an immune reaction and thus could be explained by principles derived from immunology. As he put it, "No theory of fertilization which fails to include the factor of specificity as one of the prime elements can be true.''[23] Loeb, however, simply used the egg-sperm interaction as a model to explain why tissue cells were immune to the lysins produced by their own blood cells. His interests lay elsewhere. His goal, as expressed in an address to the First International Congress of Monists in Hamburg, was to answer the basic question ''whether our present knowledge gives us any hope that ultimately life, i.e., the sum of all life phenomena, can be unequivocally explained in physico-chemical terms.'' Because he viewed fertilization in these terms, he argued that

> the process of activation of the egg by the spermatozoon, which twelve years ago was shrouded in complete darkness, is today practically completely reduced to a physico-chemical explanation. . . . Individual life begins . . . with the acceleration of the rate of oxidation in the egg, and this acceleration begins after the destruction of its cortical layer.[24]

In 1910 Lillie began his investigations of fertilization by observing the interaction between egg and sperm in *Nereis,* a polychaete annelid. This early work was clearly influenced by Loeb's findings (in 1900 both joined the University of Chicago faculty: Loeb as a full professor, Lillie as an assistant professor), and thus it was not surprising that Lillie came to much the same conclusion as Loeb had over the role of the sperm. The sperm, Lillie noted, functioned both to transfer paternal qualities to the egg and to stimulate egg development. Lillie's observations on *Nereis* also suggested that two phases were involved in the stimulation process. The first phase involved complex changes in the exterior layers of the egg by means of which the head of the sperm was drawn into the egg. The second phase, a long, continuous process extending to the time of nuclear fusion, involved, Lillie suggested, the creation of ''a free oxidation in the interior of the egg.'' Lillie's work showed that, contrary to the usual opinion, the sperm did not enter into the egg by simple mechanical boring. Instead, "the presence of the sperm calls forth a reaction on the part of the egg that leads to the absorption of the former." By centrifuging the egg at various stages of the fertilization process, Lillie also showed that the failure to segment was due either to the failure of the sperm to penetrate or to the destruction of the sperm nucleus within the egg. However, since in the latter situation the nuclear material of the sperm was still present within the egg (although "existing only as so much chemical matter"), the fertilizing power of the sperm must be "in some way bound up with its organization and growth." Lillie may have essentially agreed with Loeb that "individual life begins with the acceleration of the rate of oxidation in the egg, and this acceleration begins after the destruction of its cortical layer," but he also believed that these processes required the presence of a completely formed spermatozoon.[25]

Even at this stage of his work, Lillie's approach differed from Loeb's. In the preface to his famous monograph on fertilization, which was published in 1919, Lillie noted that past work on the problem had revealed "the inevitable conflict between the strictly biological and the physico-chemical methods of analysis."[26] Describing his own work as a reconciliation of the two approaches, he urged physiologists to use the tools and methods of physics and chemistry without losing sight of the inherent biological unity of organization which characterizes living things. Fertilization was a physiological process, and thus could never be understood without reference to the sperm. On the other hand, any explanation of fertilization had also to account for the fact that eggs could sometimes develop in the absence of sperm. Lillie had clearly expressed these views in 1911, at the end of his first paper on the subject:

> The experiments on artificial parthenogenesis are sometimes regarded as involving the entire problem of fertilization. But if it be true, as many believe, that biological fertilization is fundamentally a sexual reaction, then the physico-chemical analysis of fertilization must compass the entire problem of sex, which is much wider than the problem of parthenogenesis. . . . From the zoological point of view, at least, parthenogenesis and fertilization are not interchangeable functions. There is a factor present in fertilization which is absent in parthenogenesis, and the latter is never the exclusive mode of reproduction among animals. The biological analysis of fertilization therefore involves problems that do not occur in the physico-chemical analysis of parthenogenesis.[27]

Lillie argued that Loeb had failed to explain the role of sperm in fertilization. As he pointed out in a 1913 lecture to the Zoology Club of the University of Chicago, the current vogue of attacking the problem through artificial parthenogenesis had turned up so many factors that were capable of invoking development that one had to view with suspicion any theory that attempted to account for these phenomena.[28] Lillie was convinced that any meaningful theory of fertilization must take into account the role of the sperm—something which Loeb's work on artificial parthenogenesis clearly had not done. "There are obviously fundamental problems of fertilization," he wrote, "that cannot be touched by methods of artificial parthenogenesis."[29]

It would be tempting to suggest that Lillie's criticism of artificial parthenogenesis and of physicochemical methods sprang from his religious background (just as Loeb's extreme physicochemical thinking seemed in tune with his left-wing political and social views). Lillie had gained his B.A. degree from the University of Toronto in 1891, having enrolled there "determined to make religion my life work." With two Scottish grandfathers, both of whom were Toronto-based Congregational ministers, and a religious mother of Loyalist stock, a young man would be expected to have such an ambition. After long and passionate discussions at Toronto about science, evolution, and theology, however, Lillie opted for a career in science. Because there were few prospects for serious graduate work in Canada, he enrolled in a summer course at Woods Hole: "[In] June 1891 I took a train to Boston, my father presented me with a fine Leitz microscope, which I took with me and used for many years." Lillie returned to Canada thereafter only for brief visits. His father retired in 1896 and moved to Berlin, where he died two years later; his mother escaped from Germany and emigrated to the United States at the outbreak of World War I.[30]

Available evidence suggests that Lillie's religious upbringing did not influence his scientific attitude; religion and Canada were the rejects of his youth. Thus, Lillie's objection was more to Loeb's experimental design than to his mechanistic principles. It was an objection that sprang from Lillie's pragmatic acceptance of a holistic physiological outlook. This outlook can be seen in his earliest writings, which reflected not his religious upbringing but the influence of his Ph.D. supervisor, Charles Whitman. Lillie met Whitman that first summer at Woods Hole and under his guidance began working on cell lineage in the fresh-water clam *Unio*. He later obtained a graduate fellowship to Clark University, where Whitman was a professor. Then, when the University of Chicago, flushed with its $10 million grant from J. D. Rockefeller, offered Whitman a lucrative post there, Lillie and Whitman's other students accompanied him. Lillie gained his Ph.D. from Chicago in 1894 and, after holding junior faculty positions at the University of Michigan and Vassar College, returned to Chicago as an assistant professor in 1900. There he remained for the rest of his life, succeeding Whitman to the chair of zoology in 1910. He also returned to Woods Hole every summer, becoming its director when Whitman resigned in 1908.

Lillie's assumptions about the nature of life and living organisms mirrored Whitman's. In 1893 Whitman had criticized the "cell dogma," as he called it, and the premise that organisms are basically colonies of cells. An embryo, Whitman argued, "is not a simple adhesion of independent cells," but is an "integral structural adhesion." "Organization precedes cell formation and regulates it," he stressed; "the organization of the egg is carried forward to the adult as an unbroken physiological unity."[31] Similarly, Lillie, following

FIGURE 9.1. Woods Hole laboratory, 1893. Among those pictured are S. Watase (*front row, far left*), F. S. Conant (*front row, third from right*), F. R. Lillie (*second row, second from left*), K. Foot (*second row, third from right*), C. O. Whitman (*second row, second from right*), G. N. Calkins (*third row, second from left*), E. G. Conklin (*third row, sixth from left*), J. Loeb (*fourth row, second from left*), and T. H. Morgan (*fourth row, third from left*). Reproduced by permission of the Woods Hole Marine Laboratory.

his work on *Unio,* argued that "the egg acts as an undivided organism controlling some events by orientation of cleavage planes or shifting of cells, and others by protoplasmic movements clearly independent of cell boundaries."[32] Other organismic concepts pervaded Lillie's lengthy paper on the embryology of *Chaetopterus.* He denied, for example, the assumption that physiological unity arose secondarily as previously independent cells began to act as one unit. Instead, he argued, "there are certain properties of the whole, constituting a principle of unity of organization, that are part of the original inheritance, and thus continuous through the cycles of the generations."[33]

Lillie, like Whitman, was not averse to thinking in teleological terms. This facet of his physiological approach is best seen in his early work on cell lineages. The problem of germ layers and cell lineages had become a contentious issue by the 1890s.[34] The basis of the germ layer theory had been laid down in the early nineteenth century. In all animals, it appeared, the embryo became differentiated into two or three layers that could be distinguished both by their location in the embryo and by the fact, so it seemed, that each gave rise to particular tissues and organs. With the advent of evolutionary thinking, Ernst Haeckel and others maintained that "ontogeny is the short and rapid recapitulation of phylogeny.'' Moreover, they postulated that the two-layered gastrula stage, which was present in all Metazoa, represented a gastraea, the ancestral form of all multicellular animals. As we have seen, the search for ancestry through homology dominated biology during the second half of the nineteenth century, and the germ layer theory proved to be a vital tool in these endeavors. Organs were assumed to be homologous if they could be derived from the same ancestral germ layer.

At Woods Hole, however, Whitman argued that this embryological search for homologies should begin with the individual blastomeres, not merely with the gastrula stage. The germ layers, and even the organs that arose from them, he argued, ''can be traced directly to special blastomeres.'' This search for the cellular origins of the germ layers provided the major impetus for Lillie's earliest work on *Unio,* for by that time the value of germ lineage studies to the general problem of homology had been undermined to some extent by the discovery that the same germ layer in different animals might not have the same cell lineage. This became a crucial issue. As Edmund Wilson expressed it in 1892, ''The 'gastrula' cannot be taken as a starting point for the investigation of comparative ontogeny unless we are certain that the two layers are everywhere homologous.''[35] A determination of whether the layers were truly homologous necessitated ''tracing out the cell lineage or cytogeny of the individual blastomeres from the beginning of development.'' This program was carried out by Lillie, who argued that ''the most striking feature is not the contradictions existing, but the wonderful agreements.'' Cells of the same lineage, he found, usually differentiated into the same tissues and organs, but discrepancies did arise, because of the future needs of particular organisms. A firm advocate of the biogenetic law, Lillie believed that any seeming irregularities reflected adaptations ''to the needs of the future larva.'' Special needs, such as the early requirements for a functional shell gland in mollusks, may alter the order of appearance and character of segmentation, Lillie argued, ''but that in no way renders invalid the value of cell lineage studies to homology.''[36]

In a series of lectures delivered at Woods Hole in 1898, Lillie reminded his audience that however useful lineage studies might be in uncovering homologies, one must not lose sight of ''the special features of the cleavage in each species, which are, I believe, so definitely adapted to the needs of the future larva as is the latter to the actual conditions of its environment.'' It was Lillie's belief that differences in the cleavage patterns of organisms had ''prospective significance'' that could not be explained simply by ''mechanical laws of cell division.''[37] The mechanistic view of life, he again argued in a lecture given twenty-one years later to the Philosophy Club at the University of Chicago, ''[is] not inconsistent with a general teleological point of view.'' After quoting such holistic thinkers as Hans Driesch, J. S. Haldane, and Lawrence Henderson, Lillie concluded this lecture as follows: ''Many doctrines classed as vitalistic do not reject experimental determinism. They, therefore, interfere in no wise with the program of science."[38] This organismic, or holistic, approach typified Lillie's work on fertilization, particularly after 1911, when Lillie became dissatisfied with the methods he had been using to investigate the penetration of the egg by sperm.

How was an egg stimulated to begin development? This was the basic question that had been posed by nineteenth-century physiologists and more recently by Loeb and others. Although Lillie had contributed to the answering of this question, by showing, for example, that Boveri's centromere theory was invalid because only the head and not the middle piece of the sperm entered the egg, he began to feel that too much emphasis had been placed on this single question. The question, he argued, had become an "inhibiting factor." Because eggs could develop without the influence of sperm, their role in fertilization had been underplayed. More than that, "the problem of specificity had been left almost entirely out of account." Proponents of contact, ionic, and osmotic theories had never posed the question, Why can an egg be stimulated only by a sperm of the same species? To an organismic physiologist like Lillie, these problems of specificity were absolutely fundamental; the mysteries of fecundation could never be explained without an understanding of egg-sperm specificity.

To attack this very basic question, Lillie began to investigate how sperm responded to egg secretions. "It occurred to me," he remarked in a lecture to the Zoology Club at Chicago, "that the spermatozoa might prove better indicators of some of the processes taking place in fertilization than the slowly reacting egg."[39] The sperm, he noted, might be able to indicate "the presence of an agent to which spermatozoa of the same species are positively chemotactic,"

> if the union of the ovum and spermatozoon after they have come into contact operates not mechanically, but through some biochemical reaction between spermatozoon and ovum, the sea-water in which eggs have been standing should contain a substance also capable of reaction with the sperm, which should be an efficient indicator for it.[40]

The results of Lillie's early experiments were extremely significant. They showed that when a suspension of active sperm of *Arbacia* and *Nereis* was added to sea water in which unfertilized eggs had been allowed to stand, not only was a specific chemotactic response noted, but, more significantly, the sperm agglutinated. The sperm, he reported, became "sticky," and some change in their membranes took place such that the sperm adhered to each other and could not be separated by shaking. Eggs, he suggested, liberate an isoagglutinin which acts only on the sperm of the same species. It was now clear to Lillie that the reaction between egg and sperm was an immune reaction; the specificity of sperm and egg had become the crucial issue. "We must consider," he wrote, "the general fact that ova and spermatozoa of the same species do behave in a specific way with reference to one another in the process of fertilization. This must have some chemical basis and on the chemical side, the only reactions that exhibit a corresponding degree of specificity are those between antigens and antibodies in the field of immunity.''[41]

At this stage Lillie must have read reports of some of the recent work in immunology, for in 1913 he proposed a theory of fertilization based on the work of the great immunologist Paul Ehrlich. It is certain that Ehrlich's Croonian lecture of 1900 was known to Lillie, and in 1912 Lillie had acquired the second edition of Ehrlich and Boldvan's *Studies in Immunity*.[42] The many marginal notations in Lillie's copy of this edition draw attention to the parallel between fertilization and Ehrlich's work. By the early twentieth century, all theories of immunity, and there were many, had been ''overshadowed by that associated with the name of Ehrlich.''[43]

It was thought at this time that there were two ways in which bacteria elicited a response from the host organism. The first was by way of poisons or toxins produced by the bacteria; the second, which was associated with inflammatory reactions, was elicited only by the actual bacterial organism and not by chemicals secreted by them. Immunity, whether natural or acquired, was associated with the formation of antitoxins to the first group of bacteria and bactericidal sera to the second. The beauty of Ehrlich's theory lay in its applicability to both groups, whereas Metchnikoff's older theory of phagocytosis had only partially explained the reactions to the second group.

In 1897 Ehrlich had discovered that, although the toxicity of bacterial toxins decreased with time, exactly the same amount of antitoxin was needed to neutralize the ''young'' and the ''old'' toxins. He suggested, therefore, that the toxins included both a ''haptophorous'' group, which combined with the antitoxin molecules, and a ''toxophorous'' group, which exerted the poisonous action on host cells. The former fixed the toxin to the cells and thereby allowed the toxophorous group to exert its effect. If, then, a series of toxins was injected into the host, the host cells would attach to the haptophorous group by their ''receptors'' or ''side chains,'' which, being necessary for the normal metabolism of the cells, would be generated anew as more and more were used up in the attachment process. Eventually, Ehrlich postulated, so many receptors would be produced in this way that many would be secreted into the serum. These cast-off receptors would then act as antitoxic agents in the immune serum. By saturating the haptophorous group of toxins, they would prevent them from attaching to the host cell, and thus the toxophorous group would be unable to function (Fig. 9.2 [1–6]).

Ehrlich explained the immune reaction in the second group of bacteria along much the same lines. It was known that the immune serum of guinea pigs would dissolve the red blood cells of rabbits, but if heated to 55°C for 30 minutes, the serum lost this capability. On the other hand, when the serum of a nonimmunized guinea pig was added to the heated serum, the hemolytic activity returned. These results suggested to Ehrlich that two distinct bodies were also involved in this activity: a ''complement,'' which was susceptible to heat and present in all serum, and a stable "immune body," which was present in the immune serum. He visualized the immune body as being analogous to the haptophorous group of the bacterial toxins in that it had two receptor sites, one of which linked to the red blood cells and the other to the complement. The complement, on the other hand, was analogous to the toxophorous group in that it lysed the cells only through the intermediation of the immune body. Thus, when red blood cells or other foreign bodies such as bacteria were passed into nonimmune sera, they would link to body-cell receptors. However, since these receptors performed a normal function in the body, they would be produced in excess and become liberated into the serum as immune bodies. Any red blood cells or bacteria entering this immune serum would then be destroyed as they linked with the immune bodies and complements. The key to immunity in both groups of bacteria lay in the activity of a double-sited haptophorous group, or "amboceptor" molecule. As James Ritchie, professor of pathology at Oxford, summarized the theory in 1902:

> The methods by which bacteria are dealt with in the body are similar to those which obtain when many kinds of foreign cells gain an entrance into the latter. The development of artificial immunity against such bacteria depends on the latter being introduced either in a form not strong enough to cause death or, if virulent, not in sufficient numbers to cause death. In either case, the affected animal probably resists infection because it can develop in its body or already possesses a substance—immune body—which attached itself to the bacterial protoplasm and, in virtue of this attachment, permits another body—the complement—which exists normally in the animal's body to act on the bacteria with a fatal result to the latter.[44]

It would have taken little imagination for Lillie to realize that a parallel existed between the action of immune sera on bacteria and the action of a sperm on an egg. He would have noted, in particular, the discovery made by Karl Landsteiner and others that when spermatic fluid is injected into a guinea pig, the serum acquires the capacity to immobilize fresh sperm. The parallel with bactericidal activity became even clearer when it was discovered that this property, which was destroyed by heating the serum, returned when nonimmune serum was subsequently added. "The essential conclusion," Lillie wrote in 1914, "is that fertilization is a reaction between three bodies of which one is borne by the sperm and one by the egg; the third body, which is secreted by the egg, reacts with both the others. The spermatozoon functions especially as an activator of the third body, which I propose to name 'Fertilizin'; the latter, when activated, enters into certain reactions in the cortex of the egg which leads to membrane formation."[45] Fertilizin, in other words, was a typical "amboceptor" molecule with two reactive sites: a spermophile group that reacted with the sperm, and an ovophile group that reacted with the egg (Fig. 9.3).

By means of this theory, Lillie was able to explain not only the crucial issue of egg-sperm specificity but also why polyspermy was prevented. Sperm

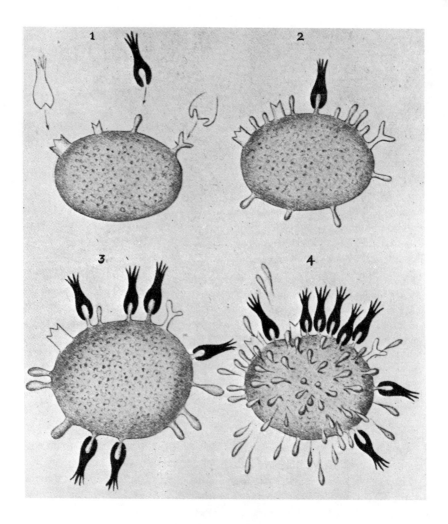

agglutination stopped, he discovered, when the eggs disintegrated. This suggested that the eggs contained within themselves a substance that could react with the spermophile side chain of fertilizin and thereby block the entry of additional sperm. As Lillie put it, "When we consider the extraordinary activity of the unfertilized egg in its secretion, and the equally extraordinary activity of the spermatozoa for it, one cannot escape the conviction that it must be a link in the normal fertilization process. When, moreover, one finds that the egg contains a more centrally located substance that can occupy the same combining group as the sperm, one seems to have before one, a view of a mechanism for preventing polyspermy."[46] The egg was thus assumed to carry egg-receptor molecules and "antifertilizin" molecules. The former attached to the ovophile group of fertilizin as soon as the fertilizin was activated

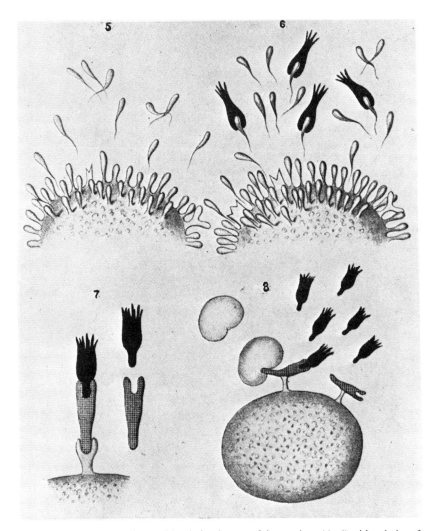

FIGURE 9.2. Paul Ehrlich's side-chain theory of immunity: (*1–4*) side chain of the cell attaches to black, toxin molecules with haptophore group; (*5*) excess side chains are liberated to become the antitoxin molecules; (*6*) free antitoxin molecules unite with toxins and thus protect the cell; (*7*) side chain of cell (*bottom*) unites with immune body (*middle*), which in turn links with the complement (*black*); (*8*) not relevant to the present discussion. Diagram reproduced from *The Collected Papers of Paul Ehrlich* (London: Pergamon Press, 1957), by permission of the publisher.

by union with the sperm. The antifertilizin molecules immediately combined with the spermophilic group of adjacent fertilizin molecules, thereby preventing the entry of additional sperm.

The assumed presence of spermophilic and ovophilic side chains on the

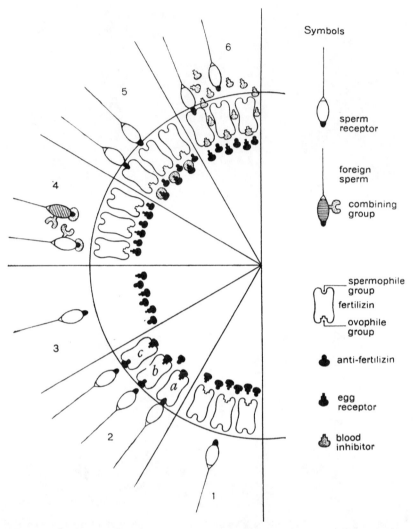

FIGURE 9.3.   Frank Lillie's theory of fertilization. Successive sectors of the egg show the mechanism of fertilization and the blocks to that mechanism as follows: (*1*) The arrangement of substances in the unfertilized egg and in the spermatozoon that are active in fertilization (see symbols). (*2*) The mechanism of normal fertilization. The sperm receptor unites with the spermophilic group of the fertilizin and the egg receptors unite with the ovophile group of the fertilizin owing to activation of the latter by the sperm (*a*). Molecules of the antifertilizin combine with the spermophilic group of the adjacent fertilizin (*b* and *c*) and thus block the way for supernumerary spermatozoa. This is the postulated mechanism for prevention of polyspermy. At the same time, molecules *b* and *c* of the fertilizin have also united with the egg receptors. (*3*), (*4*), (*5*), and (*6*) not relevant to the present discussion. Reproduced from Frank Lillie, "Studies on Fertilization," *J. Exp. Zool.* 16 (1914), by permission of Alan R. Liss, Inc., New York.

fertilizin molecules also enabled Lillie to propose a simple explanation for parthenogenesis—something that Loeb was unable to do except by undermining the role played by the sperm. The binding of the sperm with the spermophilic side chain of fertilizin, Lillie wrote,

> activates the ovophile-combining group of the fertilizin which then seizes upon egg receptors, and it is the latter union which results in membrane formation. If this were so, it is obvious that the spermatozoon is only secondarily a fertilizing agent, in the sense of initiating development, and that the egg is in reality self-fertilizing, an idea that agrees well with the facts of parthenogenesis and with the amazing multiplicity of means by which parthenogenesis may be effected. For the agents need only facilitate the union of fertilizin and egg receptor.[47]

Loeb continued to criticize Lillie's theory and to postulate a more chemical hypothesis, but Lillie's theory quickly became the standard explanation, presumably because its application was so wide.[48]

The problem of fertilization, like that of sex determination, had been redefined by basically ignoring the developmental issue. Explanations of sex determination now rested on the inheritance of sex chromosomes and ignored the puzzle of how such chromosomes could determine sex in an embryological sense. Similarly, the issue of how sperm initiated development had faded. Lillie expressed this very forcibly in his text *Problems of Fertilization:*

> It is commonly said that there are two main problems in the physiology of fertilization, viz.: the initiation of development, or activation, and biparental inheritance; but these are more properly results of fertilization. Indeed, so long as we regard fertilization primarily as a function of prospective significance in the life of the organism, we shall miss the more specific aspects of the process. Once fertilization is accomplished, development and inheritance may be left to look after themselves.[49]

TEN

The End of an Era

I end, as I began, with the plants, whose intricate life cycles had puzzled and fascinated generations of botanists. At the end of World War I, not only did Bower's beautiful theory explaining Hofmeister's alternation of generations come under attack but a "new new botany" began to question the very value of historical botany. Botany, it was argued, must join the mainstream of science; botanists must stop their tired, outmoded, and pointless investigations into ancestors and phylogeny.

In 1915 the brilliant French algologist Camille-François Sauvageau had discovered that *Saccorhiza,* a common marine kelp, alternated between the massive diploid kelp plant and a minute haploid gametophyte stage.[1] Thus, the Swedish botanist Nils Svedelius noted in a letter to Bower, "even here, the sporophyte has become prevailing during the course of development and is the absolutely dominating phase." The implication was clear. "For my part," Svedelius wrote,

I believe that if a biological explanation of the alternation of generations is to be found, we must look for it elsewhere and not in connection with the migration from water to land. For we find absolutely the same phenomenon—that of the domination of the sporophyte over the gametophyte—in the evolution of the marine algae.[2]

In his answering letter, Bower agreed "that the simple and direct views of 1890 are no longer sufficient." He still insisted, however, that "for all land plants we can see a biological reason for the elaboration of 2X [the diploid sporophyte generation]." He admitted that no reason existed for a similar life cycle in the algae, but still argued that under uniform conditions *most* algae will show a uniformity in both form and structure.[3] He was, of course, mistaken. Algae have been found to have complex and very diverse life cycles; like the land plants, many of them exhibit both a cytological (change in ploidy) and a morphological (change in form) alternation of generations.

The partial undermining of Bower's theory also reflected a change that was taking place within botany itself. By the end of World War I, botany had accommodated itself to the new outlook that had permeated some branches of zoology somewhat earlier. It was no longer popular to argue, as Bower had in 1908, that "the comparative study of plant form from the point of view of descent . . . must be pursued as in itself a substantive branch of the science."[4]

This change was particularly obvious among British botanists, whose leading figures in the years immediately prior to and after World War I were subjected to the same criticisms they had delved out to their predecessors during the 1870s. As late as 1924, Bower continued to argue in support of morphology. It was, he wrote, quoting Darwin, "the most interesting department of natural history, and it may be said to be its very soul." To Bower, recording the evolutionary history of plants still appeared to be the prime goal of botany. "A school based primarily on the study of 'process of development', and with the 'record of development' relegated to the background," he concluded, "might turn out good statisticians, but it would probably fail in converting them into historians."[5] The strains were visible, however. Bower admitted that the construction of phylogenetic trees seemed to be a hopeless quest and that "there is at present a marked reversion from that hopeful attitude." He acknowledged that phylogenetic lines always lead downward into uncharted mist, where they are "left in the air," but he urged botanists, at least for the moment, to be satisfied with such unsatisfactory answers. "It is a feeble mind that is daunted by the difficulty of the quest," he wrote.

In 1918 and 1919 this older generation of botanists came under fire in the pages of the *New Phytologist*, a journal that had been launched in 1902 by Arthur Tansley, a plant ecologist and lecturer in botany at Oxford. By that time biology had fragmented into many independent disciplines, many of them overtly experimental—genetics, biochemistry, immunology, comparative physiology, and experimental embryology—and these new disciplines had become the focus of research in vibrant new laboratories. Naturally, followers of these new disciplines regarded the older descriptive work with disdain. However, the attacks mounted in the *New Phytologist* were not aimed primarily at descriptive botany but at *morphology*. Their "revolt from morphology," to use the phrase of Garland Allen, was not so much a revolt of the experimentalists against the descriptive botanists as it was a rejection of the latter's emphasis on speculative historical and phylogenetic questions.[6] Indeed, with the advent of improved light microscopes and the electron microscope, microscopical *descriptive* studies have, if anything, enjoyed a new vogue in the twentieth century; what has changed is the raison d'être of these studies. The attacks on Bower and other occupants of established chairs in botany were attacks on their concern with uncovering the stages of plant phylogeny. Historical botany, not descriptive botany, became the central issue.

The criticism began with the December 1917 article written by Arthur Tansley; Frederick Keeble, assistant secretary to the Board of Agriculture and, from 1920 to 1927, professor of botany at Oxford; Francis W. Oliver, professor of botany at University College, London; Tansley's teacher, Frederick Blackman, reader in botany at Cambridge; and Blackman's brother, Vernon, professor of plant physiology at London's Imperial College of Science and Technology. This article which was labeled by some as the ''Tansley Manifesto,'' argued that academic botany was in a dangerous state and needed to be totally reformed. University classes in elementary botany required drastic reconstruction, its authors argued. ''It has to be frankly recognized that the study of the detailed evolution of the plant world, which has acquired a factitious importance owing to the overwhelming effect on the imagination of botanists of the doctrine of descent, has no valid claim to the dominating position which, especially, in this country, it has so long held.''[7] Such morphological botany, the article continued, ''does not attract the type of mind that wants a deeper insight into the working of plant processes.''[8] Thus, introductory botany should cater to those who wished to study physiology, ecology, biochemistry, and biophysics as well as to those who were interested in morphology and systematic botany. A year later, Walter Stiles, assistant lecturer in botany at Leeds University, argued that what needed to be taught was not botanical history but ''the science of living organisms.''[9] As Vernon Blackman put it in 1919, ''The main problem to be presented to the student seems to them to be that of living, not that of origin.''[10] Questioning an imaginary witness, ''W,'' ''the five wise men'' (the manifesto's authors) asked ironically:

> Has it ever occurred to you that the predilection for this subject . . . has had a pernicious effect on British Botany? Is it fitting that a teacher who deliberately chooses to spend the greater part of his leisure in endeavouring to interpret ''aged-dimmed tablets traced in doubtful wit'' should be entrusted with the training of men and women who wish to become botanists, gardeners and planters of waste lands?

''I submit,'' ''W'' answered, ''that it is undesirable that the study of botany should be organized on the rigid lines of ''Infantry Training.''[11]

It must have appeared to Bower that he, perhaps more than any other botanist, was at the center of the storm. The manifesto's reference to plants as ''living organisms'' seemed especially pointed his way, for in 1917, with Lang's withdrawal from a proposed Lang-Bower text entitled *Botany for Medical Students,* Bower had agreed to write *Botany of the Living Plant,* a text based on his own lectures at Glasgow. As Bower wrote to Tansley, ''Personally, it is the *living* and not the *dead* plant that I try to present every year to my elementary class—and always have done.'' He then added rather sarcastically: "It may be news to you and others that I gave a class of practical physiology with lecture experiments in 1884. Perhaps the first such in Britain."[12]

The authors of the "Tansley Manifesto" also attacked "crude Darwinian teleology." It is not enough, they argued, "to explain the appearance of a structure on the grounds of its utility." This implicit attack on Bower's teleological "amphibious alternation" was made very explicit by Vernon Blackman when he remarked that he had practically eliminated discussion of the alternation of generations in his elementary class because a detailed knowledge of it was no longer essential. Bower responded indignantly. "There are those who would eliminate the facts of alternation from elementary exposition," he wrote in 1918.

> They would pass over the great part played by the rise of the sporophyte and the suppression of the gametophyte in producing land vegetation as we see it. They would promote in its place a study of ecology. While not denying the importance of this aspect of botanical study, it seems inconsistent with the elimination of the Hofmeisterian story which rightly represented is the greatest example of practical adjustment to the surroundings. . . . The establishment of the sporophyte on land is the great text of ecology.[13]

His critics could not agree, of course; to them, Bower's arguments seemed "slipshod," "teleological," descriptive and historical, and lacked insight into the "physico-chemical causation of structure and process."[14]

Yet in Bower's *Botany of the Living Plant,* which appeared in 1919, surprisingly the alternation of generations was not the central theme; discussion of this phenomenon had been relegated to the thirty-second (and final) chapter.[15] More than half the book dealt with the basic structure and function of angiosperms—with water relations, nutrition, storage, respiration, growth and movement, and the mechanical construction of the plant body. Biochemistry was absent, but so, too, was the traditional historical perspective. Although the book received numerous excellent reviews, at least one reviewer felt that the concept of alternating generations had lost its force by virtue of having been relegated to the final chapter and "on account of [the author's] previous reticence on the subject of morphology."[16] Another reviewer pointed out that Bower's work "appears most opportunely at a time when the veneration of the more modern school of British Botanists for everything German has received a fatal setback" and at a time when "the works of continental writers will never acquire the hall-mark of super-scientific value with which they have been regarded," but he wondered, nevertheless, whether anything had really changed during the last thirty years of Bower's teaching. "Is botany an exact science," he asked, in which experiments and scientific reasoning hold sway, or "has it deteriorated to Nature Study?"[17]

The most cutting review of Bower's text, however, was that by Albert Seward, professor of botany at Cambridge and a close friend of Frederick

Blackman's. Although a paleobotanist, Seward deplored Bower's failure to discuss cell constitution and biochemistry, "topics," he wrote, "which must form the basis of any real understanding of 'The Living Plant.'" Using Bower's text as an example of the crisis besetting botanical teaching, Seward wrote in the *New Phytologist:*

> It is no more difficult to frame an account of this subject [cell biochemistry] which is intelligible and interesting to the student than to deal similarly with heredity and the alternation of generations; and it is impossible to deny that the former is even more fundamental than the two latter. Until these things are faced it is impossible to carry the basic scientific concept of Causation into the necessary foundation of the study of the plant as a living organism. . . . To avoid this task seems to abrogate the first and most vital function of the teacher of science. And it is the too common avoidance of it among even the most distinguished teachers of botany, such as the author of this book, and the consequences of that avoidance, which lead one to the conviction that the teaching of elementary botany is in need of radical reform.[18]

There was more to the debate than the revamping of botany courses, however. Deeper issues were at stake, and this accounted for the rather vitriolic manner in which the argument was joined. Arthur Hill of Kew, for example, remarked to Bower that Frederick Blackman and Tansley were "two of the most dreary of elementary teachers and Keeble is a gas bag of the first order."[19] Not only was Bower (a Yorkshireman living in what Londoners regard as the "far north") obviously reacting against the rising influence of a southern, Oxford-London-based self-styled elite, but more significantly, he was responding to the political philosophy they seemed to hold.

Britain at this time was suffering from a severe economic and social malaise, and Bower's critics thought they had a cure for it. British society, they argued, "had to become more efficient in performance and more scientific in attitude." In this era of reconstruction, argued the authors of the "Tansley Manifesto," less emphasis should be placed on pure science.

> We have to readjust our intellectual ideals as well as to reconstruct the material fabric of civilization on sounder lines. Many changes are necessary: a different curriculum and a different attitude towards knowledge and science in the schools, and a new organization of research, an immense increase in the endowment, and at the same time a large supply of better human material for scientific work. It is no use [simply] to tinker with the situation. It is of no use to say: "Oh yes, by all means let us introduce more physiology into the courses." What is wanted is not patching, but reconstruction, a new spark and a new deal.[20]

In addition, the authors of the manifesto called for the formation of a national union of scientific workers, which was actually founded as an official trade union the following year. Thomas G. Hill, lecturer in botany at University College, London, even called for the formation of a small "central committee" of botanists to direct a national course of action.[21]

To Bower and his supporters, these suggestions seemed to recall "the exhortations of preachers of other doctrines than those of natural science." Since the incorporation of science into the fabric of German universities in the previous century, academic scientists had accepted the values of those German academics: university science was "pure," undertaken for intellectual, not utilitarian, reasons, and was quite distinct from the lower forms of endeavor, applied science and technology. On the contrary, wrote R. C. McLean, I champion the cause of freedom." "I object to revolutions. . . . I would claim and defend the right of every botanist to think and practise according to his own beliefs without hindrance. . . . Away with the imposition of scheduled uniformity, it is the bane and shroud of intellect." Down with "the red-souled plotters."[22]

At this Bower exploded. The signatories of the "Transley Manifesto," he wrote, "appear to advocate immediate *Botanical Bolshevism*. . . . In order to secure their own Utopia they propose to subordinate something they admit is good in itself. That is the spirit that has ruined Russia and endangered the future of civilization."[23]

In a letter to Bower, T. Briquet, director of the Conservatoire et Jardin Botanique de Genève, wondered why "scientific periodicals are lucky in Britain to have paper enough to waste on such discussions."[24] Bower immediately forwarded a copy of this letter to Tansley, appending the remark that as editor of the *New Phytologist* he had the responsibility to stop this nonsense before more foreigners became involved. "The feeling the whole thing arouses," Tansley replied, "only convinces me the more strongly that the ventilation was very badly required." Upon receiving Tansley's letter, Bower scribbled the following note to himself:

> My reply to him was that if he were to walk stark naked down King's Parade it would cause much "feeling", but that would not prove that it would be a desirable thing for him to do.[25]

According to Tansley, Bower seemed to object to "washing dirty linen in public." "What I object to," Bower replied, "is the display of the wash, dirty or clean, before the drawing room window when you have asked a party to tea."[26]

Neither Tansley nor his supporters were, in fact, in any way left-wing radicals. However much their program may have paralleled socialist science, their main concern was to enhance their own prestige by demanding an increased role for science and for Oxbridge-London scientists in British public affairs. When the left-wing movement in English science did come to the fore in the 1930s,[27] Tansley showed his true colors. In 1941 he, the Oxford cytologist John R. Baker, and the Manchester University chemist Michael

Polanyi formed the Society for Freedom in Science. Their program echoed the norms of academic science. They attacked the Marxists' position that research should be goal-directed and that historical progress in science was mainly brought about in response to human needs, a position that had been made explicit during the Second International Congress of the History of Science and Technology held in London in 1931. In the first pamphlet produced by the Society for Freedom in Science, F. Sherwood Taylor argued that historians of science needed only to look within the internal logic of science for their answers, and that scientific curiosity alone was the driving force behind pure science. "Scientists work best in conditions of freedom" and "Science has a value (or values) apart from the benefits of its applications," were the two basic principles of the society.[28] Tansley and his supporters continued to deplore the government's indifference to science and to demand educational reform, but they now defined this reform as "a sound introduction to scientific culture," not an increased emphasis on applied science. Science, according to Tansley, who by this time had spent two years studying under Sigmund Freud, develops the mind to trust "humility, submission to the logic of established facts, dispassionate, objective investigation and respect for objective truth," and to mistrust "authoritarianism, bias, prejudice, wilfulness and self-interest." The national well-being demands that "we . . . jealously guard its freedom and independence," he wrote.[29] Bower, who had retired from Glasgow in 1925 after difficult and very moving attempts to find a replacement in a generation that had virtually been destroyed at Flanders and Picardy, would not have disagreed.[30]

Whatever Bower's private feelings may have been, and however much he may still have wished to solve the mystery of the origin of the antithetic generation, the subordinate place of the alternation-of-generations story in his text of 1919 reflected his realization that the story of Hofmeister could never again provide the basic focus of introductory botany courses. Today, I am sad to say, students of biology are almost totally unaware of the existence of Wilhelm Hofmeister and Frederick Bower. Historians of biology have not rectified this situation either; too many of them seem to suggest that the only significant work done in nineteenth-century biology is to be found in the Darwinian notebooks!

Postscript

With the exception of my brief account of Bower's postwar problems with his fellow British botanists, this study of sexual reproduction ends with Thomas Hunt Morgan's *Heredity and Sex* (1913) and Frank Lillie's important 1914 paper on fertilization. By making 1914 the cutoff point, I have omitted any reference to the sex hormones, work on which moved into high gear only after the war. Many readers, I am sure, will find this omission inexcusable, but the choice reflects my lack of interest in endocrinology and my unwillingness to add more pages to what I consider to be the optimal length of any book. Furthermore, the history of endocrinology has received little attention; I prefer to leave it that way rather than attempt to fit a hasty and ill-conceived summary into a few extra chapters.

My decision to terminate with the year 1914 is based primarily, however, on the curious fact that what general biology students now learn in high school and at the university about sexual reproduction is basically identical to the views that were propogated in 1914—plus the more modern work on sex hormones. Textbooks always lag behind contemporary work on any subject, but to lag behind by over half a century seems to call for a special explanation. Biology texts fail to mention not only the current controversy over sexual reproduction, which began in the late 1960s, but also the work of the 1920s, which exposed the oversimplicity of much of the earlier work, particularly that on sex determination.

We know today, and have known for some time, that sex determination is an extremely complex business. Contrary to what the biology textbooks tell us, sex can be determined environmentally (those nineteenth-century naturalists were right after all). In such cases, the sex of the individual is determined by the environment it encounters as a juvenile. Isolated larvae of the genus *Bonellia,* for example, develop into females whose secretions influ-

ence other larvae to form parasitic males. Likewise, as J. Bull has recently
shown, the sex of some reptiles is determined by the temperature of the nest.[1]

Of course, many other organisms, including the bulk of the vertebrates,
have genetic sex-determining systems. Such mechanisms "switch the pattern
of development onto one or [the] other of two alternative tracks."[2] The
switching mechanism can involve a single gene or multiple genes. In the latter
case, males, females, intermediate sexes, hermaphrodites, and changes of sex
are controlled by different combinations of these multiple switching genes. In
other organisms sex is controlled either by the number of chromosome sets or
by the presence of special "sex chromosomes," and in *Drosophila* it is
controlled by the balance between the number of sex chromosomes and the
number of sets of autosomes. In every case, whether genetically or environmentally controlled, the individual has the potential to become either sex;
zygotic cells are not sexually predetermined.

That is not the impression one gains from reading today's introductory
biology texts, however. Even Edmund Wilson, in the heyday of cytology,
was aware of the dangers inherent in assuming simply that sex is determined
by X and Y chromosomes. Because of his experience and expertise he was
aware of the existence of hermaphrodites, gynandromorphs, intersexes, etc.,
all of which undermined the validity of the simple X-Y assumptions: "I do not
myself consider [sex chromosomes] as sex-determinants in any exclusive
sense," he remarked in 1911, but "as one link—probably an essential
one—in a chain of factors by which sex is determined and inherited."[3] Such
warnings have fallen on deaf ears, however. The information packaged into
modern library texts remains very much the same as that contained in
Morgan's *Heredity and Sex,* which was published nearly seventy years ago.
Without exception—at least not to my knowledge—these encyclopedic texts
discuss sex determination solely in terms of the X and Y chromosomes.
William Keeton, for example, in his *Biological Science,* one of the most
widely used texts and my own particular favorite, concludes his discussion of
the X and Y chromosomes with this statement: "[In most organisms] the sex
of an individual is normally determined at the moment of fertilization and
depends on which of the two types of sperm fertilizes the egg."[4] Perhaps the
use of the word "normally" allows him to extricate himself from my criticisms, but his statement serves to perpetuate the erroneous belief that zygotes
carry information for only one sex. In Francisco Jose's *Modern Genetics,*
which was published in 1980, one would expect to find more up-to-date
information, but here, too, sex determination is linked to the different
chromosome constituents of the zygote, and the involvement of other
mechanisms is only hinted at.[5]

Part of the explanation for this emphasis on X and Y chromosomes lies in
modern biology's preoccupation with evolutionary theory. Because of this
preoccupation, not only is the genetics of sex determination stressed, but the
most puzzling and difficult question of all is rarely asked in introductory texts:

What exactly does the sperm do to the egg to initiate its development? Although some of the answers have become apparent in the last fifteen years or so, we are still told that the sperm "stimulates" the egg—a word that is equivalent to and as meaningless as the 1914 word "rejuvenate." The issue is ignored. As Lillie put it, once fertilization is accomplished, development may be left to look after itself.

Similarly, these same texts present the Weismann-like argument that sexual reproduction is the mechanism by which new genetic combinations are introduced and spread through populations to provide raw material on which natural selection can act. "Fertilization," Helena Curtis writes in her *Biology*, another voluminous and popular introductory text, "is not necessary for individual development but it is important to the species. It is the means by which the genetic information is rearranged and new combinations [are] introduced, providing the possibilities for the variation which is required if evolution is to occur."[6] Although this explanation is essentially correct, fundamental problems with it have recently come to light. Still, enough years have passed for these authors to have incorporated some of the new material into their texts (new editions seem to come out often enough for them to do it, and sex is one of the most important parts of biology to get straight).

However, the real problem with presenting the 1914 material in the guise of modern science is that it creates the impression that all the problems of sexual reproduction have been solved. We know that the sperm and egg are haploid cells that come together to form a diploid zygote; we know why half the zygotes form males and half females; we know why organisms reproduce this way. We biologists understand sex; only sociologists do not.

At issue in the standard explanation of sexual reproduction is the fundamental question, Can natural selection act at any level above that of the individual? According to Ronald Fisher, sexuality may be the only character that has evolved for the specific or group advantage rather than for the advantage of the individual.[7] Most evolutionary biologists, however, are unwilling to accept the concept of "group selection." There seems to be no mechanism by which characters that are deleterious to the individual can increase in frequency, even if advantageous to the group. As we have seen, even before the advent of population genetics, many biologists clearly were uncomfortable with the Weismannian explanation of sex and so retained the older, rejuvenation concept.

The rise of classical genetics seems to have effectively quashed these early doubts, however. In 1932 Hermann J. Muller stated categorically that "genetics has finally solved the age old problem of the reason for the existence (i.e., the function) of sexuality and sex, and that only geneticists can properly answer the question, 'Is sex necessary?'" The core of the answer, Muller noted, "was conjured up long ago by the genius of Weismann."[8] Muller's answer introduced a paradox that still remains unresolved. Although he admitted that sexless individuals have "often a temporary advantage" and that

Mendelian recombinations continually produce "disadvantageous combinations," he nevertheless argued that sexless forms "cannot keep up the pace
set by sexual beings in the evolutionary race" and that the disadvantageous
combinations "are an insurance against the day when some of them will be
needed." Implicit in his answer was the idea that although sexual reproduction may be disadvantageous to the individual, it provides advantages to the
group (the species) in that "a combination of beneficial mutations" may arise
that promotes variability and accelerates the rate of phyletic evolution—
evolution within a single line of descent.

Critics of group-selection theories naturally took issue with Muller's classical Weismannian argument, but unfortunately their concerns have not yet
found their way into introductory biology texts. Their criticisms became particularly strong following John Maynard Smith's arguments over the high
cost to individuals of reproducing sexually.[9] With no paternal investment in
the rearing of the young, as is the case in many species, the sexual female
individual makes her entire energetic investment in reproduction, but contributes only half her genetic material to the offspring. A parthenogenetic female,
on the other hand, has twice the genetic fitness of the sexual female; in the
absence of the production of males, there will be a doubling of the number of
parthenogenetic females in each generation, assuming that the parthenogenetic female in competition with a sexual female produces the same number of
eggs with the same probability of survival. Thus, as George Williams has
noted, the task of those who deny the validity of group selection "is to find a
previously unsuspected 50% advantage [of sexual reproduction] to balance the
50% cost of meiosis." As he concludes, however, "nothing remotely ap-
proaching an advantage that could balance the cost of meiosis has been
suggested. The impossibility of sex being an immediate reproductive adaptation in higher organisms would seem to be as firmly established a conclusion
as can be found in current evolutionary thought."[10]

And thus the paradox: the conclusion noted above must be wrong. "The
often assumed proposition that sexual reproduction always has a short term
disadvantage in relation to asexual is clearly untenable. It is disproved quite
simply by the existence of populations that continue to reproduce both sexually and asexually. In them there can be no net disadvantage to the sexual
process. By implication, sexual reproduction must have a (more or less)
two-fold advantage that balances the cost of meiosis."[11] The problem, however, is that no one has been able to suggest any immediate advantage that is
entirely satisfactory.[12]

Even current explanations of the long-term benefits to sexual reproduction
have been questioned. For example, Steven Stanley of Johns Hopkins University recently challenged the implicit assumption in the Weismannian interpretation that asexual species are adaptively inferior because they are unable to
evolve in any significant way.[13] In fact, he noted, many asexual species are
neither less well adapted, less variable, nor shorter-lived in geological time

than sexual species. "Contrary to most views," Suomalainen et al. noted in 1976, "parthenogenetic insects are capable of evolution. There is considerable genetic differentiation within and between populations inhabiting different regions."[14] Such data cast doubt on the basic assumption that asexual species become extinct because of their inability to adapt rapidly to changing environments. Paleontologists such as Eldredge and Gould now argue from fossil evidence that species evolved far more slowly and existed for longer periods of geological time than has been assumed by biologists. Most evolution, they contend, has been associated with the rapid multiplication of species, not with the gradual change of existing species; once formed, species change slowly. According to Stanley, therefore, the ubiquity of sex cannot be accounted for by assuming that sexual forms evolve more rapidly than asexual forms because of the rapid spread of gene mutations and combinations. In his view, sexual reproduction does not accelerate change within existing species; rather, it allows for more rapid speciation. The dominance of sexual reproduction in modern species arises, therefore, "from the fact that for most taxa normal extinction rates can only be balanced by high rates of diversification provided by the frequent formation of divergent new species." "What counts for survival," Stanley concludes, "is not rapid phyletic evolution, but rapid speciation. . . . Most asexual forms are transitory because they cannot speciate in the normal sense of the term."[15]

After studying these modern, and fascinating, controversies over the significance of sexual reproduction, one can only conclude that sex remains almost as complete an enigma today as it was three hundred years ago when those Dutch microscopists first discovered minute "animalcules" swimming about in human seminal fluid. Ninety-nine percent of the world's species reproduce sexually, and we still do not understand how that came to be. "Sexual reproduction," writes George Williams, "is analogous to a roulette game in which the player throws away half his chips at each spin. The game is fair as long as everyone behaves in this way, but if some do and some don't, the ones who keep their chips have an overwhelming advantage and will almost certainly win."[16] In truth, however, in the biological world, such players lose!

Notes

CHAPTER ONE

1. Carl Linnaeus, *Philosophia botanica* (Stockholm, 1751), par. 153. See also James Larson, "Linnaeus and the natural method," *Isis* 58 (1967): 304-20.

2. W. Watson, "An account of a treatise in Latin, entitled "Caroli Linnaei Serenissimae regiae majestatis species plantarum: With remarks," *Gentleman's Mag.* 24 (1754): 558.

3. William Withering, *A botanical arrangement of all the vegetables* (Birmingham, 1776), p. v.

4. For details of the life and work of Linnaeus, see Wilfrid Blunt, *The compleat naturalist* (London: Collins, 1971); Frans A. Stafleu, *Linnaeus and the Linnaeans* (Utrecht: Oosthoek, 1971); James Larson, *Reason and experience* (Berkeley: University of California Press, 1971).

5. Linnaeus, *Philosophia botanica,* chap. 5, sec. 146.

6. Carl Linnaeus, *Genera plantarum* (Leiden, 1737), p. 320.

7. Carl Linnaeus, "Observationes in regnum vegetabile," in *Systema naturae* (Leiden, 1735).

8. Linnaeus to Albrecht von Haller, May 1 and June 8, 1737, in James Smith, *A selection of the correspondence of Linnaeus* (London, 1821), 2: 243 and 254.

9. Linnaeus, *Genera plantarum,* 6th ed. (Leiden, 1764), p. 556.

10. John Hill, *The vegetable system* (London, 1759), pp. 198-205.

11. Johann Hedwig, "Vorlaufige Anzeige meiner Beobachtungen von den wahren Geschlectstheilen der Moose" (1761), in *Sammlung seiner zerstreuten Abhandlungen und Beobachtungen* (Leipzig, 1793).

12. Quotation from J. Reynolds Green, *A history of botany in the United Kingdom from the earliest times to the end of the nineteenth century* (London: J. M. Dent, 1914), p. 230.

13. Erasmus Darwin, *The botanic garden,* pt. 2, *The loves of the plants* (London, 1791).

14. Colin Milne, *Institutes of botany* (London, 1771), p. 198.

15. Pitton Tournefort, *Elémens de botanique* (Paris, 1694), pp. 22, 439.

16. Michel Adanson, *Familles des plantes* (Paris, 1763), p. cliv.

17. Antoine Laurent de Jussieu, *Genera plantarum* (Paris, 1789), p. 2.

18. Joseph Gaertner, *De fructibus et seminibus plantarum* (Tübingen, 1788).

19. Augustin-Pyramus de Candolle, *Théorie élémentaire de la botanique* (Paris, 1813). The word "cellular" should not be taken in the modern sense. It implied a tissue enclosing visible enclosed spaces.

20. Achille Richard, *Nouveau eléménts de botanique et de physiologie végétale*, 4th ed. (Paris, 1828), p. 400.

21. Adrien de Jussieu, *Cours élémentaire d'histoire naturelle* (Paris, 1843), p. 365.

22. Reynolds Green, *A history of botany*, p. 318.

23. John Lindley, *A synopsis of the British flora* (London, 1829), pp. vii–xi.

24. John Lindley, *An introduction to the natural system of botany* (London, 1830), p. xvii.

25. John Lindley, *The vegetable kingdom* (London, 1845), p. 5.

26. Ibid., p. 51.

27. William J. Hooker, *Flora Scotia* (London, 1821), p. 3.

28. William Harvey, *Anatomical exercises on the generation of animals*, in *The works of William Harvey*, trans. R. Willis (London: Sydenham Society, 1847), p. 462. A more extensive discussion of eighteenth-century theories of generation than is given in this chapter is presented in Jacques Roger, *Les Sciences de la vie dans la pensée française du XVIIIe siècle* (Paris: A. Colin, 1971), and Shirley Roe, *Matter, life, and generation* (Cambridge: Cambridge University Press, 1981). See also Elizabeth Gasking, *Investigations into generation, 1651–1828* (Baltimore: Johns Hopkins Press, 1967), and John Farley *The spontaneous generation controversy from Descartes to Oparin* (Baltimore: Johns Hopkins University Press, 1977). The quotation is taken from Roger *Les Sciences de la vie*, p. 267.

29. Regnier de Graaf, "On the female testes or ovaries," trans. G. W. Corner, in *Essays in biology in honour of Herbert Evans*, ed. G. W. Bartelmez (Berkeley: University of California Press, 1943), p. 128.

30. *J. Sçavans*; quotation from Roger, *Les sciences de la vie*, p. 207.

31. George Garden, "A discourse concerning the modern theory of generation," *Phil. Trans. Roy. Soc.*, 1691, p. 476.

32. Nicolas Malebranche, *De la recherche de la vérité* (Paris, 1673), bk. 1, chap. 6.

33. William Harvey, *On conception*, in *The works of William Harvey*, p. 575; *The embryological treatises of Hieronymous Fabricius of Aquapendente*, trans. H. B. Adelman (Ithaca, N.Y.: Cornell University Press, 1942), p. 190.

34. Jan Swammerdam, *Historia insectorum generalis* (1685); quoted in Iris Sandler, "The re-examination of Spallanzani's interpretation of the role of the spermatic animalcules in fertilization," *J. Hist. Biol.* 6 (1973): 196.

35. Antoni van Leeuwenhoek, "The observations of Mr. Antoni Leeuwenhoek on animalcules engendered in the semen," in *The collected letters of Antoni van Leeuwenhoek*, vol. 2 (Amsterdam: Swets and Zeitlinger, 1941), p. 279. The letter was first published in *Phil. Trans. Roy. Soc. London*, 1679; for an English translation of this letter, see Francis J. Cole, *Early theories of sexual generation* (Oxford: Clarendon Press, 1930).

36. Leeuwenhoek to N. Grew, April 25, 1679, in *Collected letters*, 3: 19.

37. Leeuwenhoek to Robert Hooke, April 5, 1680, ibid., 3: 205.

38. Leeuwenhoek to N. Grew, March 18, 1678, ibid., 2: 335.

39. Leeuwenhoek to C. Wren, Jan. 22, 1683, ibid., 4: 11.

40. Nicolas Hartsoeker, *Essai de dioptrique* (Paris, 1694), p. 231. The first publicized description of the male animalcules was that by Christiaan Huygens, to whom Hartsoeker had written describing the discovery. Huygens made no mention of Hartsoeker in this first paper (*J. Sçavans*, August 15, 1678), but gave Hartsoeker credit for the discovery two weeks later (ibid., August 29, 1678).

41. Nicholas Andry de Boisregard, *De la génération des vers dans le corps de l'homme* (Paris, 1700); William Derham, *Physico-theology; or, a demonstration of the being and attributes of God from his works of creation* (London, 1721); "Generation," in *Lexicon technicum*, 2nd ed. (London, 1710).

42. Samual Moreland, "Some new observations upon the parts and use of the flower in plants," *Phil. Trans. Roy. Soc.*, 1702, p. 1475.

43. Claude Geoffroy, "Observations sur la structure et l'usage des principales parties des fleurs," *Hist. l'Acad. Royale des Sciences, 1711* (Paris, 1777), p. 296.

44. Hill, *The vegetable system,* p. 108.

45. William Smellie, *The philosophy of natural history* (Edinburgh, 1790), p. 247. Leeuwenhoek explained this waste by referring to the difficulties of the journey: "There cannot be too great a number of adventurers, when there is so great a likelihood to miscarry." This explanation seems to have been accepted until very recent times, despite its absurdity. A bull has been estimated to ejaculate about 1 trillion sperm and man about 350 million, and yet only recently has this vast redundancy become a matter of debate (*Gamete competition in plants and animals,* ed. D. L. Mulcahy (Amsterdam: North-Holland Publishing Co., 1975).

46. *Aristotle's compleat and experienc'd midwife* (London, 1755), p. 14. This series is discussed briefly in John S. and Robin Haller, *The physician and sexuality in Victorian America* (New York: Norton, 1977), p. 93.

47. Laurence Sterne, *The life and opinions of Tristram Shandy, Gentleman* (London, 1760), chap. 2.

48. Linnaeus, *Philosophia botanica,* sec. 137.

49. Ibid., sec. 145.

50. Ibid., sec. 138.

51. Carl Linnaeus, *A dissertation on the sexes of plants,* trans. James Smith (Dublin, 1786), pp. 32-33.

52. Ibid., pp. 54-55.

53. Ibid., p. 23.

54. Pierre-Louis Maupertuis, *Vénus physique,* English trans. Simone Boas (New York: Johnson Reprint Corp., 1966), p. 41.

55. Ibid., p. 56.

56. Georges-Louis de Buffon, *Histoire naturelle, générale, et particulière,* vol. 2, *Histoire générale des animaux* (Paris, 1749), pp. 24-32.

57. Ibid., chap. 8.

58. For details of these examinations and the subsequent rebuff at the hands of Spallanzani, see Sandler, "The re-examination of Spallanzani's interpretation." Much of my account of these events is taken from her excellent paper.

59. Carl Linnaeus, *Generatio ambinega* (Uppsala, 1759), p. 5; quotation from Carlo Castellani, "Spermatozoan biology from Leeuwenhoek to Spallanzani," *J. Hist. Biol.* 6 (1973): 37-68.

60. Lazzaro Spallanzani, *De la génération de quelques animaux amphibies,* in *Oeuvres de Spallanzani,* vol. 3, p. 101. This is the French translation of Spallanzani's *Dissertazione di fisica animale e vegetabile,* vol. 2, pt. 1, published in 1780.

61. Lazzaro Spallanzani, *Opuscules de physique animale et végétale* (Paris, 1777). This is the French translation of Spallanzani's *Opuscoli di fisica animale e vegetabile* (1776). Quotation in Sandler, "The re-examination of Spallanzani's interpretation," p. 210.

62. Lazzaro Spallanzani, *De la fécondation artificielle obtenue dans quelques animaux,* in *Oeuvres de Spallanzani,* vol. 3, p. 206. This is the French translation of Spallanzani's *Dissertazione,* vol. 2, pt. 2.

63. Lazzaro Spallanzani, "Additions au mémoire sur les fécondations artificielles," in *Expériences pour servir à l'histoire de la génération des animaux et des plantes* (Paris, 1785), p. 310. The "Additions" were added by Jean Senebier to this French translation of Spallanzani's *Dissertazione,* vol. 2. They have never appeared in the original Italian edition.

64. Spallanzani, *De la fécondation artificielle,* p. 187.

65. Spallanzani, *Opuscules.* Quotation from Castellani, "Spermatozoan biology," p. 67.

66. Lazzaro Spallanzani, "Mémoire sur la génération de diverses plantes," in *Expériences pour servir à l'histoire de la génération des animaux et des plantes* (Paris, 1785), p. 339.

67. Ibid., p. 344.

68. Ibid., p. 350.

69. Ibid., p. 384.

70. Ibid., p. 410.

71. Tournefort, *Elémens de botanique,* p. 23.

72. Adanson, *Familles des plantes,* p. 121.

73. De Candolle, "Glossologie," in *Théorie élémentaire.*

74. Richard, *Nouveau éléments,* p. 401.

75. Adrien de Jussieu, *Cours élémentaire,* p. 461.

76. Smellie, *Philosophy of natural history,* p. 246.

77. Ibid., p. 255.

78. Caspar Wolff, *Theorie generationis* (Halle, 1759), p. 5. For details of Wolff's work, see Shirley Roe, "Rationalism and embryology: Caspar Friedrich Wolff's theory of epigenesis," *J. Hist. Biol.* 12 (1979): 1-43; and idem, *Matter, life, and generation.*

79. Karl Ernst von Baer, *Über Entwicklungsgeschichte der Thiere* (Königsberg, 1828), pp. 144-45.

80. Lorenz Oken, *Lehrbuch der Naturphilosophie* (Jena, 1809-11), p. 28.

81. For details of Koelreuter's work, see Herbert F. Roberts, *Plant hybridization before Mendel* (Princeton: Princeton University Press, 1929); and Robert Olby, *Origins of Mendelism* (New York: Schocken Books, 1966).

82. Kurt Sprengel, "Proper phytonomy or on the life of plants," in Augustin-Pyramus de Candolle and Kurt Sprengel, *Elements of the philosophy of plants* (Edinburgh, 1821), p. 253.

83. Kurt Sprengel, *An introduction to the study of cryptogamous plants* (London, 1807), p. 58. I am unaware of the source of this peculiar idea.

84. Ibid., pp. 261-62.

85. Quotation from John Merz, *A history of European scientific thought in the nineteenth century* (New York: Dover, 1965), p. 159.

86. For details of the rise of the German university, see ibid., chap. 2; Friedrich Paulsen, *German universities, their character and historical development* (New York, 1895); R. Steven Turner, "The growth of professional research in Prussia, 1818 to 1848: Causes and context," *Hist. Stud. Phys. Sci.* 3 (1971): 137-82. For a delightful insight into the life of a German laboratory student, see Richard Goldschmidt, *The golden age of zoology: Portraits from memory* (Seattle: University Press of Washington, 1956).

87. Turner, "Growth of research," p. 137.

88. Ibid., p. 173.

89. Quoted in David Elliston Allen, *The naturalist in Britain: A social history* (London: Allen Lane, 1976), p. 180.

90. Ibid., p. 182.

91. There are many modern examples of this split, perhaps the most obvious being those endless debates in biology departments over the content of introductory classes. In my own case, I remember vividly a biochemist remarking to me that he found it strange that Darwin was so much honored, "since he wasn't really a scientist at all." That must surely rank as one of the world's most stupid remarks.

CHAPTER TWO

1. Quotation from Savile Bradbury, *The Evolution of the microscope* (Oxford: Pergamon Press, 1967), p. 164.

2. Ibid., pp. 191 and 203.

3. Matthias Schleiden, *Principles of scientific botany* (1842), English trans. E. Lankester (New York: Johnson Reprint Corp., 1969), p. 577.

4. Jan Purkinje, "Contributions to the history of the bird's egg," in *Essays in biology in honour of Herbert Evans,* ed. G. W. Bartelmez (Berkeley: University of California Press, 1943), pp. 62-64. Karl Ernst von Baer, "On the genesis of the ovum of mammals and man," English trans. C. O'Malley, *Isis* 47 (1956): 121-53.

5. Ph. van Tieghem, "Notice sur la vie et les travaux de Jean-Baptiste Dumas," *Mem. Acad.*

Sci. Inst. Fr. 52 (1914): xliii–lxxx; and L. A. Gosse and T. Herpin, "Notice biographique sur le Dr. J.-L. Prevost," *Arch. Sci. Phys. Nat.* 15 (1850): 265–300.

6. Jean-Louis Prevost and Jean-Baptiste Dumas, "Essai sur les animalcules spermatiques de divers animaux," *Mem. Soc. Phys. Hist. Nat. Genève* 1 (1821): 195.

7. Karl Gaertner, "Notice sur les expériences concernant la fécondation de quelques végétaux," *Ann. Sci. Nat.* 10 (1827): 143.

8. Charles Brisseau de Mirbel, *Historie naturelle, générale et particulière des plantes* (Paris, 1806), 1: 52; Alphonse de Candolle, *Introduction à l'étude de la botanique* (Paris, 1835), p. 222; Achille Richard, *Nouveau éléments de botanique et de physiologie végétale,* 4th ed. (Paris, 1828), p. 184.

9. Elizabeth Gasking, *Investigations into generation, 1651–1828* (Baltimore: Johns Hopkins University Press, 1967), p. 143.

10. Jean-Louis Prevost and Jean-Baptiste Dumas, "Nouvelle théorie de la génération," *Ann. Sci. Nat.* 1 (1824): 3.

11. Ibid., pp. 4–7.

12. For details see Colin R. Austin, "Fertilization," in *Concepts of development,* ed. J. Lash and J. Whittaker (New York: Sinauer Associates, 1974).

13. Prevost and Dumas, "Nouvelle théorie," p. 286.

14. Jean-Louis Prevost and Jean-Baptiste Dumas, "Deuxieme mémoire sur la génération," *Ann. Sci. Nat.* 2 (1824): 100–121, 129–49.

15. Jean-Louis Prevost, *"De la génération chez la moule de peintres (Unio pictorum),"* ibid. 7 (1825): 454.

16. Ibid., p. 455.

17. Jean-Baptiste Dumas, "Note de M. Dumas sur la théorie de la génération," ibid. 12 (1827): 451–52.

18. Adolphe Brongniart, "Mémoire sur la génération et le développement de l'embryon dans les végétaux phanérogames," ibid., p. 159.

19. Ibid., p. 258.

20. Ibid., pp. 276–78.

21. Allen Thomson, "Generation," in *The cyclopaedia of anatomy and physiology,* ed. Robert Todd (London, 1836–39), 2: 464.

22. Karl Burdach, *Die Physiologie als Erfahrungswissenschaft,* 2nd ed. (Leipzig, 1832), vol. 1, par. 84. For details of the spontaneous-generation problem, see John Farley, *The spontaneous generation controversy from Descartes à Oparin* (Baltimore: Johns Hopkins University Press, 1977).

23. Johann Blumenbach, *An essay on generation,* English trans. A. Crichton (London, 1792), p. 9.

24. Gustav Valentin, "Über die spermatozoen des Bären," *Nova. Acta. Leopoldina* 19 (1839): 237–44; Friedrich Gerber, *Handbuch der allgemeinen Anatomie des Menschen* (Bern, 1840), p. 210; Christian Ehrenberg, *Die Infusionsthierchen als vollkommene Organismen* (Leipzig, 1838), pp. 465–68; Antoine Dugès, *Traité de physiologie comparée de l'homme et des animaux* (Paris, 1839), 3: 251; Felix Pouchet, *Théorie positive de l'ovulation spontanée et de la fécondation des mammifères et de l'espèce humaine* (Paris, 1847), p. 303.

25. Farley, *The spontaneous generation controversy.*

26. Johannes Müller, *Elements of physiology,* English trans. W. Baley (Philadelphia, 1843), p. 824; Thomson, "Generation," p. 463.

27. Pouchet, *Théorie positive de l'ovulation spontanée,* p. 350; Bory de Saint-Vincent, "Zoospermes," in *Dictionnaire classique d'histoire naturelle* (Paris, 1824), p. 737; Richard Owen, "Entozoa," in *The Cyclopaedia of Anatomy and Physiology,* ed. Todd, 2: 113; Thomson, "Generation," pp. 467, 459.

28. Felix Dujardin, "Sur les zoospermes de mammifères et sur ceux du Couchon de l'Inde en particulier," *Ann. Sci. Nat.* 8 (1837): 291.

29. John Farley, "The spontaneous generation controversy (1700–1860): The origin of parasitic worms," *J. Hist. Biol.* 5 (1972): 95–125.

30. Peter Roget, *Animal and vegetable physiology considered with reference to natural theology* (London, 1839), 2: 582.

31. Matthias Schleiden, "Contributions to Phytogenesis," in Theodor Schwann, *Microscopical researches into the accordance in the structure and growth of animals and plants,* English trans. Henry Smith (London, 1848), pp. 231 and ix.

32. Matthias Schleiden, "Some observations on the development of the organization in phanerogamous plants," *Phil. Mag.* 12 (1838): 244.

33. Louis Figuier, *Histoire des plantes* (Paris, 1865), p. 206.

34. Charles Brisseau de Mirbel, *Complément des observations sur le Marchantia polymorpha* (Paris, 1832), p. 50.

35. Hugo von Mohl, "Über Entwicklung der Sporen von *Anthoceros laevis,*" *Linneae* 13 (1839): 273–290; idem, *Morphologische Betrachtungen über das Sporangium der mit Gefässen versehenen Cryptogamen* (Tübingen, 1837); Franz Meyen, *A report on the progress of vegetable physiology during the year 1837,* English trans. W. Francis (London, 1839), p. 120.

36. William Valentine, "Observations on the development of the theca, and on the sexes of mosses," *Trans. Linn. Soc. London* 17 (1837): 480; see also ibid. 18 (1838): 499–507.

37. Schleiden, "Some observations," p. 244.

38. Schleiden, *Principles of scientific botany,* p. 165.

39. Ibid., pp. 531–32.

40. Ibid., p. 165.

41. Charles Brisseau de Mirbel and Edouard Spach, "Notes pour servir à l'histoire de l'embryogénie végétale," *Ann. Sci. Nat.* 11 (1839): 201.

42. Stephan Endlicher, *Grundzüge einer neuen Theorie der Pflanzenzeugung* (Vienna, 1838), pp. 15–20.

43. Heinrich Wydler, "Note sur la formation de l'embryon," *Ann. Sci. Nat.* 11 (1839): 142.

44. Hermann Schacht, *Entwicklungsgeschichte des Pflanzen-Embryon* (Amsterdam, 1850), p. 52. The corpuscula, described by Robert Brown in 1834, are flask-shaped organs in the "endosperm" of conifers. In modern terms they are the archegonia of the female gametophyte.

45. Hermann Schacht, "Über die Entstehung des Pflanzenkeims," *Flora* 38 (1855): 158.

46. Franz Meyen, *Neues System der Pflanzen-Physiologie* (Berlin, 1839), 3: 318.

47. Wilhelm Hofmeister, "Untersuchungen des Vorgangs bei der Befruchtung der Oenothereen," *Bot. Zeit.* 5 (1847): 786–87.

48. Ibid., p. 790.

49. Wilhelm Hofmeister, *Die Entstehung des Embryo der Phanerogamen* (Leipzig, 1849), p. 59.

50. Auguste de Saint-Hilaire, *Leçons de botanique comprenant principalement la morphologie végétale* (Paris, 1840), pp. 581–83.

51. De Mirbel and Spach, "Notes," p. 212.

52. Müller, *Elements of physiology,* pp. 815–25.

53. Schacht, "Über die Entstehung," p. 161.

54. Albert von Kölliker, *Beiträge zur Kenntniss der Geschlectsverhältnisse und der Samenflüssigkeit wirbelloser Thiere, nebst einem Versuch über Wesen und die Bedeutung der sogenannten Samenthiere* (Berlin, 1841), p. 64.

55. Ibid., pp. 72–75.

56. Ibid., pp. 78–79.

57. Ibid., p. 82.

58. Rudolph Wagner, *Lehrbuch der speciellen Physiologie* (Leipzig, 1842), p. 28. Similar views were also expressed in his earlier paper "Fragmente zur Physiologie der Zeugung," *Abh. Math. Phys. Classe K. Bayerischen Akad. Wiss.* 2 (1837): 381–414.

59. Müller, *Elements of physiology,* p. 837.

60. Von Kölliker, *Beiträge,* p. 84.

61. Theodor Bischoff, *Entwicklungsgeschichte des Kaninchen-Eies* (Brunswick, 1842), pp. 29, 33.

62. Martin Barry, "Researches in embryology, second series," *Phil. Trans. Roy Soc.,* 1839, p.315.

63. Martin Barry, "Researches in embryology, third series: A contribution to the physiology of cells," ibid., 1840, p. 555.

64. Martin Barry, "On the unity of structure in the animal kingdom," *Edinburgh New Phil. J.* 22 (1837): 116–41. For a biography of Barry, see J. B., "Memoir of the late Martin Barry," *Edinburgh Med. J.* 1 (1856): 81–91.

65. Martin Barry, "Spermatozoa observed within the mammiferous ovum," *Phil. Trans. Roy. Soc.,* 1843, p. 33.

66. Martin Barry, *Phil. Mag.* 24 (1844): 44.

67. Martin Barry, "On fibre," *Phil. Trans. Roy. Soc.,* 1842, pp. 89–135; idem, "On the corpuscles of the blood," ibid., 1841, pp. 217–68; Bischoff's reply, *Phil. Mag.* 24 (1844): 281–85.

68. Theodor Bischoff, *Entwicklungsgeschichte des Hunde-Eies* (Brunswick, 1845), p. 119.

69. Theodor Bischoff, "Theorie der Befruchtung und über die Rolle, welche die Spermatozoiden dabei spielen," *Arch. Anat. Physiol. wiss. Med.,* 1847, pp. 426–27.

70. Justus von Liebig, *Chemistry in its application to agriculture and physiology* (1840), English trans. L. Playfair (Cambridge, 1843), p. 295.

71. Ibid., p. 389.

72. Bischoff, "Theorie der Befruchtung," p. 431.

73. Ibid., pp. 433–35.

74. Rudolph Wagner and Rudolf Leuckart, "Semen," in *The cyclopaedia of anatomy and physiology,* ed. Todd, 4: 472–508; Jean-Louis de Quatrefages, "Recherches expérimentales sur les spermatozoides des hermelles de Tarets," *Ann, Sci. Nat.* 13 (1850): 111–25.

75. Bischoff, "Theorie der Befruchtung," p. 437.

76. Theodor Bischoff, *Entwicklungsgeschichte des Meerschweinchens* (Giessen, 1852), p. 14.

77. Ibid., p. 15.

78. George Newport, "On the impregnation of the ovum in the Amphibia," *Phil. Trans. Roy. Soc.,* 1851, pp. 221–23.

79. Ibid., p. 232.

80. Ibid., p. 242.

81. Henry Nelson, "The reproduction of *Ascaris mystax,*" ibid., 1852, p. 578.

82. Ibid., p. 586.

83. George Newport, "On the impregnation of the ovum in the Amphibia and on the direct agency of the spermatozoon," ibid., 1853, p. 270.

84. Theodor Bischoff, *Bestätigung des von Dr. Newport bei den Betrachiern und Dr. Barry bei den kanichen behaupteten eindringens der Spermatozoiden in das Ei* (Giessen, 1854), p. 3.

85. Ibid., p. 10.

86. Rudolf Leuckart, "Zeugung," in Rudolph Wagner, *Handwörterbuch der Physiologie,* vol. 4 (Brunswick, 1853), p. 960.

87. Jacob G. Agardh, "Observations sur la propagation des Algues," *Ann. Sci. Nat.* 6 (1836): 193–212.

88. Gustav Thuret, "Note sur l'anthère du Chara et les animalcules qu'elle renferme," ibid., 14 (1840): 65–72. *Chara,* a stonewort found at the bottom of clear lakes and streams, differs from most fresh-water algae in that it has a branching axis on which whorls of "leaves" arise.

89. Joseph Decaisne and Gustav Thuret, "Recherches sur les anthéridies et les spores de quelque *Fucus,*" ibid. 3 (1844): 5–15.

90. Gustav Thuret, "Recherches sur les zoospores des Algues et les anthéridies des Cryptogames," ibid. 14 (1850): 214–60; 16 (1851): 5–39.

91. Gustav Thuret, "Note sur la fécondation des Fucacées, "*Comptes-rendus Acad. Sci.* 26 (1853): 745.

92. Gustav Thuret, "Recherches sur la fécondation des Fucacées, suivies d'observations sur les anthéridies des Algues," *Ann. Sci. Nat.*, n.s. 2 (1854): 197–214; n.s. 3 (1855): 5–28.

93. Nathaneal Pringsheim, "Über die Befruchtung der Algen," *Bericht königl. Preuss. Akad. Wiss. Berlin*, 1855, p. 134.

94. Ibid., p. 144.

95. Ibid., p. 147.

96. Schacht, "Über die Entstehung," p. 162.

97. Hermann Schacht, "Über die Befruchtung der *Pedicularis silvatica,*" *Flora* 38 (1855): 472–73.

98. Hermann Schacht, "Den Vorgang der Befruchtung bei *Gladiolus segetum,*" *Monats. königl. Preuss. Akad. Wiss. Berlin*, 1856, p. 266. Schleiden admitted his previous error in the fourth edition of *Grundzüge*, published in 1856.

99. Julius von Sachs, *Lectures on the physiology of plants*, English trans. H. M. Ward (Oxford, 1887), p. 755.

100. William Carpenter, "Presidential address to the Microscopical Society of London, February, 1856," *Quart. J. Micro. Sci.* 4 (1856): 26.

101. Rudolf Leuckart, "Über die Micropyle und den feinern Bau der Schalalenhaut bei den Insekteneiern," *Arch. Anat. Physiol. wiss. Med.*, (1855), p. 244.

102. Austin Flint, *The physiology of man* (New York, 1875), 5: 290.

103. M. Wichura (1865), quoted in Hans Stubbe, *History of genetics* (Cambridge, Mass.: M.I.T. Press, 1972), p. 126. For details of plant hybridization studies, see Herbert F. Roberts, *Plant hybridization before Mendel* (Princeton: Princeton University Press, 1929).

104. Gregor Mendel, "Experiments in plant-hybridisation," English trans. in William Bateson, *Mendel's principles of heredity* (Cambridge: Cambridge University Press, 1930), p. 373.

105. Newport, "Impregnation of the ovum and direct agency," p. 282.

106. Ibid., pp. 288–89.

107. Georg Meissner, "Beobachtungen über das Erindringen der Samenelemente in der Dotter," *Zeit. wiss. Zool.* 6 (1855): 260–61.

108. Thomson, "Ovum," in *The cyclopaedia of anatomy and physiology*, ed. Todd, 5: 5.

109. William Kirkes, *Handbook of physiology* (London, 1848), p. 610; 3rd ed. (London. 1856), p. 648; 8th ed. (Philadelphia, 1873); 10th ed. (Philadelphia, 1881).

110. Wilhelm His, *Unsere Körperform und das physiologische Problem ihrer Entstehung* (Leipzig, 1874).

111. William Coleman, "Cell, nucleus, and inheritance: An historical study," *Proc. Amer. Phil. Soc.* 109 (1965): 134.

112. Leuckart, "Über die Micropyle," pp. 252–53.

CHAPTER THREE

1. Mary Winsor, "Barnacle larvae in the nineteenth century: A case study in taxonomic theory," *J. Hist. Med.* 24 (1969): 294–309.

2. For details of the life of Steenstrup, see C. Lützen, "Steenstrup," *Natural science: A monthly review of scientific progress* 11 (1897): 159–69; and C. F. Bricka, *Dansk biografisk lexikon, 1537–1814* (Copenhagen, 1902), pp. 326–41. Bricka's *Lexikon* includes also a biography of Johannes C. H. Reinhardt (pp. 606–9). See also P. L. Möller, "The life of H. C. Oersted," in Hans Christian Oersted, *The soul in nature*, English trans. L. and J. Horner (London, 1852).

3. J. Japetus Steenstrup, *On the alternation of generations: or, the propagation and de-*

velopment of animals through alternate generations, English trans. G. Busk (London: The Ray Society, 1845), p. 1.

4. Ibid., p. 4.

5. Ibid., p. 106.

6. In the campanularians a gonangium buds off numerous gonophores that remain attached to produce the gametes. In other forms the gonangium buds off gonophores that change into free-swimming medusae that eventually produce the gametes.

7. Steenstrup, *On the alternation of generations,* pp. 106-9.

8. Ibid., p. 109.

9. Ibid., pp. 113-14.

10. Ibid., p. 115.

11. Carl Linnaeus, *A dissertation on the sexes of plants,* English trans. James Smith (Dublin, 1786), p. 20-23.

12. Michel Guédès, "La théorie de la métamophose en morphologie végétale: Des origines à Goethe et Batsch," *Rev. Hist. Sci.* 22 (1969): 355.

13. Allen Thomson, "Ovum," in *The cyclopaedia of anatomy and physiology,* ed. Robert Todd (London, 1836-39), 5 (supp.): 1-80.

14. For details of this work, see Mary Winsor, *Starfish, jellyfish, and the order of life* (New Haven: Yale University Press, 1976), chap. 5.

15. Matthias Schleiden, *Principles of scientific botany* (1842), English trans. E. Lankester (New York: Johnson Reprint Corp., 1969), p. 102. This translation, a facsimile of the 1849 London edition, contains no translation of Schleiden's "Methodological introduction."

16. Matthias Schleiden, *Grundzüge der wissenschaftlichen Botanik* (Leipzig, 1842), preface.

17. Ibid., p. 165.

18. Ibid., pp. 185, 196.

19. Ibid., pp. 203-10.

20. Quotation from Karl von Goebel, *Wilhelm Hofmeister: The work and life of a nineteenth century botanist* (London: The Ray Society, 1926), p. 6.

21. Schleiden, *Principles of scientific botany,* p. 559.

22. Theodor Schwann, *Microscopical researches into the accordance in the structure and growth of animals and plants,* English trans. H. Smith (London, 1847), p. 39. This text also includes an English translation of Schleiden's *Beiträge zur Phytogenesis* (pp. 231-63).

23. Carl Eugen von Mercklin, *Beobachtungen an dem Prothallium der Farrnkräuter* (St. Petersburg, 1850), pp. 62-63.

24. Wilhelm Hofmeister, review of Mercklin's text, *Flora* 33 (1850): 696-701.

25. Hugo von Mohl, *Principles of the anatomy and physiology of the vegetable cell* (London, 1852).

26. Carl Nägeli, "On the nuclei, formation, and growth of vegetable cells," in *Reports and papers on botany* (London: Ray Society, 1849), p. 117.

27. Ibid., p. 122.

28. Wilhelm Hofmeister, "Untersuchungen des Vorgangs bei der Befruchtung der Oenothereen," *Bot. Zeit.* 5 (1847): 786.

29. Wilhelm Hofmeister, "Über die Entwicklung des Pollens," ibid. 6 (1848): 426.

30. Wilhelm Hofmeister, *Die Entstehung des Embryo der Phanerogamen* (Leipzig, 1849), p. 58.

31. Carl Nägeli, "Bewegliche Spiralfaden (Saamenfaden?) an Farren," *Zeit. wiss. Bot.* 1 (1844): 180-184.

32. Karl Müller, "Zur Entwickelungsgeschichte der Lycopodiaceen," *Bot. Zeit.* 4 (1846): 687.

33. Carl Nägeli, "Über die Fortpflanzung der Rhizocarpeen," *Zeit. wiss. Bot.* 3-4 (1846): 188-206.

34. Le Comte Leszczyc-Suminski, "Sur le développement des Fougères," *Ann. Sci. Nat.* 11 (1849): 121, 124-25.

35. Hofmeister, "Über die Fruchtbildung und Keimung der höheren Cryptogamen," *Bot. Zeit.* 7 (1849): 793-800; English translation in *Bot. Gaz.* (London) 2 (1850): 70-76.

36. Ibid. (English translation), p. 75.

37. George Mettenius, "Zur Fortpflanzung der Gefässkryptogamen," *Beiträge Bot.* 1 (1850): 23.

38. Alexander Braun, "The phenomenon of rejuvenescence in Nature, especially in the life and development of plants," in *Botanical and physiological memoirs*, ed. A. Henfrey (New York: Johnson Reprint Corp., 1971), p. 27.

39. Ibid., p. 51*n*.

40. Hofmeister, review of Mercklin's text, *Flora* 33 (1850): 700.

41. Arthur Henfrey, "On the reproduction and supposed existence of sexual organs in the higher cryptogamous plants," *Report of the Brit. Assoc. Adv. Sci.*, 1851, p. 120.

42. Review, *Flora* 34 (1851): 765-70.

43. Wilhelm Hofmeister, *Vergleichende Untersuchungen der Keimung, Entfaltung und Fruchtbildung hoherer Kryptogamen und der Samenbildung der Coniferen* (Leipzig, 1851), p. 139.

44. Ibid., p. 139.

45. Wilhelm Hofmeister, *On the germination, development, and fructification of the higher Cryptogamia, and the fructification of the Coniferae* (London: Ray Society, 1862), pp. 439-41.

46. Richard Owen, *Lectures on the comparative anatomy and physiology of the invertebrate animals* (London, 1843), p. 233.

47. Richard Owen, *On parthenogenesis; or, the successful production of procreating individuals from a single ovum* (London, 1849), p. 36.

48. Ibid., p. 11.

49. Ibid., pp. 14-16.

50. Ibid., p. 69.

51. Ibid., p. 56.

52. Ibid., pp. 72-75.

53. Richard Owen, "Metamorphosis and Metagenesis," *Edinburgh New Phil. J.* 50 (1851): 271.

54. Carl Theodor von Siebold, *On a true parthenogenesis in moths and bees: A contribution to the history of reproduction in animals*, English trans. W. S. Dallas (London, 1857), p. 10.

55. An exhaustive account of these events, together with a complete bibliography, is given by O. Taschenberg in "Historische Entwickelung der Lehre von der Parthenogenesis," *Abh. Nat. Gesell. Halle* 17 (1892): 367-453. My brief account is taken from this paper. A lengthier discussion can be found in Frederick Churchill, "Sex and the single organism: Biological theories of sexuality in mid-nineteenth century," *Stud. Hist. Biol.* 3 (1979): 139-78.

56. Von Siebold, "True parthenogenesis," p. 53.

57. Ibid., p. 106.

58. Ludwig Radlkofer, "On true parthenogenesis in plants," *Ann. Mag. Nat. Hist.* 20 (1857): 210.

59. Von Siebold, "True parthenogenesis," p. 10.

60. Thomas H. Huxley, "On the agamic reproduction and morphology of aphids," *Trans. Linn. Soc. London* 22 (1857): 210.

61. Ibid., p. 211.

62. Huxley, "Agamic reproduction," p. 219.

63. John Lubbock, "An account of two methods of reproduction in *Daphnia,* and of the structure of the ephippium," *Phil. Trans. Roy. Soc.*, 1857, p. 99.

64. John Lubbock, "On the ova and pseudova of insects," ibid., 1859, p. 367.

65. Rudolf Leuckart, *Zur Kenntniss des Generationswechsel und der Parthenogenesis bei den Insekten* (Frankfurt, 1858), p. 111.

66. Rudolph Wagner, "Nachtrag zu den voranstehenden Artikel "Zeugung" vom herausgeber," in *Handwörterbuch der Physiologie* (Brunswick, 1853), pp. 1004-13; idem, "*Nachtrag zum Nachtrag,*" ibid., pp. 1018a-d.

67. For details see William Coleman, "Bergmann's rule: Animal heat as a biological phenomenon," *Stud. Hist. Biol.* 3 (1979): 81.

68. Quotations from Churchill, "Sex and the single organism," p. 165.

69. Charles Darwin, "Notebooks on transmutation of species," ed. Gavin de Beer, *Bull. Brit. Mus. Nat. Hist.*, Historical Series, vol. 2, no. 2 (London, 1960), p. 41.

70. Third notebook, ibid., vol. 2, no. 4, p. 147.

71. Fourth notebook, ibid., p. 164.

72. Ibid., p. 179.

73. Charles Darwin, "On the two forms, or dimorphic condition, in the species of *Primula*, and on their remarkable sexual relations," in *Collected papers of Charles Darwin*, ed. P. Barrett (Chicago: University of Chicago Press, 1977), 2: 61.

74. Charles Darwin, *On the origin of species by means of natural selection*, 1st ed. facsim. (Cambridge, Mass.: Harvard University Press, 1964), p. 10.

75. Charles Darwin, *On the Origin of species by means of natural selection*, 6th ed. (London, 1872), p. 48.

76. Charles Darwin, *The effects of cross and self fertilization in the vegetable kingdom* (London, 1876), p. 27.

77. Charles Darwin, "On the possibility of all organic beings occasionally crossing . . .", in *Charles Darwin's Natural Selection, being the second part of his big species book written from 1856 to 1858*, ed. R. C. Stauffer (Cambridge: Cambridge University Press, 1975), p. 90-91.

78. Ibid., p. 74.

79. Darwin, *Variation of plants and animals*, p. 383.

CHAPTER FOUR

1. Charles Darwin, *The variation of plants and animals under domestication* (London, 1868), p. 359. In the second edition (1885), Darwin repeated this statement, adding that Huxley's pseudo-ova were identical to ova.

2. Ibid., p. 383.

3. Bernard Mandeville, *The fable of the bees*, facs. ed. (London: Oxford University Press, 1966), 2: 284.

4. Rudolf Leuckart, *Über den Polymorphismus der Individuen oder die Erscheinungen der Arbeitstheilung in der Natur* (Giessen, 1851), p. 34.

5. Henri Milne-Edwards, *Eléments de zoologie* (Paris, 1834), p. 210.

6. Henri Milne-Edwards, *Leçons sur la physiologie et l'anatomie comparée de l'homme et des animaux*, vol. 8 (Paris, 1863), p. 364.

7. Friedrich Engels, *The origin of the family, private property, and the state* (Chicago: C. H. Kerr, 1902), p. 79.

8. E. Lynn Linton, "What is women's work?" in *The girl of the period and other social essays* (London, 1883), p. 38.

9. James Weir, "The effect of female suffrage on posterity," *Amer. Nat.* 29 (1895): 815, 821.

10. M. Holbrook (1870), quoted in Caroll Smith-Rosenberg and Charles Rosenberg, "The female animal: Medical and biological views of woman and her role in nineteenth-century America," *J. Amer. Hist.* 60 (1973): 335.

11. William Brooks, *The law of heredity* (Baltimore, 1883), pp. 10-13.

12. Ibid., p. 69.
13. Ibid., pp. 82-84.
14. William Brooks, "The condition of women from a zoölogical point of view," *Pop. Sci. Monthly* 15 (1879): 151.
15. Ibid., pp. 154-55.
16. Ibid., p. 349.
17. Herbert Spencer, *The principles of biology* (New York, 1871), pp. 219-21.
18. Sarah Ellis, *The daughters of England* (New York, 1844), p. 101.
19. Edward J. Tilt, *Elements of health and principles of female hygiene* (Philadelphia, 1853), p. 15.
20. Isabella Beeton, *Beeton's Book of Household Management* (London, 1859), pt. 1, p. 19.
21. E. Lynn Linton, "Modern mothers," in *girl of the period*, p. 25.
22. E. Lynn Linton, "The girl of the period," ibid., pp. 2-9.
23. Robert E. Sencourt, *The life of George Meredith* (London: Chapman & Hall, 1929), p. 65. Quoted in Duncan Crow, *The Victorian woman* (London: George Allen, 1971).
24. Crow, *Victorian woman*, p. 25.
25. Annie Besant, *An autobiography* (Philadelphia, 1894), pp. 70-71.
26. Fraser Harrison, *The dark angel: Aspects of Victorian sexuality* (London: Sheldon Press, 1977), pp. 50-55.
27. Tilt, *Female hygiene*, p. 15.
28. Thomas Hardy, *Tess of the d'Urbervilles* (London, 1891), chap. 11.
29. Henry Fielding, *The history of Tom Jones* (London, 1749), bk. 6, chap. 1.
30. Ibid., bk. 5, chap. 5.
31. Quoted in Norah Gourlie, *The prince of botanists: Carl Linnaeus* (London: H. F. & G. Witherby, 1953), pp. 29-30.
32. Peter Webb, *The erotic arts* (New York: Graphic Society, 1975), p. 190.
33. Warren Roberts, *Morality and social class in eighteenth century French literature and painting* (Toronto: University of Toronto Press, 1974), pp. 21, 24.
34. Ibid., p. 120.
35. Theodore Zeldin, *France, 1848-1945: Ambition and love* (Oxford: Oxford University Press, 1979), pp. 297-99.
36. Gustave Flaubert, *Madame Bovary* (Paris, 1857); the trial proceedings are included (pp. 389-461).
37. Ibid., p. 177.
38. Ibid., p. 397.
39. Ibid., p. 408.
40. *Aristotle's master piece*, in *The works of Aristotle, the famous philosopher, in four parts* (New England, 1821), p. 26.
41. *Aristotle's compleat and experienc'd midwife*, in *The works of Aristotle*, p. 101.
42. *Aristotle's book of problems*, in *The works of Aristotle*, p. 225.
43. Quotations from John S. and Robin Haller, *The physician and sexuality in Victorian America* (New York: W. W. Norton, 1974), p. 201. The authors discuss these and other male gadgets, etc.
44. François Lallemand, *A practical treatise on the causes, symptoms, and treatment of spermatorrhea*, trans. H. McDougall (Philadelphia, 1848), p. xiii.
45. George Dangerfield, "The symptoms, pathology, causes, and treatment of spermatorrhea," *Lancet* 1 (1843): 211.
46. Lallemand, *Practical treatise*, p. 42.
47. François Lallemand, "Observations sur l'origine et le mode de développement des zoospermes," *Ann. Sci. Nat.* 15 (1841): 36-37.
48. Ibid., pp. 67-68.

49. François Lallemand, "Observations sur le rôle des zoospermes dans la génération," ibid., pp. 281-82.

50. William Acton, *The functions and disorders of the reproductive organs* (Philadelphia, 1865), p. 101.

51. Ibid., p. 113.

52. Ibid., p. 192.

53. Quotation from Bryan Strong, "Ideas of the early sex education movement in America, 1890-1920," *Hist. Educ. Quart.* 12 (1972): 130.

54. Acton, *Functions and disorders*, p. 118.

55. Ibid., p. 134.

56. Herbert Spencer, "A theory of population, deduced from the general law of animal fertility," *Westminster Rev.* 1 (1852): 477.

57. Ibid., pp. 498-99.

58. For an absorbing account of the "dark" side of Victorian life, see Steven Marcus, *The other Victorians* (New York: Basic Books, 1974). Howard Kushner, in his "Nineteenth century sexuality and the sexual revolution of the Progressive Era," *Can. Rev. Amer. Stud.* 9 (1978): 34-49, has argued that the standard view of nineteenth-century sexuality is distorted. Outpourings of warnings about sex, he argues, do not prove that Americans led repressed lives. His point is well taken, but behavior and beliefs are not necessarily related. The Victorians, bombarded on all sides by warnings about the evils and dangers of sex, could not possibly have believed otherwise, whatever their social behavior may have been.

CHAPTER FIVE

1. Robert J. Harvey-Gibson, *Outlines of the history of botany* (London: A. and C. Black, 1919), p. 137.

2. Richard Owen, *Report on the archetype and homologies of the vertebrate skeleton* (London, 1848), p. 175. For a more detailed examination of the morphological tradition, see Edward S. Russell, *Form and function* (London: J. Murray, 1916).

3. Louis Agassiz, *Essay on classification* (Cambridge, Mass.: Harvard University Press, 1962), pp. 19-21. The essay was first published in the United States in 1857 as part of Agassiz's *Contributions to the natural history of the United States*.

4. Charles Darwin, *On the origin of species*, 1st ed. facsim. (Cambridge, Mass.: Harvard University Press, 1966), pp. 435-37.

5. Quoted in William Coleman, "Morphology between type concept and descent theory," *J. Hist. Med.* 31 (1976): 162.

6. Julius von Sachs, *Textbook of botany, morphological and physiological,* English trans. A. W. Bennett and W. T. Thiselton-Dyer (Oxford, 1875), p. 831. This is a translation of the third German edition, published in 1873.

7. Ibid., p. 835.

8. Ibid., pp. 130-31.

9. Ibid., p. 842.

10. Ibid., p. 205.

11. Quotation from Russell, *Form and function*, p. 93.

12. Ernst Haeckel, *Generelle Morphologie der Organismen* (Berlin, 1866), 1: 250-51, 266.

13. Ibid., p. 266.

14. Ibid., p. 334.

15. Ibid., 2: 99-104.

16. Ibid., p. 88.

17. Ibid., pp. 105, 107.

18. Alfred Kirchoff, "Zur Lehre vom Generationswechsel im Pflanzenreich und den organologischen Analogieen der phanerogamischen und kryptogamischen Bluthe," *Bot. Zeit.* 25 (1867): 329.

19. Ibid., p. 330.

20. Ibid., p. 340.

21. Ladislav Celakovsky, "Über die allgemeine Entwickelungsgeschichte des Pflanzenreiches," *Sitz. königl. böhm. Gesell. Wiss. Prague,* 1868, p. 54. For Celakovsky's role in the Bohemian Darwinian debate, see B. Matouskova, "The beginnings of Darwinism in Bohemia," *Folia biologica* 5 (1959): 169–85.

22. Celakovsky, "Über die allgemeine Entwickelungsgeschichte," p. 51.

23. Ibid., p. 52–53.

24. Julius von Sachs, *Lehrbuch der Botanik* (Leipzig, 1868), p. 187.

25. Ibid., pp. 189–90.

26. Ibid., p. 190.

27. Von Sachs, *Textbook of botany* (trans. Bennett and Thiselton-Dyer), p. 202.

28. Ibid., p. 204.

29. Ibid., p. 205.

30. Ibid., p. 335.

31. Ibid., p. 422.

32. Ibid., p. 423.

33. Ibid., p. 202.

34. Ibid., p. 205.

35. Julius von Sachs, *Textbook of botany, morphological and physiological,* 2nd ed., trans. and ed. S. H. Vines (Oxford, 1882), p. 229. This translation is an updated version of the fourth German edition, published in 1874.

36. Ibid., pp. 232–33.

37. Ibid., p. 228.

38. Nathaneal Pringsheim, "Über die Befruchtung und den Generationswechsel der Algen," *Bericht königl. Preuss. Akad. Wiss. Berlin,* 1856, pp. 225–37.

39. Von Sachs, *Textbook of botany* (trans. and ed. Vines), p. 233.

40. Ladislav Celakovsky, "Über die verschiedenen Formen und die Bedeutung des Generationswechsel der Pflanzen," *Sitz. königl. böhm. Gesell. Wiss. Prague* 2 (1874): 22.

41. Ibid., p. 30.

42. Ibid., p. 31.

43. Ibid., p. 32.

44. Ibid., p. 34.

45. Ibid., p. 38.

46. William Farlow, "An asexual growth from the prothallus of *Pteris serrulata,*" *Quart. J. Micro. Sci.* 14 (1874): 266–72.

47. Nathaneal Pringsheim, "Über den Generationswechsel der Thallophyten und seinen Anschluss an den Generationswechsel der Moose," *Monats. königl. Preuss. Akad. Wiss. Berlin,* 1876, p. 870.

48. Nathaneal Pringsheim, "Über Sprossung der Moosefrucht und den Generationswechsel der Thallophyten," *Jahr. wiss. Bot.* 11 (1878): 4.

49. John Lindley, *A synopsis of the British flora* (London, 1829), p. x.

50. John Lindley, *A synopsis of the British flora,* 2nd ed. (London, 1835), p. vi.

51. John Lindley, *The vegetable kingdom,* 2nd ed. (London, 1847), p. 5.

52. John Lindley, *The vegetable kingdom,* 3rd ed. (London, 1853), p. 53.

53. Miles Berkeley, *Introduction to cryptogamic botany* (London, 1857), p. 558.

54. Quotation from J. Reynolds Green, *A history of botany in the United Kingdom from the earliest times to the end of the nineteenth century* (London: J. M. Dent, 1914), p. 401. The role of

Kew Gardens has been discussed more fully in Lucile Brockway, *Science and colonial expansion* (New York: Academic Press, 1979).

55. Quotation from Frederick Bower, "Kew and the empire," *London Observer*, January 22, 1931.

56. Gerald Geison, *Michael Foster and the Cambridge school of physiology* (Princeton: Princeton University Press, 1978), p. 26.

57. William Thiselton-Dyer, "Plant biology in the seventies," *Nature* 115 (1925): 710.

58. Ibid.

59. *Encyclopaedia Britannica*, 9th ed., s.v. "botany."

60. Quotation from Geison, *Michael Foster*, p. 132. Geison presents and excellent account of the introduction of the new laboratory science into Britain. For the botanical picture, see J. Reynolds Green, "The coming of the laboratory," in *History of botany in the United Kingdom*, p. 528.

61. Quotations from *Memorials: Journal and botanical correspondence of Charles C. Babington* (Cambridge, 1897), pp. xxxii and lxxvi.

62. This account of Bower's life is taken from Frederick O. Bower, *Sixty years of botany in Britain, 1875-1935* (London: Macmillan and Co., 1938). This is an excellent and very readable account of the entry of the new botany into Britain.

63. Ibid., p. 24.

64. List of students attending botany classes in South Kensington, Bower Collection, University of Glasgow Archives, catalog F1.

65. Class register for botany lectures, ibid., catalog F3.

66. Frederick O. Bower, *Address at the opening of the medical session in the University of Glasgow, Oct. 27th, 1885* (Glasgow, 1885).

67. Frederick O. Bower, "Oldest science: Early days of botany in Glasgow," *Glasgow Herald*, August 10, 12, and 20, 1926; idem, "The University of Glasgow: Memories of 40 years," ibid., October 22-November 1, 1927, both in Bower Collection, catalog F20.

68. Frederick O. Bower, "On antithetic as distinct from homologous alternation of generations in plants," *Ann. Bot.* 4 (1890): 347-70.

69. Quotation from David Allen, *The naturalist in Britain* (London: Allen Lane, 1976), p. 116.

70. Peter Roget, *Animal and vegetable physiology, considered with reference to natural theology* (London, 1839), 1: 25.

71. Charles Darwin, *On the origin of species*, facsim. ed. (Cambridge, Mass.: Harvard University Press, 1966), p. 3.

72. Herbert Spencer, *The principles of biology* (London, 1866), bk. 2, p. 4.

73. Karl von Goebel, *Organography of plants*, English trans. I. B. Balfour (Oxford: Clarendon Press, 1900), p. v.

74. Bower, "On antithetic ... generations," p. 348.

75. Celakovsky, "Über die verschiedenen Formen," p. 47.

76. Frederick O. Bower, "On apospory and allied phenomenona," *Trans. Linn. Soc. London* 2 (1887): 322-23.

77. Notes made on December 4, 1889, Bower Collection, catalog F6.

78. Walter Gardiner, "How plants maintain themselves in the struggle for existence," *Gardeners Mag.* September 28, 1889.

79. William H. Lang, lecture notes from Bower's class in elementary botany at the University of Glasgow, Bower Collection, catalog F8.

80. Bower, "On antithetic ... generations," p. 348.

81. Ibid., p. 350.

82. Ibid., p. 356.

83. Frederick O. Bower, "Presidential address to the Botanical Society of Edinburgh, November, 1894," *Trans. Proc. Bot. Soc. Edinburgh* 20 (1896): 275-84.

CHAPTER SIX

1. For reviews of this period, see William Coleman, "Cell, nucleus, and inheritance: An historical study," *Proc. Amer. Phil. Soc.* 109 (1965): 124-58. Much of this and the following chapter is derived from Alice Baxter and John Farley, "Mendel and Meiosis," *J. Hist. Biol.* 12 (1979): 137-73.

2. Arthur Lee, *The microtomist's Vade-Mecum* (Philadelphia, 1885).

3. Eduard Strasburger, "Die Befruchtung bei den Farnkräutern," *Jahr. wiss. Bot.* 7 (1869): 390-408.

4. Wilhelm His, *Unsere Körperform und das physiologische Problem ihrer Entstehung* (Leipzig, 1874), p. 152.

5. By the end of the nineteenth century, free cell-formation no longer referred exclusively to endogeny or exogeny but referred as well to the nucleus of a cell dividing several times in the absence of a corresponding division of the cell cytoplasm.

6. Eduard Strasburger, *Über Zellbildung und Zelltheilung* (Jena, 1875), p. 309.

7. Oscar Hertwig, "Beiträge zur Kenntniss der Bildung, Befruchtung und Theilung des thierischen Eies," *Morphol. Jahr.* 1 (1876): 347-452; 3 (1877): 1-86. Quotation from 3: 30.

8. Leopold Auerbach, *Zur Charakteristik und Lebensgeschichte der Zellkerne* (Breslau, 1874), p. 213.

9. Hertwig, "Beiträge zur Kenntniss der Bildung," p. 357.

10. Hermann Fol, "Sur le commencement de l'hénogénie chez divers animaux," *Arch. Zool. Expér. Gén.* 6 (1877): 145-69. A more extensive treatment appeared in idem, "Recherches sur la fécondation et le commencement de l'hénogénie chez divers animaux," *Mem. Soc. Phys. Hist. Nat. Genève* 26 (1879): 92-397.

11. Strasburger, *Über Zellbildung und Zelltheilung* (1875), pp. 308-9.

12. Walther Flemming, "Beiträge zur Kenntniss der Zelle und ihrer Lebenserscheinungen, Theil II," *Arch. mikro. Anat.* 18 (1880): 151-259. English translation by L. Piternick, *J. Cell. Biol.* 25 (1965): 1-69.

13. Walther Flemming, "Beiträge ... Theil III," *Arch. mikro. Anat.* 20 (1881): 1-86.

14. Eduard Strasburger, *Über Zellbildung und Zelltheilung*, 3rd ed. (Jena, 1880), p. 321.

15. Eduard Strasburger, "Über den Theilungsvorgang der Zellkerne und das Verhältniss der Kerntheilung zur Zelltheilung," *Arch. mikro. Anat.* 21 (1882): 476-590.

16. Ibid., p. 498.

17. Eduard Strasburger, "Die Controversen der indirecten Kerntheilung," ibid. 23 (1884): 301.

18. Eduard Strasburger, "Über Befruchtung und Zelltheilung," *Jena. Zeit. Naturwiss.* 11 (1877): 508.

19. Eduard Strasburger, "Über den Befruchtungsvorgang," *Sitz. nied. Gesell. Nat. Heil. Bonn,* in *Verhandlung nat. Vereines Preuss. Rhein. Westfalens* 39 (1882): 185.

20. Eduard Strasburger, *Neue Untersuchungen über den Befruchtungsvorgang bei den Phanerogamen als Grunglage für eine Theorie der Zeugung* (Jena, 1884), p. 86.

21. Eduard Strasburger, et al., *Lehrbuch der Botanik* (Jena, 1898), English trans. H. C. Porter (London, 1898), p. 277.

22. François Maupas, "Le rejeunissement caryogamique chez les cilies," *Arch. Zool. Expér. Gén.* 7 (1889): 149-517.

23. Strasburger, *Neue Untersuchungen,* p. 133.

24. Edouard van Beneden, "De la distinction originelle du testicule et de l'ovaire, caractère sexuel des deus feuillets primordiaux de l'embryon, hermaphroditism morphologique de toute individualité animale: Essai d'une théorie de la fécondation," *Bull. Acad. Roy. Sci. Belg.* 37 (1874): 530-95.

25. Edouard van Beneden, "Recherches sur la maturation de l'oeuf et la fécondation," *Arch. Biol.* 4 (1883): 610-20.

26. Edouard van Beneden and Adolphe Neyt, "Nouvelles recherches sur la fécondation et la division mitosique chez l'*Ascaride mégalocephale,*" *Bull. Acad. Roy. Sci. Belg.,* n.s. 14 (1887): 238.

27. Van Beneden, "Recherches sur la maturation," p. 611.

28. Edouard van Beneden, "Contributions à l'histoire de la vésicule germinative et du premier noyau embryonnaire," *Bull. Acad. Roy. Sci. Belg.* 41 (1876): 79.

29. Van Beneden, "Recherches sur la maturation," p. 526.

30. Ibid., p. 527.

31. Wilhelm Roux, *Über die Bedeutung der Kerntheilungsfiguren: Eine hypothetische Erörterung* (Leipzig, 1883).

32. Walther Flemming, "Neue Beiträge zur Kenntniss der Zelle," *Arch. mikro. Anat.* 29 (1887): 389-463.

33. August Weismann, "The continuity of the germ-plasm as the foundation of a theory of heredity," in *Essays upon heredity and kindred biological problems* (Oxford, 1889), pp. 162-248.

34. Ibid., p. 188.

35. Ibid., p. 168.

36. Ibid., p. 214.

37. August Weismann, "On the number of polar bodies and their significance in heredity," ibid., pp. 335-84.

38. Ibid., p. 360.

39. Ibid., p. 355.

40. Theodor Boveri, *Zellenstudien I: Die Bildung der Richtungskörper bei Ascaris megalocephala und Ascaris lumbricoides* (Jena, 1887).

41. Oscar Hertwig, "Vergleich der Ei-und Samenbildung bei Nematoden: Eine Grundlage für celluläre Streitfragen," *Arch. mikro. Anat.* 36 (1890): 61.

42. Ibid., p. 71.

43. Theodor Boveri, "Zellenstudien III: Über das Verhalten der chromatischen Kernsubstanz bei der Bildung der Richtungskörper und bei der Befruchtung," *Jena. Zeit. Naturwiss.* 24 (1890): p. 374.

44. Ibid., p. 385.

45. Hertwig, "Ei-und Samenbildung," p. 92.

46. August Weismann, "Amphimixis, or the essential meaning of conjugation and sexual reproduction" (1891), in *Essays upon heredity,* 2: 121.

47. Ibid., p. 122.

48. Ibid., p. 136.

49. Ibid., p. 132. Frederick Churchill has discussed the genesis of these ideas in "August Weismann and a break from tradition," *J. Hist. Biol.* 1 (1968): 91-112. One break involved the reemergence of sex: from an "ill-defined blending of two stuffs" to a "unique combination of material worthy of investigation in its own right." This aspect of sex was elaborated on in much greater depth in Churchill's "Sex and the single organism: Biological theories of sexuality in mid-nineteenth century," *Stud. Hist. Biol.* 3 (1979): 139-78, and forms the major focus of my work.

50. Hertwig, "Ei-und Samenbildung," pp. 62-71.

51. Boveri, "Zellenstudien III," p. 62.

52. Leon Guignard, "Nouvelles études sur la fécondation," *Ann. Sci. Nat. (Bot.)* 14 (1891): 170-76, 255.

53. Leon Guignard, "Le développement du pollen et la réduction chromatique dans le *Naias major,*" *Arch. Anat. Micro.* 2 (1898): 505.

54. August Brauer, "Zur Kenntniss der Spermatogenese von *Ascaris megalocephala,*" *Arch. mikro. Anat.* 42 (1893): 207.

55. John Farmer, "On spore formation and nuclear division in the Hepaticae," *Ann. Bot.* 9

(1895): 518; Friedrich Meves, "Über die Entwicklung der männlichen Geschlectszellen von *Salamandra maculosa,"* Arch. mikro. Anat. 48 (1897): 1-83.

56. Otto vom Rath, "Zur Kenntniss der Spermatogenese von *Gryllotalpa vulgaris* mit besonderer Berucksichtigung der Frage der Reductionstheilung," *Arch. mikro. Anat.* 40 (1892): 102-32; Valentin Haecker, "Die Vorstadien der Eireifung," ibid. 45 (1895): 200-273.

57. Johannes Rückert, "Die Chromatinreduktion bei der Reifung der Sexualzellen," *Ergebnisse der Anatomie und Entwicklungsgeschichte* 3 (1893): 517-83; idem, "Zur Eireifung bei Copeopoden," *Anat. Hefte* 4 (1894): 263-351.

58. Rückert, "Chromatinreduktion," p. 576.

59. Theodor Boveri, *Zellenstudien II. Die Befruchtung und Theilung des Eies von Ascaris megalocephala* (Jena, 1888), p. 5.

60. Boveri, "Zellenstudien III," p. 63.

61. Eduard Strasburger, "The periodic reduction of the number of chromosomes in the life-history of living organisms," *Ann. Bot.* 8 (1894): 302.

62. Valentin Haecker, "The reduction of the chromosomes in the sexual cells as described by botanists: A reply to Prof. Strasburger," ibid. 9 (1895): 96.

63. Rückert, "Chromatinreduktion," p. 580.

64. Wilhelm Roux, "Beiträge zur Entwicklungsmechanik des Embryo," *Zeit. Biol.* 21 (1885): 414.

65. Gilbert Bourne, "Epigenesis or evolution," *Sci. Prog.* 1 (1894): 112.

66. Strasburger, "Periodic reduction," pp. 306-7.

67. Ibid., p. 304.

68. Oscar Hertwig, *The biological problem of today: Preformation or epigenesis?* trans. P. C. Mitchell (New York, 1894), p. 11. The English edition has recently been reprinted by Dabor Science Publications, Dabor Services, New York.

69. Carl Correns, "G. Mendel's Regel über das Verhalten der Nachkommenschaft der Rassenbastarde," *Deut. Bot. Gesell. Berlin* 18 (1900): 163.

70. Strasburger, "Periodic reduction," p. 310. This English translation omits the words "nor anywhere else," which appear in the original German article in *Biol. Centralblatt* 14 (1894): 851.

CHAPTER SEVEN

1. Wilhelm Roux, "The problems, methods, and scope of developmental mechanics," *Woods Hole Biological Lectures* (Boston, 1894), p. 149.

2. Ibid., p. 170.

3. Edmund Wilson, "Amphioxus and the mosaic theory of development," *J. Morph.* 8 (1893): 610.

4. Ibid., p. 615.

5. Henry Crampton, "Experimental studies on Gastropod development with an appendix by Edmund B. Wilson," *Arch. Entwick.* 3 (1896): German summary, p. 16. In previous experiments by Roux such incomplete development occurred only in a two-celled blastomere that was still attached to the killed blastomere.

6. Ibid., p. 20.

7. Edmund Wilson, "Experiments on cleavage and localization in the Nemertine egg," *Arch. Entwick.* 16 (1903): 448.

8. Ibid., pp. 457-58.

9. Edmund Wilson, *The cell in development and inheritance,* 2nd ed. (New York: Macmillan, 1900), p. 425.

10. Edmund Wilson, "The problem of development," *Science* 21 (1905): 289-92. For details of Wilson's work, see Alice Baxter, "Edmund B. Wilson as a preformationist: Some reasons for his acceptance of the chromosome theory," *J. Hist. Biol,* 9 (1976): 29-57.

11. Edmund Wilson, "Mendel's principles of heredity and the maturation of germ cells," ibid. 16 (1902): 992.

12. Walter Sutton, "On the morphology of the chromosome group in *Brachystola magna*," *Biol. Bull.* 4 (1902): 24–39.

13. Thomas Montgomery, "The spermatogenesis in *Pentatoma* up to the formation of the spermatid," *Zool. Jahr.* 12 (1898): 1–88.

14. Thomas Montgomery, "A study of the chromosomes of the germ cells of Metazoa," *Trans. Amer. Phil. Soc.*, 20 (1901): 154–236.

15. Ibid., pp. 222–23, gives these observational reasons.

16. Ibid., p. 223.

17. Theodor Boveri, *Ergebnisse über die Konstitution der chromatischen Substanz des Zellkerns* (Jena, 1904), p. 43.

18. Ibid., p. 44.

19. Theodor Boveri, "On multipolar mitoses as a means of analysis of the cell nucleus" (English trans.), in *Foundations of Experimental Embryology*, ed. B. H. Willier and J. M. Oppenheimer (Englewood Cliffs, N.J.: Prentice Hall, 1964), p. 77.

20. Ibid., p. 85.

21. Ibid., p. 84.

22. Boveri, *Ergebnisse,* p. 70. This famous paper is an expanded version of an earlier study: "Über die Konstitution der chromatischen Kernsubstanz," *Verhand. Deut. zool. Gesell. Leipzig* 13 (1903): 10–33.

23. Ibid., p. 71.

24. Boveri, "Multipolar mitoses," p. 89.

25. Sutton, "Chromosome group in *Brachystola*," p. 39.

26. Walter Sutton, "The chromosomes in heredity," *Biol. Bull.* 4 (1903): 231–51.

27. Boveri, *Ergebnisse,* p. 116.

28. Eduard Strasburger, "Über cytoplasmastructuren, Kern und Zelltheilung," *Jahr. wiss. Bot.* 30 (1897): 375–405.

29. Eduard Strasburger and David Mottier, "Über den zweiten Theilungs schritt in Pollen-mutterzellen," *Berichte Deut. bot. Gesell.* 15 (1897): 331.

30. Strasburger, *Über Reduktionstheilung, Spindelbildung, Centrosomen und Cilienbildner im Pflanzenreich* (Jena: G. Fisher, 1900), p. 86.

31. David Mottier, *Fecundation in plants* (Washington, D.C., 1904), p. 56. See also idem, "The behaviour of the chromosomes in the spore–mother cells of higher plants and the homology of the pollen and embryo sac mother cells," *Bot. Gazette* 35 (1903): 250–82.

32. R. P. Gregory, "The reduction division in ferns," *Proc. Roy. Soc. London* 73 (1904): 86–92; D. Tretjakoff, "Die Spermatogenese bei *Ascaris megalocephala,*" *Arch. mikro. Anat.* 65 (1905): 383–438.

33. Valentin Haecker, "Die Chromosomen als angenommene Vererbungstrager," *Ergebnisse Fortschritte Zool.* 1 (1907): 69.

34. John Farmer and J. E. S. Moore, "New investigations into the reduction phenomena of animals and plants—Preliminary communication," *Proc. Roy. Soc. London,* 72 (1903): 104–8.

35. Eduard Strasburger, "Über Reduktionstheilung," *Sitz. königl. Preuss. Akad. Wiss. Berlin* 18 (1904): 596.

36. Hugo de Vries, *The mutation theory,* trans. J. B. Farmer and A. D. Darbishire (Chicago, 1910), pp. 645–47.

37. Hugo de Vries, "Fertilization and hybridization," in *Intracellular pangenesis, including a paper on fertilization and hybridization,* trans. C. S. Gager (Chicago: Open Court, 1910), p. 224.

38. Ibid., p. 227.

39. Ibid., p. 237.

40. Ibid., p. 263.

41. Ibid., foreword by Strasburger.

42. Strasburger, "Über Reduktionstheilung," p. 612.

43. Eduard Strasburger, "Über Befruchtung," *Bot. Zeit.* 59 (1901): 352.

44. Eduard Strasburger, "The minute structure of cells in relation to heredity," in *Darwin and modern science*, ed. A. C. Seward (Cambridge: Cambridge University Press, 1909), p. 110. Strasburger appears to be one of the few scientists who would change his theories if faced with seemingly incompatible facts. At the Darwin-Wallace celebration of the Linnaean Society (held fifty years after the famous papers of the two men were presented to the society in 1858), where he was awarded a gold medal, Strasburger noted in a brief address: "When Darwin believed he had erred, he never hesitated to admit it. Here again he showed the mettle of really great men, whose intellectual wealth easily supports a loss, whereas indigent minds cling anxiously to their presumed property." Historians of science will agree with me, I think, that few scientists fit Strasburger's model of "really great men." As Max Planck wrote in his *Wissenschaftliche Selbstbiographie* (Leipzig: J. Barth, 1955), p. 22: "A new scientific truth does not become accepted by way of convincing and enlightening the opposition. Rather the opposition dies out and the rising generation becomes well acquainted with the new truth from the start."

45. Charles Allen, "Nuclear division in the pollen–mother cells of *Lilium canadensis*," *Ann. Bot.* 19 (1905): 242.

46. Ibid., pp. 249-52.

47. John Farmer and J. E. S. Moore, "On the maiotic phase (reduction division) in animals and plants," *Quart. J. Micro. Sci.* 48 (1905): 494-96.

48. M. M. Rhoades, "Meiosis," in *The cell*, vol. 3, *Meiosis and mitosis*, ed. Jean Brachet and Alfred Mirsky (New York: Academic Press, 1961), p. 4.

49. Hans von Winiwarter, "Recherches sur l'ovogenese et l'organ oogenese de l'ovaire des mammifères (lapin et homme)," *Arch. Biol.* 17 (1901): 33-199.

50. Edmund Wilson, *The cell in development and inheritance*, 3rd ed. (New York: Macmillan, 1925), p. 503.

51. William Cannon, "A cytological basis for Mendelian laws," *Bull. Torrey Bot. Club* 29 (1902): 661.

52. Strasburger, "Über Reduktionstheilung," p. 587.

53. Frederick O. Bower to G. MacMillan, March 10, 1918, Bower Collection, University of Glasgow Archives, catalog E154.

54. Isabel, Marchioness of Ailsa, to Frederick O. Bower, November 19, 1919, ibid., catalog E52.

55. F. O. Bower, J. G. Kerr, and W. E. Agar, *Lectures on sex and heredity* (London: Macmillan, 1919), p. 7.

56. August Weismann, *Vorträge über Deszendenztheorie*, 2nd ed. (Jena, 1904), chap. 17; 3rd ed. (Jena, 1913), pp. 41-47.

57. Charles Brown-Séquard, "Des Effets produits chez l'homme par des injections sous-cutanées d'un liquide retiré des testicles frais de cobaye et de chien," *Compte-rendus Soc. Biol.* 1 (1889): 415-21. For details of Brown-Séquard's work and the beginnings of endocrinology, see Merriley Borell, "Brown-Séquards' organotherapy and its appearance in America at the end of the nineteenth century," *Bull. Hist. Med.* 50 (1976): 309-20; and "Organotherapy, British physiology, and discovery of the internal secretions," *J. Hist. Biol.* 9 (1976): 235-68.

58. François Maupas, "Recherches expérimentales sur la multiplication des infusoires ciliés," *Arch. Zool. Expér. Gén.* 6 (1888): 261.

59. Charles Minot, "On the formation of the germinal layers and the phenomena of impregnation among animals," *Proc. Boston Soc. Nat. Hist.* 19 (1877): 170.

60. Charles Minot, "Growth as a function of cells," ibid. 20 (1879): 190.

61. Gary Calkins, "Degeneration in *Paramoecium* and so-called 'rejuvenescence' without conjugation," *Science* 15 (1902): 526; idem, "Studies on the life history of Protozoa: I. The life cycle of *Paramoecium caudatum*," *Arch Entwick.* 15 (1902): 139-86; Gary Calkins and C. C.

Lieb, "Studies on the life history of Protozoa: II. The effect of stimuli on the life-cycle of *Paramoecium caudatum,*" *Arch. Protistenkunde* 1 (1902): 355-71 (quotation from p. 370).

62. Gary Calkins, "Studies on the life history of Protozoa: III. The six hundred twentieth generation of *P. caudatum,*" *Biol. Bull.* 3 (1902): 202.

63. Gary Calkins, "Studies in the life history of Protozoa: IV. Death of the A series," *J. Exp. Zool.* 1 (1904): 447. Calkins reached the same conclusion in "The protozoan life-cycle," *Biol. Bull.* 11 (1906): 229-44.

64. Lorande Woodruff, "The life cycle of *Paramoecium* when subjected to varied environment," *Amer. Nat.* 42 (1908): 520-26; idem, "Further studies in the life cycle of *Paramoecium,*" *Biol. Bull.* 17 (1909): 287-308; idem, "Two thousand generations of *Paramoecium,*" *Arch. Protistenkunde* 21 (1911): 263-66.

65. Charles Minot, *The problem of age, growth, and death* (London: John Murray, 1908); quotation from idem, "The problem of age, growth and death," *Pop. Sci. Monthly* 71 (1910): 377.

66. Herbert Jennings, "Age, death, and conjugation in the light of the work on lower organisms," *Pop. Sci. Monthly* 80 (1912): 577.

67. Herbert Jennings, "The effect of conjugation in *Paramoecium,*" *J. Exp. Zool.* 14 (1913): 279-391; table on p. 362.

68. Ibid., p. 391; Jennings, "Age, death, and conjugation," p. 573.

69. John Maynard Smith, "The origin and maintenance of sex," in *Group selection,* ed. George C. Williams (Chicago: Aldine-Atherton, 1971), p. 163.

CHAPTER EIGHT

1. Eduard Strasburger, "Die Ontogenie der Zelle seit 1875," *Progressus Rei Botanica* 1 (1907): 136.

2. Charles E. Overton, "On the reduction of the chromosomes in the nuclei of plants," *Ann. Bot.* 7 (1893): 139-43.

3. Eduard Strasburger, *Über Kern und Zelltheilung im Pflanzenreiche* (Jena, 1888), p. 36.

4. Ibid., p. 38.

5. Ibid., p. 51.

6. Ibid., p. 57.

7. Ibid., p. 238.

8. Leon Guignard, "Nouvelles études sur la fécondation," *Ann. Sci. Nat. (Bot.)* 14 (1891): 257.

9. Eduard Strasburger, *Schärmsporen, Gameten, pflanzliche Spermatozoiden und das Wesen der Befruchtung* (Jena, 1892), p. 148.

10. Eduard Strasburger, "The periodic reduction of the number of the chromosomes in the life-history of living organisms," *Ann. Bot.* 8 (1894): 295. This paper was presented to the Oxford meeting of the British Association for the Advancement of Science in 1894.

11. Ibid., p. 302.

12. Ibid., p. 290.

13. Ibid., pp. 288-89.

14. Ibid., p. 311.

15. The law was implicit in the work of Celakovsky and others but to my knowledge this was the first explicit statement linking the law to the alternation of generations.

16. Arthur Marshall, "Development of animals," *Report of the Brit. Assoc. Adv. Sci.,* 1890, p. 827; Dukinfield H. Scott, "Present position of morphological botany," ibid., 1896, p. 992. Stephen Gould's recent study of the biogenetic law, *Ontogeny and phylogeny* (Cambridge, Mass.: Harvard University Press, 1977), makes no mention of the alternation of generations in plants.

17. Eduard Strasburger, "Über die Bedeutung phylogenetischer Methoden für die Erforschung lebender Wesen," *Jena. Zeit. Naturwiss.* 8 (1874): 56-80.

18. Strasburger's address in *The Darwin-Wallace Celebration* (London: Linnaean Society, 1908), p. 23.

19. Strasburger, "Periodic reduction," p. 281-84.

20. Frederick O. Bower, "On antithetic as distinct from homologous alternation of generations in plants," *Ann. Bot.* 4 (1890): 348.

21. Frederick O. Bower, "Presidential address to the Botanical Society of Edinburgh," *Trans. Proc. Bot. Soc. Edinburgh* 20 (1896): 282-83.

22. Scott, "Morphological botany," p. 999.

23. Ibid., p. 992.

24. Ibid., p. 997.

25. Frederick O. Bower, "Presidential address to the British Association for the Advancement of Science," ibid., 1898; reprinted in *Nature* 59 (1898): 69.

26. Scott, "Morphological botany," p. 1000.

27. Ibid., p. 1001. For a modern reevaluation of this question, see George L. Stebbins and C. J. C. Hill, "Did multicellular plants invade the land?" *Amer. Nat.* 115 (1980): 342-53. The authors argue that plants invaded land as unicellular, undifferentiated forms.

28. Ibid., p. 1003.

29. Bower, "Presidential address" (1898), p. 89.

30. David Mottier, "Nuclear and cell division in *Dictyota dichotoma,*" *Ann. Bot.* 14 (1900): 163-92; J. Lloyd Williams, "Studies in the Dictyotaceae: I. The cytology of the tetrasporangium and the germinating tetraspore," ibid. 18 (1904): 141-60; idem, "Studies in the Dictyotaceae: II. The cytology of the gametophyte generation," ibid., pp. 183-204; J. J. Wolfe, "Cytological studies on *Nemalion,*" ibid. 23 (1904): 607-30.

31. Charles Allen, "Die Keimung der Zygote bei *Coleochaete,*" *Berichte Deut. bot. Gesell.* 23 (1905): 290.

32. Eduard Strasburger, "Typische und allotypische Kerntheilung," *Jahr. wiss. Bot.* 42 (1906): 62.

33. Frederick O. Bower, *The origin of a land flora: A theory based upon the facts of alternation* (London: Macmillan, 1908), pp. 83-84.

34. Ibid., p. 44.

35. Ibid., p. 52.

36. Ibid., p. 84.

37. Theodor Boveri, "Über den Anteil des Spermatozoon an der Teilung des Eies," *Sitz. Gesell. Morphol. Physiol. München* 3 (1887): 155.

38. Edmund Wilson, *The cell in development and inheritance* (New York, 1896), p. 140.

39. Ibid., p. 109. Wilson made much the same remarks in the second edition, published in 1900.

40. François Maupas, "Sur le déterminisme de la sexualité chez l'*Hydatina senta,*" *Comptes-rendus Acad. Sci.* 113 (1891): 388.

41. An account of these various theories is given in Patrick Geddes and J. A. Thomson, *The evolution of sex* (London, 1889).

42. Maupas, "Déterminisme de la sexualité," pp. 388-90.

43. S. Watasé, "On the phenomenon of sex differentiation," *J. Morphol.* 6 (1892): 485.

44. Chas Riley, "Controlling sex in butterflies," *Amer. Nat.* 7 (1873): 518.

45. Lucien Cuénot, "Sur la détermination du sexe chez les animaux," *Bull. Sci. Fr. Belg.* 32 (1899): 471. For details about Cuénot, see Camille Limoges, "Natural selection, phagocytosis, and preadaptation: Lucien Cuénot, 1887-1914," *J. Hist. Med.* 31 (1976): 176-214.

46. Ibid., p. 515.

47. William Bateson and E. R. Saunders, "Experimental studies in the physiology of heredity," *Evolution Committee of the Royal Society of London: Report No. 1,* 1902, p. 138.

48. William Castle, "The heredity of sex," *Bull. Mus. Comp. Zool.* 40 (1903): 191.

49. Ibid., p. 194.

50. Ibid., p. 195.

51. Thomas Hunt Morgan, "Recent theories in regard to the determination of sex," *Pop. Sci. Monthly* 64 (1903): 111-12.

52. Ibid., p. 114.

53. C. E. McClung, "The accessory chromosome—sex determinant?" *Biol. Bull.* 3 (1902): 72.

54. Ibid., p. 80.

55. Morgan, "Recent theories," p. 116.

56. William Castle, "Mendel's law of heredity," *Proc. Amer. Acad. Arts Sci.* 38 (1903): 545.

57. Edmund Wilson, "The chromosomes in relation to the determination of sex in insects," *Science* 22 (1905): 500.

58. Edmund Wilson, "Studies on chromosomes: III. The sexual differences of the chromosome groups in Hemiptera, with some considerations of the determination and inheritance of sex," *J. Exp. Zool.* 3 (1906): 26.

59. Ibid., p. 29.

60. Nettie Stevens, "Studies in spermatogenesis, with especial reference to the 'accessory chromosome,'" Carnegie Institution Report no. 36 (Washington, D.C., September, 1905), p. 13. For details of her work, see Stephen Brush, "Nettie M. Stevens and the discovery of sex determination by chromosomes," *Isis* 69 (1978): 163-72.

61. Wilson, "Determination of sex in insects," p. 502.

62. Wilson, "Studies on chromosomes: III," pp. 34-35.

63. Ibid., p. 39.

64. Edmund Wilson, "A new theory of sex production," *Science* 23 (1906): 189-91.

65. Nettie Stevens, "Studies in spermatogenesis: Part II," Carnegie Institution Report no. 36, pt. 2 (Washington, D.C., October 1906), p. 56. Brush maintains that Stevens "was still ahead of Wilson in realizing the significance of their discovery," whereas, of course, Wilson was far more aware of the difficulties inherent in the Mendelian scheme.

66. Edmund Wilson, "Sex determination in relation to fertilization and parthenogenesis," *Science* 25 (1907): 376.

67. Robert Harper, "Sex-determining factors in plants," ibid., p. 381.

68. Thomas Hunt Morgan, "Sex-determining factors in animals," ibid., p. 384.

69. Wilson, "Studies on chromosomes: III," p. 28.

70. Carl Correns, "Die Bestimmung und Vererbung des Geschlechtes," *Arch. Rassen-Gesell. Biol.* (not seen).

71. Edmund Wilson, "Studies on chromosomes: IV. The 'accessory' chromosome in *Syromastes* and *Pyrrochoris,* with a comparative review of the types of sexual differences of the chromosome groups," *J. Exp. Zool.* 6 (1909): 96.

72. Reginald C. Punnett and W. Bateson, "The heredity of sex," *Science* 27 (1908): 785-87.

73. Thomas Hunt Morgan, F. Payne, and Ethel Brown, "A method to test the hypothesis of selective fertilization," *Biol. Bull.* 18 (1910): 76-78.

74. Edmund Wilson, "Recent researches on the determination and heredity of sex," *Science* 29 (1909): 62. Elie and Emil Marchals, "Aposporie et sexualité chez les mousses," *Bull. Acad. Roy. Belg.,* 1907, pp. 765-89.

75. Nettie Stevens, "A study of the germ cells of certain Diptera, with reference to the heterochromosomes and the phenomena of synapsis," *J. Exp. Zool.* 5 (1908): 359.

76. William Johannsen, "The genotype conception of heredity," *Amer. Nat.* 45 (1911): 129-59.

77. Edmund Wilson, "Some aspects of cytology in relation to the study of genetics," *Amer. Nat.* 46 (1912): 60.

78. Wilson, "Recent researches," p. 62.

79. Ibid., pp. 63-64.

80. Ibid., p. 64.

81. Garland Allen, *Thomas Hunt Morgan: The man and his science* (Princeton: Princeton University Press, 1978), p. 143.

82. Thomas Hunt Morgan, "Sex determination and parthenogenesis in Phylloxerans and Aphids," *Science* 29 (1909): 234-37.

83. Thomas Hunt Morgan, "A biological and cytological study of sex determination in Phylloxerans and Aphids," *J. Exp. Zool.* 7 (1909): 306.

84. William Castle, "A Mendelian view of sex heredity," *Science* 29 (1909): 395-400.

85. Thomas Hunt Morgan, "The application of the conception of pure lines to sex-limited inheritance and to sexual dimorphism," *Amer. Nat.* 45 (1911): 65.

86. Thomas Hunt Morgan, "Sex-limited inheritance in *Drosophila*," *Science* 32 (1910): 120-22.

87. Morgan, "Application of pure lines," pp. 67-69.

88. Ibid., pp. 70-71.

89. Thomas Montgomery, "Are particular chromosomes determinants?" *Biol. Bull.* 19 (1910): 15.

90. Ibid., pp. 12-13.

91. Information from Harvey Jordan, "Recent literature touching the question of sex-determination," *Amer. Nat.* 44 (1910): 245-52. Russo's paper not seen.

92. For details of Goldschmidt's later work, see G. Allen, "Opposition to the Mendelian-chromosome theory: The physiological and developmental genetics of Richard Goldschmidt," *J. Hist. Biol.* 7 (1974): 4972.

93. Maurice Caullery, "Le Problème du déterminisme du sexe," *Biol. Rev. Sci. Med.* 3 (1913): 193-202; idem, *Les Problèmes de la sexualité* (Paris: Flammarion, 1913).

94. Thomas Hunt Morgan, *Heredity and sex* (New York: Columbia University Press, 1913), pp. 43-44.

95. Ibid., p. 46.

96. Ibid., pp. 83-84.

97. Edmund Wilson, "The sex-chromosomes," *Arch. mikro. Anat.* 77 (1911): 265-67.

CHAPTER NINE

1. Edmund Wilson, "Recent researches on the determination and heredity of sex," *Science* 29 (1909): 70.

2. Jacques Loeb, "The biological problems of today: Physiology," ibid. 7 (1898): 154-56.

3. Hermann von Helmholtz, "Aim and progress of physical science," in *Popular lectures on scientific subjects,* English trans. E. Atkinson (New York, 1873), p. 384. For details of Loeb's life and work, see Donald Fleming's introduction to *The mechanistic conception of life,* ed. D. Fleming (Cambridge, Mass.: Harvard University Press, 1964).

4. Thomas Hunt Morgan, "The action of salt-solutions on the unfertilized and fertilized eggs of *Arbacia* and other animals," *Arch. Entwick.* 8 (1899): 527.

5. Jacques Loeb, "On the nature of the process of fertilization and the artificial production of normal larvae (plutei) from the unfertilized eggs of the sea urchin," *Amer. J. Physiol.* 3 (1899): 135-38.

6. Ibid., p. 138.

7. Loeb, "On the nature of the process of fertilization" (1899), in *Mechanistic conception of life,* ed. Fleming, p. 115.

8. Jacques Loeb, "Further experiments on artificial parthenogenesis and the nature of the process of fertilization," *Amer. J. Physiol.* 4 (1900): 181.

9. Jacques Loeb, "On the artificial production of normal larvae from the unfertilized eggs of the sea urchin (*Arbacia*)," *Amer. J. Physiol.* 3 (1900): 468.

10. Jacques Loeb, "Artificial parthenogenesis in annelids," *Science* 12 (1900): 170; idem, "Experiments on artificial parthenogenesis in annelids (*Chaetopterus*) and the nature of the process of fertilization," *Amer. J. Physiol.* 4 (1901): 423–59.

11. *Harper's Weekly*, December 13, 1902, p. 1936; and Garrett Serviss, "Artificial creation of life," *Cosmopolitan* 39 (1905): 459–68. Quotation from Carl Snyder, "The mysteries of life and mind: Dr. Loeb's researches and discoveries," *Fortnightly Rev.* 77 (1902): 1016.

12. Loeb, "Experiments on annelids (*Chaetopterus*)," p. 456.

13. J.-B. Piéri, "Un nouveau ferment soluble: L'ovulase," *Arch. Zool. Expér. Gén.* 7 (1899): 29–30; Raphael Dubois, "Sur la spermase et l'ovulose," *Comptes-rendus Soc Biol.* 52 (1900): 197–99.

14. William Gies, "Do spermatozoa contain enzymes having the power of causing development of mature ova?" *Amer. J. Physiol.* 6 (1901): 53–76.

15. Loeb, "Artificial production of normal larvae," p. 469.

16. Victor Hensen, *Physiologie der Zeugung* (Leipzig, 1881), p. 126.

17. Yves Delage, *Les Théories de la fécondation* (Jena: Gustav Fischer, 1902), p. 14.

18. Ibid., p. 17.

19. Oscar Hertwig, "Kritische Betrachtungen über neuere Erklärungsversuche auf dem Gebiete der Befruchtungslehre," *Sitz. Preuss. Akad. Wiss.* 17 (1905): 375.

20. Jacques Loeb, "On the nature of formative stimulation (artificial parthenogenesis)," in *Mechanistic conception of life*, ed. Fleming, p. 121.

21. Ibid., p. 125.

22. Ibid., p. 128.

23. Frank Lillie, *Problems of fertilization* (Chicago: University of Chicago Press, 1919), p. 26.

24. Jacques Loeb, "The Mechanistic conception of life" (1911), in *Mechanistic conception of life*, ed. Fleming, pp. 5–6, 16.

25. Frank Lillie, "Function of the spermatozoon in fertilization from observations on *Nereis*," *Science* 31 (1910): 836; idem, "Studies on Fertilization in *Nereis* I and II," *J. Morphol.* 22 (1911): 389.

26. Lillie, *Problems of fertilization*, p. vii.

27. Frank Lillie, "Studies on fertilization in *Nereis*," *J. Morphol.* 22 (1911): 389.

28. Frank Lillie, "Fertilization" (lecture to Zoology Club, University of Chicago, October 1913), F. R. Lillie Papers, Marine Biological Laboratories, Woods Hole, Mass. (hereafter cited as Lillie Papers).

29. Lillie, *Problems of fertilization*, p. 22.

30. Frank Lillie, "My early life" (a biography written for Lillie's children in 1944), Lillie Papers.

31. Charles O. Whitman, "The inadequacy of the cell-theory of development," *J. Morphol.* 8 (1893): 649, 658.

32. Frank Lillie, "The organization of the egg in *Unio*, based on a study of its maturation, fertilization, and cleavage," ibid. 17 (1901): 265.

33. Frank Lillie, "Observations and experiments concerning the elementary phenomena of embryonic development in *Chaetopterus*," *J. Exp. Zool.* 3 (1906): 251.

34. For excellent accounts of cell-lineage work, see Alice Baxter, "Edmund Beecher Wilson and the problem of development" (Ph.D. diss., Yale University, 1974), pp. 12–19; and Jane Maienschein, "Cell lineage, ancestral reminiscence, and the biogenetic law," *J. Hist. Biol.* 11 (1978): 129–58.

35. Edmund Wilson, quoted in Baxter, "Edmund Beecher Wilson," p. 90.

36. Frank Lillie, "The embryology of the Unionidae: A study in cell-lineage," *J. Morphol.* 10 (1895): 38–39.

37. Frank Lillie, "Adaptation in cleavage," *Biological lectures at Woods Hole, 1897-1898* (Boston, 1899), pp. 43-56.

38. Frank Lillie, "The mechanistic view of vital phenomena" (lecture to Philosophy Club, University of Chicago, December 1919), Lillie Papers.

39. Lillie, "Fertilization," Lillie Papers.

40. Frank Lillie, "Studies on fertilization: V. The behaviour of the spermatozoa of *Nereis* and *Arbacia*, with special reference to egg-extractives," *J. Exp. Zool.* 14 (1913): 549.

41. Ibid., p. 565.

42. Paul Ehrlich, "On immunity, with special reference to cell-life" (Croonian Lecture, Royal Society of London, 1900), reprinted in *Collected papers of Paul Ehrlich*, vol. 2, ed. F. Himmelweit (New York: Permagon Press, 1957), pp. 178-95.

43. James Ritchie, "A review of current theories regarding immunity," *J. Hyg.* 2 (1902): 221.

44. Ibid., p. 261.

45. Frank Lillie, "Studies on fertilization: VI. The mechanism of fertilization in *Arbacia*," *J. Exp. Zool.* 16 (1914): 524.

46. Ibid., p. 549.

47. Ibid., p. 563.

48. Jacques Loeb, "On some non-specific factors for the entrance of the spermatozoon into the egg," *Science* 40 (1914): 316-18; idem, "Cluster formation of spermatozoa caused by specific substances from eggs," *J. Exp. Zool.* 17 (1914): 123-40; Frank Lillie, "Recent theories of fertilization and parthenogenesis," *Trans. Illinois Acad. Sci.* 7 (1914): 1-10; idem, "Sperm agglutination and fertilization," *Biol. Bull.* 28 (1915): 18-33.

49. Lillie, *Problems of fertilization*, p. 129.

CHAPTER TEN

1. Camille-François Sauvageau, "Sur la sexualité heterogamie d'une Laminaire (*Saccorhiza bulbosa*)," *Comptes-rendus Acad. Sci.* 161 (1915): 769-99.

2. Nils Svedelius to Frederick O. Bower, November 3, 1916, Bower Collection, University of Glasgow Archives, catalog B326.

3. Bower to Svedelius, November 25, 1916, ibid., catalog B327.

4. Frederick O. Bower, *Origin of land flora* (London: Macmillan, 1908), p. 8.

5. Frederick O. Bower, "Remarks on the present outlook on descent," *Proc. Roy. Soc. Edinburgh* 44 (1924): 1-7.

6. Garland Allen, *Life science in the twentieth century* (Cambridge: Cambridge University Press, 1978). See also the special section on American morphology in *J. Hist. Biol.* 14 (1981): 83-191.

7. Arthur Tansley et al., "The reconstruction of elementary botanical teaching," *New Phytol.* 16 (1917): 247-48.

8. Ibid., p. 243.

9. Walter Stiles, ibid. 17 (1918): 251-57.

10. Vernon Blackman, ibid. 18 (1919): 51.

11. T. F. W. M., "The reconstruction of elementary botany teaching: The examination of a witness," ibid. 17 (1918): 3-8.

12. Frederick O. Bower to Arthur Tansley, January 31, 1918, Bower Collection, catalog C514.

13. Manuscript notes in Bower's handwriting (n.d.), ibid., catalog C521.

14. Arthur Tansley, "Postscript," *New Phytol.* 18 (1919): 108-10.

15. Frederick O. Bower, *Botany of the living plant* (London: Macmillan, 1919).

16. Review of *Botany of the living plant*, *Oxford Mag.*, June 13, 1919.

17. Review of *Botany of the living plant*, *J. Bot.*, August 1919.

18. Albert Seward, review of *Botany of the living plant*, *New Phytol.* 18 (1919): 259–61.

19. Arthur Hill to Frederick O. Bower, June 27, 1918, Bower Collection, catalog C234.

20. Tansley et al., "Reconstruction of botanical teaching," p. 250.

21. Thomas Hill, "Some practical suggestions," *New Phytol.* 17 (1918): 9–12.

22. R. C. McLean, "A plea for freedom," ibid., pp. 54–56.

23. Frederick O. Bower, "Botanical Bolshevism," ibid., pp. 105–7.

24. T. Briquet to Frederick O. Bower, July 10, 1918, Bower Collection, catalog C522.

25. Arthur Tansley to Frederick O. Bower, July 25, 1918, ibid., catalog C524.

26. Bower to Tansley and Tansley to Bower, July and August 1918, ibid., catalogs C526, C527.

27. This movement has been examined by Paul Werskey in *The visible college: The collective biography of British scientific socialists of the 1930's* (New York: Holt, Rinehart & Winston, 1978).

28. F. Sherwood Taylor, *Is the progress of science controlled by the material wants of man?* Society for Freedom in Science, Occasional Pamphlet no. 1, April 1945.

29. Arthur Tansley, *The values of science to humanity,* Herbert Spencer Lecture, Oxford University, June 1942 (London: George Allen, 1942), p. 31.

30. Bower, for whom I have formed a deep respect, lived a long life, passing away in 1948. His *Botany of the living plant* could still be used with profit by modern students, and, indeed, only a few years ago was still used as a reference text by English botany students. In 1938, thirteen years after his retirement, Bower published the delightful *Sixty years of botany in Britain.* The book was actually suggested to Bower by Duncan S. Johnson of The Johns Hopkins University, who in a letter of February 1923 (Bower Collection, catalog D128) asked Bower whether he had "ever thought of telling American botanists something of [his] experiences and companions while in the Wurzburg laboratory of Sachs? These must have been very stimulating days and there are few left now to tell from actual experience. . . ." Since Bower underlined this passage, we can assume that Johnson put the idea into Bower's head.

POSTSCRIPT

1. J. Bull, "Sex determination in reptiles," *Quart. Rev. Biol.* 55 (1980): 3–21.

2. M. J. D. White, *Animal cytology and evolution,* 3rd ed. (Cambridge: Cambridge University Press, 1973), p. 575.

3. Edmund Wilson, "The sex-chromosomes," *Arch. mikro. Anat.* 77 (1911): 260.

4. William Keeton, *Biological science* (New York: W. W. Norton, 1972), p. 488.

5. Francisco Jose, *Modern genetics* (Menlo Park, Calif.: Cummings Publishing Co., 1980).

6. Helena Curtis, *Biology* (New York: Worth, 1975), p. 514.

7. Ronald R. Fisher, *The genetical theory of natural selection.* (New York: Dover, 1958), p. 50.

8. Hermann J. Muller, "Some genetic aspects of sex," *Amer. Nat.* 66 (1932): 118–32.

9. John Maynard Smith, "The origin and maintenance of sex," in *Group selection,* ed. George C. Williams (Chicago: Aldine, Atherton, 1971), pp. 163–75.

10. George C. Williams, *Sex and evolution* (Princeton: Princeton University Press, 1975), p. 11.

11. George C. Williams and J. B. Mitton, "Why reproduce sexually?" *J. Theor. Biol.* 39 (1973): 553.

12. Other important studies of this paradox include George C. Williams, *Adaptation and natural selection* (Princeton: Princeton University Press, 1966): John Maynard Smith, "Evolution in sexual and asexual populations," *Amer. Nat.* 102 (1968): 469–73; idem, "What use is sex?" *J. Theor. Biol.* 30 (1971): 319–35; E. O. Wilson, "The origin of sex," *Science,* 188 (1975): 139–40; David Barash, "What does sex really cost?" *Amer. Nat.* 110 (1976): 894–97; and John Maynard Smith, *The evolution of sex* (Cambridge: Cambridge University Press, 1978).

13. Steven Stanley, "Clades versus clones in evolution: Why we have sex," *Science* 190 (1975): 382-83.

14. Quoted in Steven Stanley, *Macroevolution: Pattern and process* (New York: Freeman, 1979), p. 219.

15. Stanley, *Macroevolution*, pp. 226-27.

16. Williams, Introduction to *Group Selection*, p. 13.

Index

The Johns Hopkins University Press

GAMETES & SPORES

This book was composed in Times Roman text and display
type by The Composing Room of Michigan from a design
by Lisa S. Mirski. It was printed on S. D. Warren's
50-lb. Sebago Eggshell paper and bound in Kivar 5 by
the Maple Press Company. The manuscript
was edited by Penny Moudrianakis.